CONTAMINATED SOILS, SEDIMENTS AND WATER
Volume 10:
Successes and Challenges

CONTAMINATED SOILS, SEDIMENTS AND WATER
Volume 10:
Successes and Challenges

Edited by

Edward J. Calabrese
Paul T. Kostecki
James Dragun

 Springer

Library of Congress Cataloging-in-Publication Data

ISBN-10: 0-387-28322-6   ISBN-13: 978-0-387-28322-7   Printed on acid-free paper.
e-ISBN-10: 0-387-28324-2   e-ISBN-13: 978-0-387-28324-1

Printed in the United States of America.

9 8 7 6 5 4 3 2 1   SPIN 11415152

springeronline.com

Contents

PART V: RADIONUCLIDES

PART VI: REGULATORY

PART VII: REMEDIATION

PART IX: SITE ASSESSMENT

Foreword

Every spring, the University of Massachusetts – Amherst welcomes all "Soils Conference" Scientific Advisory Board members with open arms as we begin the planning process responsible for bringing you quality conferences year after year. With this "homecoming" of sorts comes the promise of reaching across the table and interacting with a wide spectrum of stakeholders, each of them bringing their unique perspective in support of a successful Conference in the fall.

This year marks the 20th anniversary of what started as a couple of thoughtful scientists interested in developing partnerships that together could fuel the environmental cleanup dialogue. Since the passage of the Superfund Law, regulators, academia and industry have come to realize that models that depend exclusively on "command and control" mandates as the operative underpinning limit our collective ability to bring hazardous waste sites to productive re-use. It is with this concern in mind that the Massachusetts Department of Environmental Protection privatized its cleanup program in 1993, spurring the close-out of over 20,000 sites and spills across the Commonwealth to date, in a manner that is both protective of human health and the environment while also flexible and responsive to varied site uses and redevelopment goals.

So we gather together again, this year, to hear our collective stories and share success and challenges just as we share stories at a family gathering. Take a read through the stories contained in these proceedings, Volume 10 of the Contaminated Soils, Sediments and Water. This jewel of a volume contains a valuable collection of successes (and challenges) in the areas of environmental fate, heavy metals, modeling, MtBE and oxygenates, regulatory, remediation, risk assessment, site assessment and sampling methodology. As you can see, there is something for everybody. Most importantly, in our minds at least, is the embodiment of how, as a community, we have worked together toward the optimization of established approaches as well as embracing departures from traditional regulatory models in order to address the challenges posed by emerging unregulated constituents that threaten our natural resources.

It is with great joy and pride that we write this Foreword, an affirmation of our commitment to this international, one-of-a-kind conference. A conference that over the last 20 years has taken into account where we all have been – public and private sector alike – so we can responsibly chart where we need to go as stewards of the environment.

Millie Garcia-Surette, MPH and Janine Commerford, LSP
Massachusetts Department of Environmental Protection

Contributing Authors

John L. Albrecht, Metcalf & Eddy, Inc., 860 North Main Street Extension, Wallingford, CT 06392

Claudio Alimonti, Universita di Roma "La Sapienza", Via Eudossiana, 18, 00184 Rome, Italy

George Alther, Biomin, Inc., PO Box 20028, Ferndale, MI 48220

Dora Angelova, Bulgarian Academy of Science, Geological Institute, Acad. G. Bonchev Str., Block 24, Sofia 1113, Bulgaria

Christian Bélanger, Biogenie S.R.D.C. Inc., 350, rue Franquet, entrée 10, Sainte-Foy, Québec, Canada G1P 4P3

Rob Breault, U.S. Geological Survey, 10 Bearfoot Road Northborough, MA 01532

Thomas E. Bridge, Emporia State University, Emporia, Kansas

James P. Byrne, US EPA Region 1, Brownfields Team, Work Assignment Manager, 1 Congress Street, Suite 1100 (HIO), Boston, MA 02114-2023

Ruth Chang, Sr. Hazardous Substances Scientist, Department of Toxic Substances Control, Hazardous Materials Laboratory, 700 Heinz Avenue, Suite 100, Berkeley, CA 94710

Charles Chrosniak, George Mason University, Chemistry Department, Fairfax, VA 22030

Cecilia Clavet, University of Maine, Department of Resource Economics and Policy, Orono, ME

Scott R. Compston, Malcolm Pirnie, Inc., 15 Cornell Road, Latham, NY 12110

Matthew Cooke, U.S. Geological Survey, 10 Bearfoot Road Northborough, MA 01532

Scott DeHainaut, CH2M HILL, 318 East Inner Road, Otis ANG Base, MA 02542

Charles Denton, Varnum, Riddering, Schmidt and Howlett LLP, Bridgewater Place, PO Box 352, Grand Rapids, MI 49501

Paul DiBenedetto, George Mason University, Chemistry Department, Fairfax University, VA 22030

Wilson Doctor, Department of the Navy, 1230 Columbia Street, Suite 1100, San Diego, CA

Larry Dudus, Tetra Tech FW, Inc., 1230 Columbia Street, Suite 500, San Diego, CA

Dale W Evans, Metrix, Inc., 17301 West Colfax Ave, Suite 170, Golden, CO 80401

Nile Fellows, Minnesota Pollution Agency, 520 Lafayette Rd, St. Paul, MN 55155

Rose H. Forbes, Air Force Center for Environmental Excellence, 322 East Inner Road, Otis-ANG Base, MA 02542-5028

Timothy Fortman, Pacific Northwest National Lab, Marine Sciences Lab, 1529 West Sequim Bay Rd, Sequim, WA 98382

Martin P.N. Gent, Connecticut Agricultural Experiment Station (CAES), Dept of Forestry and Horticulture, 123 Huntington Street, New Haven, CT 06504

John Glass, CH2M HILL, 13921 Park Center Road, Suite 600, Herndon, VA 20171-5416

Dennis Goldman, Tetra Tech FW, Inc., 1230 Columbia Street, Suite 500, San Diego, CA 92101

Joshua Goldowitz, Rochester Institute of Technology, Civil Engineering Technology, Environmental Management & Safety Department, 78 Lomb Memorial Drive, Rochester, NY 14623

Ilana S. Goldowitz, Cornell University, Plant Science Dept, Plant Sciences Building, Ithaca, NY 14853

Lucas A. Hellerich, Metcalf & Eddy, Inc., 860 North Main Street Extension, Wallingford, CT 06392

Robert Howe, Tetra Tech, Inc., 4940 Pearl East Circle, Suite 100, Boulder, CO 80301

Michael Huesemann, Pacific Northwest National Laboratory, Marine Sciences Lab, 1529 West Sequim Bay Rd, Sequim, WA 98382

Meer Husain, Kansas Dept. of Health and Env., 130 S. Market, 6th Floor, Wichita, KS 67202

Krasimir I. Ivanov, Univeristy of Agriculture, Dept of Chemistry, Mendeleev Street 12, Plovdiv, 4000, Bulgaria

Q. Jaradat, Chemistry Department, Faculty of Science, Mutah University, Al-Karak, Jordan

Mark D. Kauffman, ENSR International, 2 Technology Park Drive, Westford, MA 01886

Dhanuja Kumar, Department of Environmental and Civil Engineering, Texas A&M University-Kingsville, MSC 213, Kingsville, TX 78363

John C. LaChance, Terra Therm, 356 Broad St., Fitchburg, MA 01420

Daniele Lausdei, Golder Associates S.r.l., Via Mesina, 25, 00198 Rome, Italy

Christopher Lawless, Johnson Wright, Inc., 3687 Mt. Diablo Blvd, Suite 330, Lafayette, CA 94549

Gregory V. Lowry, Carnegie Mellon University, Department of Civil & Environmental Engineering, 119 Porter Hall, Pittsburgh, PA 15213-3890

David Ludwig, Blasland, Bouck & Lee, Inc., 326 First Street, Suite 200, Annapolis, MD 21403-2678

Edward Lynch, Wisconsin Department of Natual Resources (RR/3), 101 S. Webster Street, Masison, WI 53701

Ron Marnicio, Tetra Tech FW, Inc., 133 Federal Street, 6th Floor, Boston, MA, 617-457-8262

A. Marquette, Lousiana State University, Department of Chemical Engineering

A. Massadeh, Jordan University of Science and Technology (JUST), Dept of Medicinal Chemistry and Pharmcognosy, Faculty of Pharmacy, PO Box 3030, Irbid 22110 Jordan

Nicolas Moreau, Biogenie Corporation, 2085 Quaker Pointe Dr, Quakertown, PA 18951

Douglas Mose, Center for Basic and Applied Sciences, Center for Basic and Applied Sciences, 20099 Camp Road, Culpepper, VA 22701

P.J. Murphy, Carnegie Mellon University, Department of Civil and Environmental Engineering, 119 Porter Hall, Pittsburgh, PA 15213-389

George Mushrush, Center for Basic and Applied Science, 20099 Camp Road, Culpepper, VA 22701

Stuart Nagourney, NJDEP, Office of Quality Assurance, PO Box 424, Trenton, NJ 08625

Bruce R. Nelson, Malcolm Pirnie, Inc., 43 British American Boulevard, Latham, NY 12110

Ian T. Osgerby, US Army Corp of Engineers New England District, 696 Virginia Road, Concord, MA 01742

Zaki Parrish, Connecticut Agricultural Experiment Station, Dept of Soil and Water, 123 Huntington Street, New Haven, CT 06504

Timothy S. Patterson, Environment Canada, National Water Research Institute, 867 Lakeshore Road, PO Box 5050, Burlington, Ontario, L7R 4A6 Canada

John Peckenham, University of Maine, 102 Norman Smith Hall, Orono, ME 04469

Karen Pelto, Riverways Programs, Dept of Fish and Game, 251 Causeway Street, Suite 400, Boston, MA 02114

Brent Peyton, Chemical Engineering Department, Washington State University, Pullman, WA 99164

Michael Pouliot, Biogenie Corporation, 2085 Quaker Pointe Dr, Quakertown, PA 18951

Srilakhmi Ramaraju, Dept. of Environmental and Civil Engineering, MSC 213, Texas A&M University-Kingsville, Kingsville, TX 78363

Danny D. Reible, The University of Texas at Austin, Bettie Margaret Smith Chair of Environmental Health Engineering, Civil Engineering Department, 1 University Station C1700, Austin, TX 78712-0283

Ronald Richards, Stone and Webster Massachusetts, Shaw Environmental and Infrastructure, 100 Technology Center Drive, Stoughton, MA 02072-4705

Robert Riley, Pacific Northwest National Lab, Marine Sciences Lab, 1529 West Sequim Bay Rd, Sequim, WA 98382

Carl Rodzewich, Biogenie Corporation, 2085 Quaker Point Dr, Quakertown, PA 18951

Jonathan Rubin, University of Maine, Senator Margaret Chase Smith Center for Public Policy, Orono, ME

George Saiway, Chemistry Department, George Mason University, Fairfax, VA 22030

Cristen Sardano, Stone and Webster Massachusetts, Shaw Environmental and Infrastructure, 100 technology Center Drive, Stoughton, MA

Denise M. Savageau, Town of Greenwich, Conservation Director, 101 Field Point Road, Greenwich, CT 06830

Steven D. Schroeder, RMT, Inc., 100 Verdae Boulevard, Greenville, SC 29607

Richard C. Schwenger, Noranda, Inc., PO Box 200, Bathurst, NB E2A 3Z2 Canada

Fiorella Simoni, George Mason University, Chemistry Department, Fairfax, VA 22030

Michael Sklash, Dragun Corporation, 30445 Northwestern Highway, Suite 260, Farmington Hills, MI 48334

Carl Tammi, ENSR International, 2 Technology Park Drive, Westford, MA 01886

LeeAnn M.H. Thomas, Canadian Pacific Railway, 501 Marquette Avenue, Suite 804, Minneapolis, MN 55402

Christopher Thompson, Pacific Northwest National Laboratory, Marine Sciences Lab, 1529 West Sequim Bay Rd, Sequim, WA 98392

C. Brian Trask, Illinois State Geological Survey, Environmental Site Assessments Section, 615 E. Peabody Dr., Champaign, IL 61820-6964

Andrea M. Traviglia, ENSR International, 2 Technology Park Drive, Westford, MA 01886

Stephen P. Truchon, Blasland, Bouck & Lee, Inc., 326 First Street, Suite 200, Annapolis, MD 21403-2678

Michael Truex, Pacific Northwest National Lab, Marine Sciences Lab, 1529 West Sequim Bay Rd, Sequim, WA 98382

Lester Tyrala, Stone and Webster Massachusetts, Shaw Environmental and Infrastructure, 100 Technology Center Drive, Stoughton MA

Venkatesh Uddameri, Dept. of Environmental and Civil Engineering, MSC 213, Texas A&M University-Kingsville, Kingsville, TX 78363

Scott A. Underhill, Malcolm Pirnie, Inc., 15 Cornell Road, Latham, NY 12110

Mark Velleux, Colorado State University, Dept of Civil Engineering, A211 Engineering Research Center, Fort Collins, CO 80523-1372

James H. Vernon, ENSR International, 2 Technology Park Drive, Westford, MA 01886

Andrew R. Vitolins, Malcolm Pirnie, Inc., 15 Cornell Road, Latham, NY 12110

Zheming Wang, Pacific Northwest National Lab, Marine Sciences Lab, 1529 West Sequim Bay Rd, Sequim, WA 98382

Barbara Weir, Metcalf & Eddy, Inc., 701 Edgewater Drive, Wakefield, MA 01880

Jason C. White, The Connecticut Agricultural Experiment Station, Department of Soil and Water, 123 Huntington Street, New Haven, CT 06504

Alborz Wozniak, Johnson Wright, 3687 Mt Diablo Blvd, Suite 330, Lafayette, CA 94549

Kathy Yager, US EPA, Technology Innovation Office, 11 Technology Drive, Chelmsford, MA 01863

Penka S. Zaprjanova, Institute of Tobacco and Tobacco Products, Markovo, 4108, Bulgaria

Alex J. Zeman, Environment Canada, National Water Research Institute, 867 Lakeshore Road, PO Box 5050, Burlington, Ontario, L7R 4A6 Canada

John Zupkus, Massachusetts Department of Environmental Protection, BWCC, NE Regional Office, 1 Winter Street, Boston, MA 02108

Acknowledgments

We wish to thank all agencies, organizations and companies that sponsored the conference. Without their generosity and assistance, the conference and this book would not have been possible.

Benefactors

ENSR International
Environmental Remediation and Financial Services, LLC
MA DEP
Massachusetts Riverways Program
Regenesis Bioremediation Products
US EPA, ORD, National Risk Management Research Laboratory

Sponsors

AMEC Earth and Environmental Inc.
American Petroleum Institute
BBL
U.S. Army Engineer Research and Development Center
US EPA, Office of Superfund Remediation and Technology
Innovation (OSRTI)

Supporters

3M
Adventus Americas
Alpha Analytical Labs
DPRA, Inc.
EA Engineering, Science & Technology

Geovation Technologies, Inc.
K-V Associates
LSP Association
New York DEC
Security, Emergency Preparedness and Response Institute (SEPRI)
Tighe & Bond, Inc.

In addition, we express our deepest appreciation to the members of the Scientific Advisory Boards. The tremendous success of the conference has been result of the dedication and hard work of our board members.

Scientific Advisory Board

Nader Al-Awadhi	*Kuwait Institute for Scientific Research*
Akram N. Alshawabkeh	*Northeastern University*
Ernest C. Ashley	*Camp, Dresser & McKee, Inc.*
Alan J.M. Baker	*The University of Melbourne*
Ralph S. Baker	*Terra Therm, Inc.*
Ramon M. Barnes	*University Research Institute for Analytical Chemistry*
Michael Battle	*EA Engineering, Science & Technology*
Bruce Bauman	*American Petroleum Institute*
Mark Begley	*Environmental Management Commission*
Gary Bigham	*Exponent Environmental Group*
Scott R. Blaha	*GE Global Research Center*
Carol de Groot Bois	*Bois Consulting Company*
Clifford Bruell	*University of Massachusetts Lowell*
Peter Cagnetta	*SIAC*
Barbara Callahan	*University Research*
Robert H. Clemens	*AMEC Earth & Environmental, Inc.*
Andrew Coleman	*Electric Power Research Institute*
Janine Commerford	*MA Department of Environmental Protection*
Abhijit V. Deshmukh	*University of Massachusetts Amherst*
Mary Donohue	*Environmental Forensics*
James Dragun	*The Dragun Corporation*
John W. Duggan	*Wentworth Institute of Technology*
Mohamed Elnabarawy	*3M (retired)*
Timothy E. Fannin	*US Fish and Wildlife Service*
Samuel P. Farnsworth	*AMEC Earth & Environmental, Inc.*
Linda Fiedler	*US EPA*
Kevin T. Finneran	*GeoSyntec Consultants*

John Fitzgerald	*MA Department of Environmental Protection*
Millie Garcia-Surette	*MA Department of Environmental Protection, South East Region*
Connie Gaudet	*Environment Canada, Soil & Sediment Quality Section*
Steve Goodwin	*Associate Dean, University of Massachusetts Amherst*
Michael Gorski	*MA Department of Environmental Protection*
Peter R. Guest	*Parsons*
Eric Hince	*Geovation Technologies, Inc.*
Ihor Hlohowskyj	*Argonne National Laboratory*
Duane B. Huggett	*Pfizer, Inc.*
Barry L. Johnson	*Assistant Surgeon General (ret.)*
Evan Johnson	*Tighe & Bond*
William B. Kerfoot	*Kerfoot Technologies, Inc.*
Stephen S. Koenigsberg	*Regenesis Bioremediation Products*
Bill Kucharski	*Ecology & Environment, Inc.*
Cindy Langlois	*Journal of Children's Health*
Steven C. Lewis	*University of Medicine & Dentistry of New Jersey*
Dave Ludwig	*Blasland, Bouck & Lee, Inc.*
Ronald J. Marnicio	*TetraTech FW, Inc.*
Rick McCullough	*MA Turnpike Authority*
Chris Mitchell	*ENSR*
Robert Morrison	*DPRA, Inc.*
Ellen Moyer	*Greenvironment, LLC*
Willard Murray	
Lee Newman	*University of South Carolina*
Gopal Pathak	*Birla Institute of Technology*
Brad Penney	
Frank Peduto	*Spectra Environmental Group*
Ioana G. Petrisor	*DPRA, Inc.*
David Reckhow	*University of Massachusetts Amherst*
Corinne E. Schultz	*RED Technologies, LLC*
Nicholas P. Skoulis	*Arch Chemicals, Inc.*
Frank Sweet	*ENSR International*
Christopher Teaf	*Florida State University*
James C. Todaro	*Alpha Analytical Labs*
Allen D. Uhler	*Newfields -Environmental Forensics Practice, LLC*
Mark Vigneri	*Environmental Remediation and Financial Services, LLC*
A. Dallas Wait	*Gradient Corporation*

Richard Waterman   *EA Engineering, Science, and Technology, Inc.*
Jason C. White   *The Connecticut Agricultural Experiment Station*
Katie Winogroszki   *3M*
Peter Woodman   *Risk Management Incorporated*
Baoshan Xing,   *University of Massachusetts*
Ed Zillioux   *FPL Environmental Services*

Federal Advisory Board

An appreciated acknowledgement to **Denise Leonard**, Conference Coordinator, for her time and energy contributed throughout the year to the organization of The Annual International Conference on Soils, Sediment & Water at the University of Massachusetts, Amherst. Also acknowledgement to her assistant, **Holly Dodge,** for her support over the last year in her contribution to the compilation of this book and other aspects pertaining to the conference.

About the Editors

Edward J. Calabrese is a board certified toxicologist and professor of toxicology at the University of Massachusetts School of Public Health at Amherst. Dr. Calabrese has researched extensively in the area of host factors affecting susceptibility to pollutants and has authored more than three hundred papers in scholarly journals, as well as twenty-four books, including: Principles of Animal Extrapolation; Nutrition and Environmental Health, Vols. 1 and 2; Ecogenetic: Safe Drinking Water Act: Amendments. Regulations, and Standards; Soils Contaminated by Petroleum: Environmental and Public Health Effects; Petroleum Contaminated Soils, Vols. 1, 2 and 3; Ozone Risk Communication and Management; Hydrocarbon Contaminated Soils, Vols. 1, 2, 3, 4 and 5; Hydrocarbon Contaminated Soils and Groundwater, Vols. 1, 2, 3, and 4; Multiple Chemical Interactions; Air Toxics and Risk Assessment; Alcohol Interactions with Drugs and Chemicals; Regulating Drinking Water Quality; Biological Effects of Low Level Exposures to Chemicals and Radiation; Contaminated Soils; Diesel Fuel Contamination; Risk Assessment and Environmental Fate Methodologies; Principles and Practices for Petroleum Contaminated Soils, Vols. 1, 2, 3, 4, and 5; Contaminated Soils, Vol. 1; and Performing Ecological Risk Assessments. He has been a member of the U.S. National Academy of Sciences and NATO Countries Safe Drinking Water Committees, and the Board of Scientific Counselors for the Agency for Toxic Substances and Disease Registry (ATSDR). Dr. Calabrese also serves as Director of the Northeast Regional Environmental Public Health Center at the University of Massachusetts, Chairman of the BELLE Advisory Committee and Director of the International Hormesis Society.

Paul T. Kostecki, Vice Provost for Research Affairs, University of Massachusetts at Amherst and Associate Director, Northeast Regional Environmental Public Health Center, School of Public Health, University of Massachusetts at Amherst, received his Ph.D. from the School of Natural Resources at the University of Michigan in 1980. He has been involved with human and ecological risk assessment and risk management research for the last 13 years. Dr. Kostecki has co-authored and co-edited over fifty articles and sixteen books on environmental assessment and cleanup including: remedial Technologies for Leaking Underground Storage Tanks; Soils Contaminated by Petroleum Products; Petroleum Contaminated Soils, Vols. 1, 2, and 3; Hydrocarbon Contaminated Soils and Groundwater, Vols. 1, 2, 3 and 4; Hydrocarbon Contaminated Soils, Vols. 1, 2, 3, 4 and 5; Principles ad Practices for Petroleum Contaminated Soils; Principles and Practices for Diesel Contaminated Soils, Vols. 1, 2, 3, 4 and 5; SESOIL in Environmental Fate and Risk modeling; Contaminated Soils, Vol. 1; and Risk Assessment and Environmental Fate Methodologies. Dr. Kostecki also serves as Associate Editor for the Journal of Soil Contamination, Chairman of the Scientific Advisory Board for Soil and Groundwater Cleanup Magazine, as well as an editorial board member for the journal Human and Ecological Risk Assessment.

In addition, Dr. Kostecki serves as Executive Director for the Association for the Environmental Health of Soils (AEHS). He is a member of the Navy's National Hydrocarbon Test Site Advisory Board and a member of the Steering Committee for the Total Petroleum Hydrocarbon Criteria Working Group and the Association of American Railroads Environmental Engineering and Operations Subcommittee.

James Dragun, as a soil chemist (Ph.D. Penn State University in Agronomy), has a rich background in the fate of hazardous materials in soil systems and groundwater. He has assessed the migration and degradation of chemicals and waste of national concern in soil-groundwater systems such as dioxin, PBB, Radionuclides at Three Mile Island Nuclear Power Plant, PCB and petroleum spills, organics and inorganics at the Stringfellow Acid Pits, pesticides in San Joaquin Valley groundwater, and solvents in Silicon Valley groundwater. In addition, he has analyzed engineering controls designed to prevent the leakage of chemicals and wastes from landfills, surface impoundments, deepwell injection systems, underground storage tanks, land treatment systems, manufacturing and processing facilities, and hazardous waste sites. He has analyzed the chemical integrity and reactivity of materials used to treat and store hazardous and nonhazardous wastes. He has served as an expert reviewer of over 40 projects and programs involving the siting, design, construction, performance, and failure mechanisms of landfills, land treatment systems, surface impoundments, and waste piles. In addition, he has authored exposure assessments for over 100 chemicals and wastes.

Widely recognized for his expertise, Dr. Dragun was appointed the primary technical advisor on exposure assessment to the Interagency Testing Committee, a consortium of 14 federal agencies that selects chemicals for potential regulatory control. He directs the Association of Official Analytical Chemist's development of standard methods to measure the migration and degradation of chemicals and wastes, and has authored test methods that are used today by environmental laboratories in North America, Europe, and Asia. His counsel and scientific findings have been disseminated and utilized by 24 nations including Japan, Canada, the United Kingdom, Australia, West Germany, Switzerland, Italy, France, Spain, Scandinavia, and the Netherlands.

Dr. Dragun is a member of Phi Kappa Phi and Sigma Xi, both international honorary scientific societies, and was awarded the U.S. EPA Bronze Medal for distinguished service in 1980.

PART 1: ENVIRONMENTAL FATE

CHAPTER 1

SLOW DESORPTION OF PHENANTHRENE FROM SILICA PARTICLES: INFLUENCE OF PORE SIZE, PORE WATER, AND AGING TIME

Michael H. Huesemann[1], Timothy J. Fortman[1], Robert G. Riley[1], Christopher J. Thompson[1], Zheming Wang[1], Michael J. Truex[1], and Brent Peyton[2]

[1]*Marine Sciences Laboratory, Pacific Northwest National Laboratory, 1529 West Sequim Bay Rd, Sequim, WA 98382;* [2]*Chemical Engineering Department, Washington State University, Pullman, WA 99164*

Abstract: When micro-porous and meso-porous silica particles were exposed to aqueous phenanthrene solutions for various durations it was observed that sorbed-phase phenanthrene concentrations increased with aging time only for meso-porous but not micro-porous silicas. Desorption equilibrium was reached almost instantaneously for the micro-porous particles while both the rate and extent of desorption decreased with increasing aging time for the meso-porous silicas. These findings indicate that phenanthrene can be sequestered within the internal pore-space of meso-porous silicas while the internal surfaces of micro-porous silicas are not accessible to phenanthrene sorption, possibly due to the presence of physi- or chemi-sorbed water that may sterically hinder the diffusion of phenanthrene inside water-filled micro-pores. By contrast, the internal surfaces of these micro-porous silicas are accessible to phenanthrene when incorporation methods are employed which assure that pores are devoid of physi-sorbed water. Consequently, when phenanthrene was incorporated into these particles using either supercritical CO_2 or via solvent soaking, the aqueous desorption kinetics were extremely slow indicating effective sequestration of phenanthrene inside micro-porous particles. Finally, a two-

compartment conceptual model is used to interpret the experimental findings
and the implications for contaminant fate and transport are discussed.

Key words: Contaminant Fate and Transport, Contaminant Sequestration, Desorption
Kinetics, Aging Methodology, Porous Silica, Phenanthrene, Two-
Compartment Model.

1.   INTRODUCTION

The remediation of aged hydrophobic contaminants in soils and
sediments has in many cases been complicated by the extremely slow or
incomplete release of these compounds from the mineral particles. It has
been postulated that the slow desorption and related sequestration of these
hydrophobic contaminants is most likely caused by several mechanisms such
as the slow diffusion within either hard or soft organic matter domains or by
sorption-retarded and sterically hindered diffusion in small mineral pores
(Alexander, 1995; Huesemann, 1997; Hatzinger and Alexander, 1997; Luthy
et al., 1997; Pignatello, 1990; Pignatello and Xing, 1996; Steinberg et al.,
1987; Weber and Huang, 1996; Weber et al., 1998; Xing and Pignatello,
1997).

Considering that most naturally occurring soils and sediments contain
significant amounts of organic matter, it is not surprising that most research
has focused on elucidating the nature of contaminant sequestration in the
various organic matter (OM) phases. In fact, it has been suggested by
Cornellissen et al. (1998) that the presence of OM is more important for
slow desorption than mineral micropores in soils and sediments with more
than 0.1-0.5% OM. As a result, comparatively little contaminant
sequestration research has been carried out to evaluate the role of mineral
micropores in the absence of OM (Nam and Alexander, 1998; Huang et al.,
1996; Farrell and Reinhard, 1994a, 1994b; Alvarez-Cohen et al., 1993;
Werth and Reinhard, 1997a, 1997b; McMillan and Werth, 1999).

Huang et al. (1996) studied the aqueous sorption and desorption of
phenanthrene in meso-porous silica gels (40Å, 100Å, and 150Å) and found
that little or no phenanthrene sorption occurred on internal pore-surfaces.
These investigators hypothesized that the presence of physi-sorbed water in
silica pores results in the size-exclusion of phenanthrene from the interior
pore space. They therefore concluded that the use of models that invoke
solute diffusion in meso- and micro-porous mineral structures as a
significant rate-limiting factor for sorption by soils and sediments is highly
questionable. Nam and Alexander (1998) measured the biodegradation
kinetics of phenanthrene that had been incorporated onto non-porous and

meso-porous (25Å, 60Å, and 150Å) silica particles via aqueous sorption. Since no significant difference in biodegradation rates between non-porous and porous silicas was observed, these investigators also concluded that the internal surfaces of these porous beads sorb little or no phenanthrene.

It is the objective of this research to further elucidate the various factors that affect slow desorption and sequestration of hydrophobic contaminants in mineral micro- and meso-pores in the absence of organic matter. Specifically, we are interested in how the pore-diameter, the presence (or absence) of water during the phenanthrene incorporation process, and the aging time influence the aqueous desorption kinetics of phenanthrene from silica particles.

2. MATERIAL AND METHODS

2.1 Silica Particles

The types of silica particles used in this study are listed in Table 1. Four batches of meso-porous silica particles ranging in size from 1 –10µ and median pore diameter (based on pore-volume) from 18Å to 76Å were synthesized using techniques similar to those described by Bruinsma et al. (1998) and Beck et al. (1992). Cetyltrimethylammonium chloride was used in combination with tetraethoxysilane to prepare the particles in batches 1 and 2. A cetyltrimethylammonium hydroxide/cetyltrimethylammonium chloride mixture in combination with sodium aluminate and mesitylene was employed to prepare the silicas in batches 3 and 4. The synthesized particles were calcined by heating using a temperature ramp from 20°C to 540°C under a nitrogen purge. Prior to use in the experiments, the cooled particles were ground lightly with a mortar and pestle to break up large aggregates.

Table 1. Physical and Chemical Properties of Silica Particles

Particle Type	Particle Diameter (µ)	Median Pore Diameter (Å)	Surface Area (m²/g)	TOC (%, w/w)
Batch #1	1 – 10	18	825	0.0031
Batch #2	1 – 10	21	755	0.0096
Batch #3	1 – 10	76	858	0.0047
Batch #4	1 – 10	66	845	0.0068
Davisil	250 –500	202	314	0.0078
Spheriglass	2	NA	2	ND

NA = Not Applicable, ND = Not Determined

Davisil silica gel with a median pore diameter of 202Å and a particle size range of 250-500µ was purchased from Supelco, Bellefonte, PA. Finally, non-porous silica beads (i.e., spheriglass solid spheres) with a mean particle size of 2μ were purchased from Potters Industries, Inc., Carlstadt, NJ.

The total organic carbon (TOC) content of the silica particles was determined by placing an aliquot into a platinum crucible and heating it at 550°C for 16.5 hours. The carbon dioxide that was released as a result of this oxidation process was catalytically converted to methane which was subsequently analyzed by gas chromatography. The BET surface area and the pore-diameter distribution (based on pore-volume) of the silica particles was determined using a Micrometrics Surface Area Analyzer (Model 2010 Micrometrics Instrument Corp., Norcross, GA) according to procedures given in the operating manual (Micrometrics, 1995).

2.2   Hydration of Silica Particles

Preliminary aqueous sorption experiments involving dry silica particles from batches 1 and 2 indicated that phenanthrene sorption processes are significantly affected by changes in the silica surface chemistry that occur slowly when dry silica is exposed to water. In order to eliminate this confounding factor, all silica particles used in aqueous sorption experiments were pre-wetted in de-ionized water for one week. After the wetting period, the supernatant was carefully removed, and the sorption experiments were initiated by adding aqueous phenanthrene solution as described in more detail below.

In addition, all silica particles that were loaded with phenanthrene using non-aqueous methods (see details below) were also hydrated prior to phenanthrene loading to avoid the unusual aqueous desorption kinetics that are due to changes in silica surface chemistry. Five to ten grams of silica particles were equilibrated with 150 mL of de-ionized water over a period of 3 to 4 days at room temperature. The equilibrated particles were then filtered and dried under house vacuum in a dessicator containing Drierite for 5 to 6 days. The loss of water was monitored during the drying process. Drying was terminated when the weight of silica closely approximated its original starting weight. Additional water was then removed by subjecting the silica particles to high vacuum (4 to 5 $X10^{-6}$ torr) for a period of 5 to 6 days. This procedure is known to remove all physi-sorbed water while chemi-sorbed water remains on the silica surfaces (Young, 1958).

2.3   Phenanthrene Sorption

2.3.1   Preparation of Phenanthrene Stock Solution

The phenanthrene stock solution was prepared as follows. 100 mg of ultrapure (99.5%+) phenanthrene (Aldrich Chemical Company) was placed into a small polyethylene bag (1.5" X 1.5", 4 mil). After the addition of 10 mL hexane (95% pure, Burdick and Jackson Chemical Company), the polyethylene bag was heat-sealed. The bag was slowly inverted until all phenanthrene crystals were dissolved. The hexane was evaporated by placing the bag into a hood for 2-3 days.

The bag was then transferred to an amber glass bottle (ca. 3.8 L) filled with a buffered (pH 7) solution containing 5 mg/L sodium bicarbonate (NaHCO$_3$) and 100 mg/L sodium azide (NaN$_3$) dissolved in de-ionized water (Huang et al., 1996). The bottle was sparged with nitrogen, capped, and then placed on a magnetic stirrer. The submerged polyethylene bag was mixed within the bottle until the aqueous phenanthrene concentration reached after 9 days equilibrium levels at 874 ug/L, which is close to the maximum reported solubility of this compound (Mackay et al., 1992). Aliquots of this phenanthrene stock solution were used in all sorption experiments.

This particular procedure was developed to assure that the aqueous solution is truly free of phenanthrene crystals that have been known to negatively affect the reproducibility of sorption and desorption experiments. In addition, in this method the use of solvents (e.g., methanol) that are commonly used to dissolve phenanthrene prior to the spiking of water was also avoided, thereby eliminating any potential negative influences that a co-solvent may have on sorption kinetics or equilibria.

2.3.2   Sorption Experiments

All sorption experiments were carried out in 30 mL amber centrifuge glass tubes with screw caps and Teflon-lined silicone septa. Prior to use, the glass tubes were ashed at 450°C for 4 hours to remove any potential organic materials that may interfere with phenanthrene sorption to silica particles. To initiate sorption experiments, 20 mL of the phenanthrene stock solution (874 ug/L) was added to 0.2 grams of silica particles (18Å, 76Å, and 202Å) that had been placed inside the glass tube. Thus, the water-to-solids ratio was equal to 100 in all sorption experiments.

During the sorption studies, the centrifuge tubes were tightly capped, covered with paper towels to protect against potential photo-oxidation of phenanthrene by fluorescent laboratory lights, and placed on a modified rock roller (Model NF-1, Lortone Inc.) @100 to 250 rpm for mixing. At specified sampling times, the tubes were taken from the rollers and centrifuged at 4000 rpm (2960g) for 5 minutes. A supernatant sample (0.1mL) was taken from each tube and analyzed for phenanthrene as outlined in the Phenanthrene Analysis section. The glass tubes were again tightly capped and placed back on the roller until the next sampling event. For the "time zero" measurements, the glass tubes were briefly mixed manually (i.e., they were not put on the roller) and placed immediately in the centrifuge. In this case, the total time for mixing, centrifugation, and sampling took ca. 15 minutes.

A detailed mass balance calculation was carried out for each tube to determine the sorbed-phase phenanthrene concentration at termination of the sorption experiments. Thus, the mass of sorbed phenanthrene was computed as the initial mass of phenanthrene added to each tube minus any phenanthrene that was either removed via sampling or remained dissolved in the supernatant. An acetonitrile extraction of tubes and septa used in sorption experiments indicated that the mass of phenanthrene sorbed to glass walls or septa is neglible (< 0.1 ug per tube, or equivalently <5% (wt) of the initial mass of phenanthrene). In addition, control experiments carried out with tubes containing no silica particles confirmed that the observed decreases in aqueous phenanthrene concentrations are due to sorption and are not caused by biodegradation, photo-oxidation, or volatilization.

2.4   Incorporation of Phenanthrene into Silica Particles Using Non-Aqueous Methods

In the aqueous sorption methods outlined above, the pores of all silica particles were filled with water. In order to determine whether the presence of water has any significant effect on phenanthrene sequestration, we used the following three different "non-aqueous" methods to incorporate phenanthrene into the internal pore space of the silica particles in the absence of pore water. (Note: As outlined above, all silica particles were subjected to a specific hydration procedure that assured the elimination of all physi-sorbed water from the pores.)

2.4.1   Incorporation of Phenanthrene into Silica Particles Using Supercritical Carbon Dioxide

Phenanthrene (Aldrich, zone-refined) was used for all supercritical fluid (SCF) loading experiments. The SCF system consisted of a Dionex model SFE-703 supercritical extraction instrument that was modified to circulate supercritical carbon dioxide in a closed loop (Riley et al., 2001). Included in the closed loop system was a high-pressure stainless steel vessel (10 mL,

Keystone Scientific) used to dissolve the phenanthrene in supercritical carbon dioxide. A second vessel (10 mL) in the system contained the silica particles. An Eldex model B-100-S HPLC pump was used to circulate the supercritical solution through the closed loop system and a Shimadzu UV-2401PC spectrophotometer equipped with a custom-mounted high-pressure flow cell (Shimadsu SPD–M6A) was employed to monitor real-time changes in phenanthrene concentrations during loading.

The general procedure for the SCF incorporation of phenanthrene into the silica particles was as follows. Before starting the experiment, the two high-pressure vessels were removed from the system and loaded with appropriate amounts of silica (substrate vessel) and phenanthrene (sorbate vessel). After re-installing the vessels, the SFE-703 oven chamber was maintained at 30°C and the system was pressurized at 300 atm (4409 psi) with SFE-grade carbon dioxide. The supercritical carbon dioxide was then pumped through the sorbate vessel until all phenanthrene had been dissolved in CO_2 as indicated by a stabilized UV absorbance reading. Following baseline stabilization, valves were switched to allow the phenanthrene containing supercritical CO_2 to contact the silica particles in the substrate vessel. The solution was pumped through the substrate vessel for four hours. This contact time was long enough to ensure that phenanthrene in the circulating supercritical CO_2 reached a steady-state concentration as indicated by a stabilized UV absorbance measurement. The circulating pump was then turned off, a valve was switched to depressurize the system, and the loaded silica particles were removed for use in aqueous desorption experiments. Using these procedures, rehydrated 21Å, 66Å and 202Å silica particles were loaded with phenanthrene resulting in final solid-phase concentrations of 2.9 ug/g, 5.7 ug/g, and 2.0 ug/g, respectively.

2.4.2   Incorporation of Phenanthrene into Silica Particles Using Solvent Soaking

Approximately 0.8 grams of rehydrated 21Å, 66Å and 202Å particles were each fully submerged in 10 mL methylene chloride containing 20 ug, 40 ug and 14 ug dissolved phenanthrene, respectively. The resulting slurry was mixed on a shaker table (@ 100 rpm) for four hours. A gentle stream of nitrogen was then used to evaporate the solvent while stirring the slurry periodically with a spatula until a constant weight was reached (ca. 3 hours). A subsample of the phenanthrene loaded particles was taken and analyzed for phenanthrene as outlined below. The solid-phase phenanthrene concentrations for the 21Å, 66Å and 202Å particles were 8.5 ug/g, 11.0 ug/g, and 1.1 ug/g, respectively (Note: g dry weight). All silica particles were placed into a freezer (-20 °C for 5 days) to immobilize the phenanthrene until the initiation of desorption experiments.

2.4.3   Incorporation of Phenanthrene into Silica Particles Using Solvent Spiking with Aging

This method involves the spiking of a small volume of solvent containing phenanthrene onto silica particles and the subsequent addition of water for moisture adjustment. It should be recognized that this phenanthrene incorporation procedure is a hybrid between a non-aqueous spiking procedure and an aqueous sorption experiment. It is most likely that phenanthrene is deposited on the outside surfaces of the silica particles during the spiking procedure while the subsequent addition of moisture will not only fill the pores with water but cause the dissolution of phenanthrene which in turn enables the diffusion along pores and aqueous sorption on silica surfaces. Despite the mechanistic complexity of this phenanthrene incorporation procedure, it was decided to evaluate this method because it is the "aging" technique that is most commonly reported in the literature (Chung and Alexander, 1998; Nam and Alexander, 1998; Hatzinger and Alexander, 1995; Kelsey and Alexander; 1997).

Approximately 1 gram of rehydrated 21Å, 66Å and 202Å particles were each spiked with 10uL methylene chloride containing 3 ug, 6 ug, and 2 ug dissolved phenanthrene, respectively. In addition, ca. 1 gram of dry 18Å particles were spiked with 10uL methylene chloride containing 110ug dissolved phenanthrene. All spiked silica particles were mixed every 30 minutes with a spatula for a total duration of four hours. A gentle stream of nitrogen was then used to evaporate the solvent while stirring the slurry periodically until a constant weight was reached (ca. 30 minutes). At this point, an aqueous solution containing 2% (wt) sodium azide was added to the spiked silica particles in order to adjust the moisture content to ca. 80% of the field capacity. The silica particles were subsequently mixed with a spatula for ca. 15 minutes and then transferred to an amber glass jar. The jar was tightly capped and stored in the dark at room temperature until the initiation of desorption experiments. The 21Å, 66Å and 202Å particles were aged in this manner for 61 days whereas the 18Å particles were aged for 100 days.

At the end of the aging period, the solid-phase phenanthrene concentrations were determined as outlined below. For the 18Å, 21Å, 66Å and 202Å particles, the phenanthrene concentrations were found to be 89 ug/g, 1.0 ug/g, 2.8 ug/g, and 0.9 ug/g, respectively. The corresponding moisture contents (g water/g moist silica) for these particles were 0.65, 0.66, 0.62, and 0.55, respectively. Finally, non-porous silica beads were spiked with phenanthrene in a manner similar to the procedures employed for the 18Å particles (see above for details). However, these glass beads were not aged prior to use in desorption experiments. The phenanthrene concentration for these dry glass beads was 72 ug/g.

In order to determine the state of phenanthrene on the silica particles, fluorescence excitation and emission spectra were measured as described elsewhere (Wang et al., 2001). Results indicated that phenanthrene at these low solid-phase concentrations is present as a monomer (i.e., not crystalline) in all particles where phenanthrene was incorporated using non-aqueous methods.

2.5 Phenanthrene Desorption

In all cases where phenanthrene had been incorporated into silica particles via aqueous sorption, the glass tubes were centrifuged and most of the supernatant was carefully removed. Desorption was initiated by adding a known volume of buffered (pH 7) background solution containing 5 mg/L sodium bicarbonate ($NaHCO_3$) and 100 mg/L sodium azide (NaN_3) to each tube. The initial liquid volume in all desorption tubes was ca. 21 mL which translates into a water-to-solids ratio of 105 in all desorption experiments involving particles aged by aqueous sorption.

For the 21Å, 66Å and 202Å silica particles where phenanthrene had been incorporated using supercritical carbon dioxide, solvent soaking, or solvent spiking with long-term aging, all desorption experiments were initiated by adding 20 mL of buffered background solution to 0.3 grams of silica particles (Note: As outlined earlier, the particles loaded via SCF CO_2 and solvent soaking were dry while the long-term aged particles were moist). For the 100-day aged 18Å silica particles and spiked non-porous glass beads, desorption experiments were initiated by adding 20 mL of buffered background solution to 0.1 grams of (moist) silica particles or (dry) glass beads.

In all desorption experiments, the procedures for mixing, centrifugation, sampling, and analyses were similar to those described for the sorption studies (see details above). Relative phenanthrene concentrations were calculated by dividing the measured aqueous-phase phenanthrene concentration by the maximum achievable phenanthrene concentration that would result if all phenanthrene present in the tube at the beginning of the desorption experiment were to be dissolved in the aqueous phase.

In order to check the phenanthrene mass balance at termination of the desorption experiments, the contents of the glass tube were filtered using is porous glass frit to recover all of the silica particles which were then subsequently extracted and analyzed for phenanthrene according to the methods outlined in the Phenanthrene Analysis Section. A subsample of the wet silica particles was dried at 105°C for 24 hours and the dry weight of silica (M_{sd}) was determined gravimetrically. The volume of interstitial water (V_{iw}) was calculated using the difference in the dry and wet weight masses of the silica. The concentration of phenanthrene (C_{sd}) on dry silica was then calculated according to:

$$C_{sd} = \frac{M_p - C_{sup} \cdot V_{iw}}{M_{sd}} \qquad (1)$$

where M_p is the total mass of phenanthrene extracted from the subsample and C_{sup} is the aqueous phenanthrene concentration in the last supernatant sample (i.e., it is assumed that the phenanthrene concentration in this sample is similar to the one in the interstitial water). For most desorption experiments, the mass balance was greater than 90%.

2.6   Phenanthrene Analysis

For the analysis of aqueous-phase phenanthrene, a 0.5 mL sample of the supernatant is diluted with 0.5 mL HPLC grade acetonitrile (ACN). For the analysis of sorbed-phase phenanthrene, ca. 50 mg of silica is extracted via sonication for 40 minutes in 10 mL ACN. A 0.5 mL aliquot of the extract is then forced through a 0.2 μm Teflon syringe filter and diluted with 0.5 mL deionized water. For both types of analyses, a 100 uL aliquot of the resulting 50:50 ACN:H₂O mixture is injected using an autosampler (Spectra Physics AS 3000 HPLC Autosampler) for subsequent HPLC analysis. The HPLC instrumentation consisted of a Perkin-Elmer biocompatible Model 250 binary HPLC pump, a Perkin-Elmer Model LC-101 column oven, a Supelco Supelcosil LC-PAH HPLC column, and a Waters Model 474 fluorescence detector.

The HPLC was operated using a 75:25 ACN:H₂O solvent at a flow rate of 1.5 mL/min. The column temperature was set at 30 °C. Under these conditions, the retention time of phenanthrene was approximately 6 minutes. The fluorescence detector's excitation wavelength was set at 260 nm and the resulting signal was measured at an emission wavelength of 380 nm. Based on calibration results using known phenanthrene standards in 50:50 ACN:H₂O, this method yields a linear response for phenanthrene concentrations ranging from 0.4 ppb to 500 ppb. For aqueous samples containing phenanthrene concentrations greater than 500 ppb, the sample was further diluted with ACN to assure that phenanthrene quantification occurred within the linear range of the detector response.

3.   RESULTS

3.1   Aqueous Sorption of Phenanthrene on Silica Particles

As shown in Figure 1, the sorbed-phase phenanthrene concentration in the 202Å silica particles increases with the length of exposure to the aqueous phenanthrene solution. Thus, the sorbed-phase phenanthrene concentration was approximately 5- and 10-times greater in the 24-day and 48-day exposed particles, respectively, than in the silica particles that were subjected to aqueous sorption for only 1 hour. Similarly, 48-day-long sorption treatment increased sorbed-phase phenanthrene concentrations on the 76Å particles ca. three-fold compared to 24-day exposure. However, for these particles there was no significant change in sorbed-phase concentrations between 1 hour and 24-day-long sorption treatment. Finally, no increase in sorbed-phase phenanthrene concentrations as a function of sorption duration was observed for the 18Å particles.

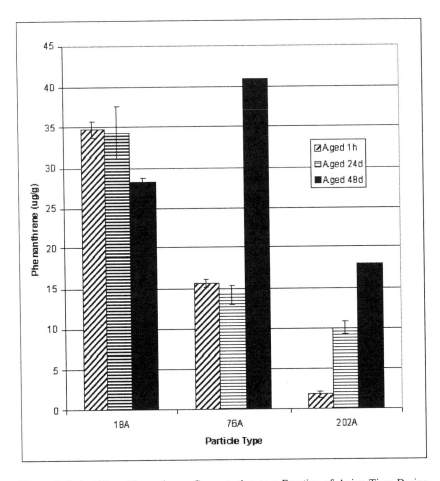

Figure 1. Sorbed-Phase Phenanthrene Concentration as a Function of Aging Time During Aqueous Sorption for 18 Å, 76Å, and 202Å Silica Particles. The Error Bars Show the High and Low Concentration Values Obtained from Two Replicate Sorption Ex periments.

3.2   Desorption of Phenanthrene from Particles Loaded By Aqueous Sorption

As shown in Figure 2, the rate and extent of phenanthrene desorption in 202Å particles is strongly influenced by the length of sorption treatment. The shorter the sorption duration, the greater the fraction of phenanthrene desorbed within the first few hours. For example, close to 100% of all sorbed-phase phenanthrene is released in the particles that had been subjected to sorption for 1 hour while only around 25% is desorbed within 24 hours for particles that had been loaded with phenanthrene for 48 days. In addition, the rate at which apparent desorption equilibrium is reached is also affected by the duration of sorption loading. The longer the sorption exposure, the longer it takes to reach the final aqueous phenanthrene concentration plateau that is indicative of equilibrium conditions. For example, the 1-hour exposed 202Å particles reached desorption equilibrium almost instantaneously while the 48-day exposed particles still had not reached a concentration plateau after more than 800 hours.

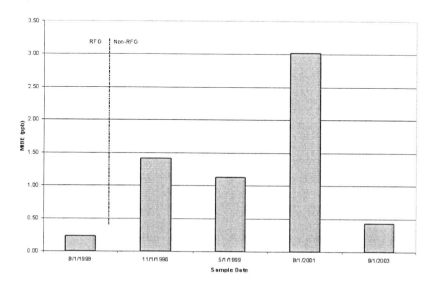

Figure 2. Aqueous Desorption of Phenanthrene from 202Å Silica Particles that had been Subjected to Aqueous Sorption for Various Durations. The Error Bars Show the High and Low Concentration Values Obtained from Two Replicate Desorption Experiments.

The 76Å particles exhibit a similar desorption behavior as the 202Å particles. As shown in Figure 3, an increase in sorption duration resulted in smaller fractions of phenanthrene to be released in the aqueous phase and longer times for phenanthrene to reach desorption equilibrium. However, compared to the 202Å particles, the 76Å particles reached equilibrium much faster. For example, the 48-day exposed 76Å particles reached equilibrium after ca. 50 hours while the corresponding 202Å particles took more than 800 hours.

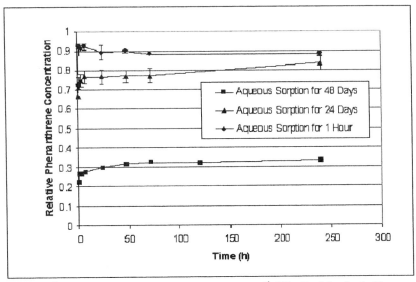

Figure 3. Aqueous Desorption of Phenanthrene from 76Å Silica Particles that had been Subjected to Aqueous Sorption for Various Durations. The Error Bars Show the High and Low Concentration Values Obtained from Two Replicate Desorption Experiments.

Finally, the 18Å particles exhibited a somewhat attenuated version of the desorption behavior that was observed for the 76Å and 202Å silicas (Figure 4). While increased sorption time also resulted in a smaller fraction of phenanthrene to be released in the aqueous supernatant, the difference between the 1-hour, 24-day and 48-day exposed particles is almost insignificant. In addition, apparent desorption equilibrium was reached in these 18Å particles at approximately the same rate in less than 24 hours for all three sorption exposure times.

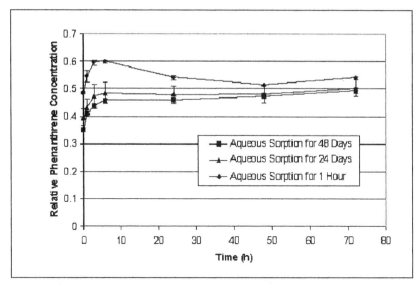

Figure 4. Aqueous Desorption of Phenanthrene from 18Å Silica Particles that had been Subjected to Aqueous Sorption for Various Durations. Th e Error Bars Show the High and Low Concentration Values Obtained from Two Replicate Desorption Experiments.

While the fraction of phenanthrene that is released within the first few hours varies greatly as a function of sorption exposure time for the 76Å and 202Å particles (Figures 2 and 3), it is interesting to note that the *mass* of initially released phenanthrene appears to be almost independent of the sorption exposure time. For example, while the fraction of initially released phenanthrene for the 76Å particles ranges from ca. 25% to 90%, the mass of phenanthrene released within the first few hours was in the range from 9.1 to 11.5 ug/g. Similarly, the initially released mass of phenanthrene is almost identical in the 24-day and 48-day sorption exposed 202Å particles, namely 3.4 and 3.6 ug/g. However, the initially released mass of phenanthrene from the 1 hour 202Å particles was slightly less, namely 1.8 ug/g.

Assuming that a true desorption equilibrium has been reached, it is possible to calculate the desorption K_d values (mL/g) according the following equation (ASTM, 1988; Scharzenbach et al., 1993; McDonald and Evangelou, 1997):

$$K_d = \frac{V}{M}\left(\frac{1}{f} - 1\right) \qquad (2)$$

where f is the fraction of phenanthrene desorbed at equilibrium and V/M is the supernatant volume to silica mass ratio employed in the desorption experiment.

Since the V/M ratio is the same (i.e., 105) in all aqueous desorption treatments, it is possible to use equation (2) to estimate and compare the different desorption equilibrium partition coefficients (K_d) from the respective "f" values that are observed in each experiment. For example, since the f value for the 18Å particles is around 0.5 for all three sorption times (see Figure 4), it follows that the length of sorption exposure does not affect the magnitude of the desorption equilibrium partition coefficient which is around 90 mL/g in all cases. By contrast, the desorption K_d value increases with the length of sorption duration for both 76Å and 202Å particles. For instance, the desorption K_d values are 14 mL/g, 21 mL/g and 221 mL/g for 76Å particles that had been subjected to sorption treatment for 1 hour, 24 days and 48 days, respectively.

3.3   Desorption of Phenanthrene from Particles Loaded by Non-Aqueous Methods

Silica particles that had been loaded with phenanthrene by three different non-aqueous methods (i.e., SCF-loading, solvent soaking, or solvent spiking with aging) exhibited markedly different desorption behavior than those silicas that were subjected to aqueous sorption treatment. It is particularly noteworthy that apparent desorption equilibrium was reached within only 6 hours for both 66Å and 202Å particles that had been loaded with phenanthrene using any of the three non-aqueous incorporation procedures (data not shown). By contrast, the desorption kinetics of phenanthrene released from 21Å particles that had been loaded using either SCF CO_2 or via solvent soaking was extremely slow. As shown in Figure 5, phenanthrene is released fast initially but equilibrium has not been reached in either experiment even after hundreds of hours of desorption treatment. In addition, the SCF-loaded particles release a smaller fraction of phenanthrene initially (within the first 24 hours) than those that were loaded via solvent soaking.

As shown in Figure 6, the 18Å particles that were spiked with phenanthrene and aged for 100 days reached desorption equilibrium (f=0.4) relatively fast: within only 50 hours. (Note: Similar results were also obtained using 21Å silicas that were aged for 61 days– data not shown). In order to evaluate whether the release of phenanthrene from the porous 18Å particles is controlled by the kinetics of phenanthrene dissolution, non-porous silica beads that had been spiked with phenanthrene were also subjected to aqueous desorption. Within only 6 hours, almost 100% of the sorbed-phase phenanthrene was released from the non-porous particles indicating that phenanthrene dissolution processes are much faster than the release rate of phenanthrene from porous 18Å particles where pore diffusion may limit the desorption kinetics.

For both the 18Å porous and non-porous particles, the desorption equilibrium partition coefficient K_d was calculated as the ratio of the measured sorbed-phase and aqueous phase phenanthrene concentrations. The average K_d value (obtained from quadruplicate concentration data) for the porous 18Å silica was found to be 51.8 mL/g (stdev = 5.3 mL/g) while it was only 1.08 mL/g (stdev = 0.8 mL/g) for the non-porous particles. Considering that the greater K_d value in porous silica particles could be related to the potentially larger surface area (SA) available for sorption compared to the non-porous bead (see also Table 1), the surface-area specific desorption equilibrium coefficients K_d^s ($=K_d/SA$) were also computed. It was found that the K_d^s value for the non-porous particles ($=0.54$ mL/m^2) was almost one order of magnitude greater than the respective value for the porous 18Å particles ($=0.063$ mL/m^2).

4. DISCUSSION

In order to facilitate the interpretation of the above experimental findings, we propose the following conceptual model:

1. The silica particles can be viewed as consisting of two compartments or domains. The "fast" compartment (F) comprises all silica surfaces that are easily accessible to phenanthrene molecules while the "slow" compartment (S) contains the remaining surfaces that are more difficult to reach (Schrap et al., 1994; Kan et al., 1994).

2. The F domain is the fraction of the silica particle that is characterized by pores large enough to enable the relatively fast transport, either via convection or unhindered diffusion, of phenanthrene to sorption sites on the silica surface.

3. By contrast, the S domain is characterized by pores whose diameter is so small as to cause sterically hindered or restricted diffusion thereby drastically reducing the effective diffusion coefficient for phenanthrene.

4. During aqueous sorption, the F-compartment fills up very fast while it may take a long time to saturate the less readily accessible sorption sites in the S-compartment. During aqueous desorption, the F- compartment empties quickly until a new "local" desorption equilibrium is reached while any phenanthrene located inside the S-compartment is only very slowly released.

5. The fraction of phenanthrene residing in the S-compartment (relative to the F domain) can be increased by lengthening the aging time or by using aging methods that increase the diffusion rate in the small pores of the S domain. The greater the fraction of phenanthrene that is "sequestered" in the S compartment, the smaller percentage is released initially and the longer it takes to reach desorption equilibrium.

With the help of this simplified conceptual two-compartment model, it is now possible to interpret the experimental data as follows. During aqueous sorption of phenanthrene onto 18Å particles, it appears that only the F-compartment was accessible to the sorbate molecules since increasing sorption exposure time did not increase the sorbed-phase phenanthrene concentration (Figure 1). Considering the relatively small molecular diameter of phenanthrene (i.e, ca. 10Å (*14*)), it is surprising that pores which have considerably larger diameters (i.e., ca. 18Å) would be inaccessible via diffusion. In fact, theoretical models (Brusseau et al., 1991; Farrell and Reinhard, 1994b; Satterfield et al., 1973) predict that the pore diffusion coefficient (D_p) for phenanthrene would be somewhat but not excessively reduced relative to the aqueous diffusion coefficient (D_a) at this particular molecular diameter to pore size ratio (i.e., if 10 Å /18Å = 1.8, then $D_p \approx$ 0.025 D_a). As was hypothesized earlier by Huang et al. (1996), it is more likely that the presence of chemi-sorbed and physi-sorbed water on pore wall surfaces effectively reduces the diameter of the pore to such an extent as to severely limit or completely block the diffusion of phenanthrene. For example, depending on the number of layers of physi-sorbed water molecules, it is possible that water films on these pore walls have thicknesses ranging from at least 9Å to more than 30Å (Huang et al., 1996). Consequently, the effective diameter available for phenanthrene diffusion may be reduced by at least 18Å if not more than 60Å relative to the actual diameter of the dry pore. Thus, in the presence of water, 18Å pores are not accessible to phenanthrene and only the F-compartment – consisting of the readily accessible surface areas of large macro- (>500 Å) or meso-pores (20Å -500Å) - is available for phenanthrene sorption.

By contrast, the sorbed-phase phenanthrene concentration increases with sorption exposure time in the 76Å and 202Å particles (Figure 1) indicating that effective pore diameters in these particles are large enough to allow slow diffusion of phenanthrene into the pore-spaces of the S-compartment. Apparently, the effective diffusion rate is so small that it takes many weeks to fill up the S-compartment. When these 76Å and 202Å particles are exposed to clean water to initiate desorption (Figure 2 and 3), the fraction of phenanthrene present in the F compartment is released from the silica within a few hours. Since the mass of initially released phenanthrene was found to be within a narrow range independently of the sorption exposure time, it can be concluded that the F compartment has a fixed sorption capacity which, in turn, is related to the limited pore surface area that is accessible to phenanthrene sorption.

The fraction of initially released phenanthrene is inversely related to the sorption exposure time, i.e., the longer the sorption duration, the larger fraction of phenanthrene that is "sequestered" in the S-compartment, and the less is released during the initial fast desorption phase. Similarly, the greater the fraction of phenanthrene that is located in the S-compartment, the longer it takes to be completely released from this domain and reach apparent desorption equilibrium. In addition, the desorption equilibrium partition coefficient K_d predictably increases with increasing phenanthrene sequestration in the S-compartment. Finally, since no phenanthrene entered the S-compartment in the 18Å particles, the same desorption equilibrium was reached very quickly within a few hours for all three sorption exposure times (Figure 4).

In contrast to the aqueous sorption experiments, phenanthrene apparently was able to be sequestered in the S-compartment of dry 21Å particles that were loaded using either supercritical CO_2 or via solvent soaking. The extremely slow desorption of phenanthrene from these particles is evidence of phenanthrene sequestration (Figure 5). Because of the absence of water during the phenanthrene incorporation process, the pores are not restricted by water layers and thus their diameters are large enough to allow for the relatively unhindered transport of phenanthrene inside the S-compartment. Upon contact with water at the initiation of desorption treatment, the water that is entering the pores is apparently not restricting the slow transfer of phenanthrene out of the particles. It may be possible that the presence of hydrophobic phenanthrene on the silica surfaces discourages the chemi- and physi-sorption of water which otherwise would sterically hinder the diffusion of this sorbate. Since the fraction of phenanthrene released initially from the F-compartment is greater for the solvent-soaked particles compared to the SCF-loaded ones, it follows that supercritical CO_2 enables the greater sequestration of phenanthrene in the S-compartment, possibly due to the increased diffusion coefficient of phenanthrene in supercritcal CO_2 compared to other solvents or water (Lee and Markides, 1990).

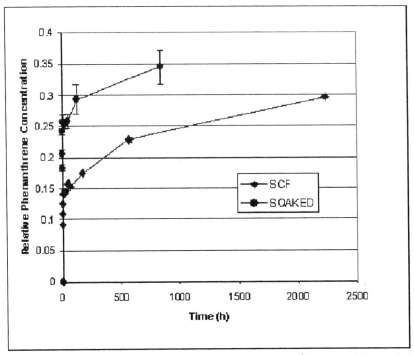

Figure 5. Long-term Aqueous Desorption of Phenanthrene from 21Å Silica Particles that had been Aged Using either SCF CO$_2$ or Via Solvent Soaking. The Error Bars Show the High and Low Concentration Values Obtained from Two Replicate Desorption Experiments.

It is interesting to note that phenanthrene desorption equilibrium was reached comparatively fast--within only 50 hours in 18Å particles that had been spiked with phenanthrene and aged after moisture adjustment for 100 days (Figure 6). This observation would indicate that even during the 100-day aging period, mostly the silica surfaces in the F-compartment and very few in the S-compartment, if any, were accessible to phenanthrene. This finding is not surprising in view of the fact that these 18Å particles were not dry but had been wetted with water prior to initiating the 100-day aging process. Consequently, the rigid water layers inside the pores sterically hindered the diffusion of phenanthrene that had been deposited, probably in the form of small micro-crystals, on the outside of the silica particles. In short, the same processes that obstructed the entry of phenanthrene into the S-compartment of the 18Å particles during aqueous sorption (Figure 1) also hindered the diffusion of phenanthrene into the smaller pores during the 100-day aging process.

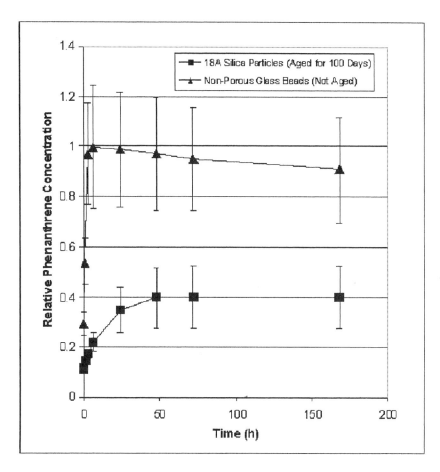

Figure 6. Aqueous Desorption of Phenanthrene from Aged 18Å Silica Particles and Non-Porous Glass Beads. The Plotted Data Represent t he Average of Measurements Taken from Four Replicate Desorption Experiments.

Another indication that only a relatively small fraction of the surface area of the porous 18Å particles was accessible to phenanthrene during the 100-day aging process can be inferred by comparing the surface-area specific desorption equilibrium coefficients (K_d^s) of both the porous 18Å silicas and their respective non-porous counterparts. In the non-porous silica beads, all of the surface area should have been accessible to phenanthrene sorption and the respective K_d^s value was 0.54 mL/m^2. Assuming that sorption processes are similar on the porous and non-porous silica surfaces, the significantly smaller K_d^s value of only 0.063 mL/m^2 that was observed in the porous particles would indicate that only a small fraction of the surface area, possibly that present in the F-compartment, was accessible to phenanthrene.

Contrary to earlier reports by Huang et al. (1996) and Nam and Alexander (1998), our experimental findings indicate that internal pore surfaces of silica particles are accessible to phenanthrene sorption. Consequently, it can be concluded that even in absence of organic matter, the small pores that are present in soil mineral particles could conceivably cause significant contaminant sequestration which in turn affects the bioavailability, biodegradability, and toxicity of these hydrophobic soil pollutants. Most likely, contaminant sequestration in small mineral pores could be particularly relevant in soils or sediments characterized by very low organic matter contents and a long contaminant aging history. However, more research is needed to determine the environmental importance of contaminant sequestration in mineral pores relative to the more commonly observed organic matter sorption phenomena in aged soils or sediments.

ACKNOWLEDGEMENTS

This research was supported by the Environmental Technology Partnerships (ETP) Program, Office of Biological and Environmental Research (BER), Office of Energy Research, U.S. Department of Energy. Pacific Northwest National Laboratory is operated for the U.S. Department of Energy by Battelle under Contract DE-AC06-76RLO 1830.

REFERENCES

Alexander, M. 1995. How Toxic are Toxic Chemicals in Soil? *Environ. Sci. Technol.* 29, 2713-2717.

Alvarez-Cohen, L., McCarty, P.L., and Roberts, P.V. 1993. Sorption of Trichloroethylene onto a Zeolite Accompanied by Methanotrophic Biotransformation. *Environ. Sci. Technol.* 27, 2141-2148.

ASTM Method E1195-87. 1988. *Standard Test Method for Determining a Sorption Constant (K_{oc}) for an Organic Chemical in Soil and Sediments* . April 1988.

Beck, J.S., Vartuli, J.C., Roth, W.J., Leonowicz, M.E., Kresge, C.T., Schmitt, K.D., Chu, C.T.-W., Loson, D.H., Sheppard, E.W., McCullen, S.B., Higgins, J.B., and Schlenker, J.L. 1992. A New Family of Mesoporous Molecular Sieves Prepared with Liquid Crystal Templates. *J. Am. Chem. Soc.* 114, 10834-10843.

Bruinsma, P.J., Kim, A.Y., Liu, J., and Baskaran, S. 1998. Mesoporous Silica Synthesized by Solvent Evaporation: Spun Fibers and Spray -dried Hollow Spheres. *Chem. Mater.* 9,2507.

Brusseau, M.L., Jessup, R.E., and Rao, P.S. 1991. Nonequilibrium Sorption of Organic Chemicals: Elucidation of Rate -Limiting Processes. *Environ. Sci. Technol.* 25,134-142.

Chung, N., and Alexander, M. 1998. Differences in Sequestration and Bioavailability of Organic Compounds Aged in Dissimilar Soils. *Environ. Sci. Technol.* , 32, 855-860.

Cornelissen, G., Van Noort, P.C.M., and Govers, H.A.J. 1998. Mechanism of Slow Desorption of Organic Compounds from Sediments: A Study Using Model Sorbents. *Environ. Sci. Technol.* 32,3124-3131.

Farrell, J., and Reinhard, M. 1994a. Desorption of Halogenated Organics from Model Solids, Sediments, and Soil under Unsaturated Conditions, 1. Isotherms. *Environ. Sci. Technol.* 28, 53-62.

Farrell, J., and Reinhard, M. 1994b. Desorption of Halogenated Organics from Model Solids, Sediments, and Soil under Unsaturated Conditions, 2. Kinetics. *Environ. Sci. Technol.* 28, 63-72.

Hatzinger, P.B., and Alexander, M. 1995. Effect of Aging of Chemicals in Soil on Their Biodegradability and Extractability. *Environ. Sci. Technol.* 29, 537-545.

Hatzinger, P.B., and Alexander, M. 1997. Biodegradation of Organic Compounds Sequestered in Organic Solids or in Nanopores Within Silica Particles, *Environ. Tox. Chem.* 16, 2215-2221.

Huang, W., Schlautman, M.A., and Weber, W.J. 1996. A Distributed Reactivity Model for Sorption by Soils and Sediments. 5. The Influence of Near-Surface Characteristics in Mineral Domains. *Environ. Sci. Technol.* 30, 2993-3000.

Huesemann, M.H. 1997. Incomplete Hydrocarbon Biodegradation in Contaminated Soils: Limitations in Bioavailability or Inherent Recalcitrance? *Biorem. J.* 1(1), 27-39.

Kan, A.T., Fu, G., and Tompson, M.B. 1994. Adsorption/Desorption Hysteresis in Organic Pollutant and Soil/Sediment Interaction. *Environ. Sci. Technol.* 28, 859-867.

Kelsey, J. W., and Alexander, M. 1997. Declining Bioavailability and Inappropriate Estimation of Risk of Persistent Compounds. *Env. Tox. Chem.* 16(3), 582-585.

Lee, M.L., and Markides, K.E., editors, *Analytical Supercritical Fluid Chromatography and Extraction*, Chromatography Conferences, Inc., Provo, UT, 1990.

Luthy, R.G., Aiken, G.R., Brusseau, M.L., Cunningham, S. D., Gschwend, P.M., Pignatello, J.J., Reinhard, M., Traina, S.J., Weber, W.W., and Westall, J.C. 1997. Sequestration of Hydrophobic Organic Contaminants by Geosorbents. *Environ. Sci. Technol.* 31, 3341-3347.

Nam, K., and Alexander, M. 1998. Role of Nanoporosity and Hydrophobicity in Sequestration and Bioavailability: Test with Model Solids. *Environ. Sci. Technol.* 32, 71-74.

Mackay, D., Shui, W.Y., and Ma, K.C., *Illustrated Handbook of Physical-Chemical Properties and Environmental Fate for Organic Chemicals* , Volume II, Lewis Publishers, 1992, p. 133.

McDonald, L.M., and Evangelou, V.P. 1997. Optimal Solid-to-Solution Ratios for Organic Chemical Sorption Experiments. *Soil Sci. Soc. Am. J.*, 61, 1655-1659.

McMillan, S.A., and Werth, C.J. 1999. Counter-Diffusion of Isotopically Labeled Trichloroethylene in Silica Gel and Geosorbent Micropores: Model Development. *Environ. Sci. Technol.* , 33, 2178-2185.

Micrometrics: ASAP 2010, *Accelerated Surface Area and Porosimetry System Operator's Manual*, Norcross, GA, 1995.

Pignatello, J. 1990. Slowly Reversible Sorption of Aliphatic Halocarbons in Soils. I. Formation of Residual Fractions. *Environ. Tox. Chem.* 9, 1107-1115.

Pignatello, J.J., and Xing, B. 1996. Mechanisms of Slow Sorption of Organic Chemicals to Natural Particles. *Environ. Sci. Technol.* 30, 1-11.

Riley, R.G., Thompson, C.J., Huesemann, M.H., Wang, Z., Peyton, B., Fortman, T., Truex, M.J., and Parker, K.E. 2001. Artificial Aging of Phenanthrene in Porous Silicas Using Supercritical Carbon Dioxide. *Environ. Sci. Technol.* . 35, 3707-3712.

Satterfield, C.N., Colton, C.K., and Pitcher, W.H. 1973. Restricted Diffusion in Liquids within Fine Pores. *AIChE J.*, 19(3), 628-635.

Schrap, S.M., Sleijpen, G.L.G., Seinen, W., an d Opperhuizen, A. 1994. Sorption Kinetics of Chlorinated Hydrophobic Organic Chemicals, Part II: Desorption Experiments. *Environ. Sci. Poll. Res.* 1, 81-92.

Schwarzenbach, R.P., Gschwend, P.M. and Imboden, D.M. *Environmental Organic Chemistry*, John Wiley and Sons, Inc., New York, NY, 1993, p. 260.

Steinberg, S.M., Pignatello, J.J., and Sawhney, B.L. 1987. Persistence of 1,2 Dibromomethane in Soils: Entrapment in Intraparticle Micropores. *Environ. Sci. Technol.* 21, 1201-1208.

Wang, Z., Friedrich, D.M.. Be versluis, M.R., Hemmer, S.L., Joly, A.G., Huesemann, M.H., Truex, M.J., Riley, R.G., and Thompson, C.J. 2001. A Spectroscopic Study of Phenanthrene Adsorption on Porous Silica . *Environ. Sci. Technol.* 35, 2710-2716.

Weber, W.J., and Huang, W. 1996. A Distributed Reactivity Model for Sorption by Soils and Sediments. 4. Intraparticle Heterogeneity and Phase -Distribution Relationships under Nonequilibrium Conditions. *Environ. Sci. Technol.* 30, 881-888.

Weber, W.J., Huang, W., and Yu, H. 1998. Hysteresis in the Sorption and Desorption of Hydrophobic Organic Contaminants by Soils and Sediments − 2. Effects of Soil Organic Matter Heterogeneity. *J. Contam. Hydrol.* 31, 149-165.

Werth, C.J., and Reinhard, M. 1997a. Effects of Temperature on Trichloroethylene Desorption from Silica Gel and Natural Sediments. 1. Isotherms. *Environ. Sci. Technol.* 31, 689-696.

Werth, C.J., and Reinhard, M. 1997b. Effects of Temperature on Trichloroethylene Desorption from Silica Gel and Natural Sediments. 2. Kinetics. *Environ. Sci. Technol.* 31, 697-703.

Xing B., and Pignatello, J.J. 1997. Dual -Mode Sorption of Low -Polarity Compounds in Glassy Poly-Vinyl-Chloride and Soil Organic Matter. *Environ. Sci. Technol.* 31, 792-799.

Young, G.J. 1958. Interaction of Water Vapor with Silica Surfaces. *J. Colloid Chem.* 13, 67-85.

PART II: HEAVY METALS

CHAPTER 2

UNDERSTANDING THE CAUSES OF AND THE PERMANENT SOLUTIONS FOR GROUNDWATER ARSENIC POISONING IN BANGLADESH

Meer T. Husain[1] and Thomas E. Bridge[2]

[1]*Kansas Dept. of Health and Env., 130 S. Market, 6th Floor, Wichita, KS 67202;* [2]*Emporia State University, Emporia, Kansas*

Abstract: The groundwater arsenic poisoning in Bangladesh is the largest disaster in the history of human civilization: more than 100 million people have been drinking arsenic-poisoned water on a daily basis. A large number of scientists believe that the groundwater arsenic poisoning in Bangladesh is a natural disaster, that the poisoning has been present for thousands of years, and that reduction of ancient soil with ferric hydroxide-bearing arsenic is the main mechanism for the mobilization for the mobilization of arsenic into groundwater. However, historical groundwater use data from the dug wells and the tube wells, historical medical data, arsenic toxicological data, hydrological, hydro geological and geochemical parameters reject the reduction hypothesis an d suggest that the groundwater arsenic poisoning in Bangladesh is a recent, man-made disaster and that exposure and oxidation of arsenic minerals previously below the water table is probably the principal mechanism for releasing arsenic into groundwater.

The oxidation of arsenic-bearing minerals present in the Bengal delta sediments is responsible for the release of arsenic oxides in solution to the groundwater. The subsequent migration of this arsenic-contaminated groundwater through the upper layers of deltaic sediments is the principal cause of arsenic poisoning in Bangladesh.

Arsenic-bearing minerals of several kinds are associated with the organic -rich sediments present in deltaic environments. Available sources for arsenic are the ocean, coal beds in India , and mountains to the north. Minerals formed in these reducing environments below the groundwater table would be stable unless they were exposed to oxidizing -environments. The groundwater table is lowered by increased irrigation during the dry season and in the cone of depression formed by pumping -tube wells and irrigation wells drilled below the zone of fluctuation. The arsenic minerals in the newly exposed sediments oxidize and release the arsenic when the water table recovers and exposes the oxidized minerals to a reducing -environment.

Increased irrigation did become necessary during India 's 30 years of unilateral diversion of river water from the Ganges, Tista and 28+ common rivers of Bangladesh and India which cut the normal flow of the 30+ rivers during the dry season. The solution to the arsenic problem is to restore the natural river flow of the Ganges, Tista and other common rivers of Bangladesh and India. This would restore the groundwater level to a level that existed in Bangladesh prior to the construction and commission of Farakka Barrage in 1975 .

Other man-made environmental disasters created by the Farakka, Tista and other barrages/dams constructed in the common rivers of Bangladesh and India would also be solved if these barrages were removed and a normal flow restored. The riverbeds could then be dredged and groundwater produced at a safe yield rate. A comprehensive plan not only for water supplies but associated waste disposal should be worked out for all of Bangladesh. Individual units within the plan could t hen be developed on the bases of need and tied into the overall plan as it develops.

1.   INTRODUCTION

Bangladesh is located in one of the major environmentally endangered areas of the world. Prior to 1975 the country had never faced an environmental crisis of the present magnitude. Thousands of people are currently suffering from numerous arsenic-related diseases. More than 100 million people are being poisoned by groundwater arsenic contamination. Eighty million more are at risk.

Scientists from around the world have been working on the groundwater arsenic poisoning in Bangladesh for some time, but no one has come up with convincing mechanisms for the process and source of the arsenic poisoning. The Bangladesh Department of Public Health and Engineering (DPHE) and the British Geological Survey (BGS) have jointly conducted an investigation and have collected the most data addressing this problem.

The source of the arsenic contamination and how long it has been present in groundwater should be determined in order to remedy the problem and prevent future occurrences.

Recently the presence of arsenic in groundwater has been reported in Assam, Bihar, Tripura, Uttar Pradesh, and Jharkhand of India and Terai areas of Nepal. Immediate action should be undertaken to find the source and the cause of the problem before the situation get worse.

2.   SOURCES OF ARSENIC IN BANGLADESH

(Nickson et.al.,1998a, 1999b), (DPHE/BGS/DFID,1999a, 2001b) and others attribute the arsenic to the reduction of arsenic in oxyhydroxides that were present in sediments washed into valleys cut by rivers when sea-level was lowered during the last glacial maximum (18 ka BP). During glacial maximum, these rivers had base levels some 100 meters lower than in interglacial times. The original sediments had been deposited during Pleistocene-Holocene time and were oxidized and flushed during the low-stand of sea level during this last glacial maximum. The sediments in-filling the valleys as the sea level rose during glacial melting were the characteristic weathered red brown Pleistocene-Holocene sediments.

Nickson et.al. further state that, "the As derives from reductive dissolution of Fe oxyhydroxide and release of its sorbed As. The Fe oxyhydroxide exists in the aquifer as dispersed phases, such as coatings on sedimentary grains. Recalculated to pure FeOOH, As concentrations in this phase reach 517 ppm. Reduction of the Fe is driven by microbial metabolism of sedimentary organic matter, which is present in concentrations as high as 6% C. Arsenic released by oxidation of pyrite, as water levels are drawn down and air enters the aquifer, contributes negligibly to the problem of As pollution."

In 1999, DPHE/BGS/DFID investigators collected about 21 sediment and soil samples, and based on their analysis have suggested that arsenic-bearing minerals were not present in the Bangladesh's sediments. The DPHE/BGS accepted the Nickson et. al., explanation of the source of arsenic poisoning in Bangladesh. In their report, DPHE/BGS/DFID stated that, "The 'pyrite oxidation' hypothesis proposed by scientists from West Bengal is therefore unlikely to be a major process, and the 'oxyhydroxide reduction' hypothesis (Nickson, R. et. al., 1998 in Nature; v395:338) is probably the main cause of arsenic mobilization in groundwater. It is difficult to account for the low sulfate concentrations if arsenic had been released by oxidation of pyrite. Moreover, mineralogical examination suggests that the small amount of pyrite present in the sediments have been precipitated since burial."

(Das, et al., 1996) conducted a geochemical survey in the six districts of west Bengal bordering the western part of Bangladesh. These districts are Mulda, Murshidabad, Bardhaman, Nadia, North 24-Pargana and South-24 pargana. They did a subsurface investigation, some laboratory analysis, and observed the presence of arsenic-rich pyrite minerals in the sediments. They stated that the source of arsenic in groundwater and in the soil is from pyrite minerals containing arsenic.

The climate of Bangladesh is conducive to the formation of laterite-type soils from which most of the elements have been leached out leaving behind only the most insoluble oxides such as aluminum hydroxide (gibsite) and ferric oxides and hydroxides. The minerals present in saturated zone below the water table could be similar to minerals found in some marshes. Drainage of some tidal marshes or the exposure of acid-firming underclays results in acid sulfate soils (cat clays) that contain pyrite, jarosite, mackinawite, and alunite (Dost, 1973: Iverson and Hallberg, 1976) Some of the minerals groups present include {Beudantite group (Sr, Be, Ca, Al, Pb) FeO_3 $(AsO_4,SO_4)(OH)_6$ Jarosite [K Fe3$(SO_4, AsO_4)2(OH)_6$]} Alunite Group: AB3$(XO_4)(OH)_6$ Apatite Group.

Arsenic present in trace amounts in the groundwater it would be concentrated in some of these minerals and released when the water table is lowered exposing this layer to oxidation. An extensive sampling and analysis of the iron-hydroxide zones at the interface of water table and reduced zone would reveal the presence of these arsenic-bearing minerals (Bridge and Husain 2000a, 2000b, 2001c).

3.   OFFSITE SOURCES OF ARSENIC CONTAMINATION IN BANGLADESH

Bangladesh is located down gradient from West Bengal.   The groundwater flow directions of major aquifers in the six districts of West Bengal are to the south and southeasterly direction towards Bangladesh. Being located down gradient, Bangladesh is probably receiving arsenic-contaminated water from West Bengal (Bridge and Husain, 2000).

On June 19, 2004, the New Delhi based *Times News Network* reported that, "Arsenic, that had seeped into the underground freshwater of eight West Bengal districts in 1998, is spreading once again. A study published earlier this month in London-based Royal Society of Chemistry's *Journal for Environmental Monitoring*, by a team of scientists of Kolkata-based Jadavpur University's School for Environmental Studies (SES), said that the killer arsenic has trickled into aquifers in entire Brahmaputra and Gangetic belt, comprising north-eastern states-particularly Assam and Tripura, West Bengal, parts of Uttar Pradesh, Bihar and Jharkhand states"

The Tripura State is located to the east and Assam State is located to the north north east of Bangladesh. High concentrations of arsenic were reported in the down-gradient regions, (Sylhet and Brhamanbaria districts in Bangladesh) to the Tripura and Assam states of India. The arsenic contaminants have not yet migrated from Assam and Tripura states of India to Bangladesh, the continued harvesting of river water in the upstream territory of India and over-pumping of groundwater will enhance migration of arsenic contaminants from the adjoining states of India to Bangladesh.

4. REDUCTION MECHANISM

(Smedley, 2003) stated that, "Several Bengali workers have proposed that the As contamination of groundwater in Bangladesh is due to the oxidation pyrite in the aquifers, brought about by over-abstraction of groundwater during the last few decades as result of increased irrigation demands (e.g. Das et.al, 1996; Mandal; et.al., 1996). While this is a plausible mechanism for As release, the strongly reducing groundwater chemistry with low dissolved SO_4 concentrations and the lack of evidence for significant seasonal groundwater drawdown in the worst-affected areas of Bangladesh, make the pyrite oxidation hypothesis unconvincing. Although sulphide minerals have been identified, albeit rarely, in some alluvial sediments from Bangladesh, this does not provide evidence for pyrite oxidation as an important mechanism of arsenic release. Indeed, sulphide minerals are an expected product of sulfate reduction under the strongly reducing conditions. Alternative mechanisms, including reductive dissolution of iron oxides (Nickson et.al 1998) and coupled reduction and desorption of As from iron oxides (BGS and DPHE, 2001) have been proposed as the dominant mechanisms driving the As mobilisation under the reducing aquifer conditions in Bangladesh. Phyllosilicate minerals (notably chlorite, biotite and Al hydroxide) have been suggested as playing an additional role in the cycling of As in the Bangladesh aquifers (Breit et al., 2001)."

The process described by Nickson and McArthur et al., and Smedly and Kinniburgh et.al., (DPHE/BGS/DFID investigators) seems to require that the arsenic remains in solution for thousands of years and that movement of groundwater through the delta sediments did not flush the arsenic from the system. They also state that the arsenic correlates well with the dissolved iron in the groundwater. Arsenic substitutes freely for sulfur in iron sulfides (pyrites or marcasite) and the chemical analyses of water from wells contaminated with arsenic do not rule out an iron sulfide source.

The arsenic contamination, with one or two exceptions, is restricted to shallow aquifers (< 80 meters). The arsenic contamination is not uniform in distribution; some wells have high concentrations and others have low concentrations. Some wells that are relatively free of arsenic and were used for domestic use have become contaminated with arsenic. These observations suggest that environmental changes have occurred recently and near the surface. This also suggests a non-uniform distribution of source material. A non-uniform distribution of organic mater in the sediments with arsenic pyrites would be expected.

In order to examine the both reduction and oxidation mechanism for the mobilization of arsenic into groundwater in Bangladesh and West Bengal of India, we developed the following questions and requested the proponents of reduction hypothesis to answer the following questions:

1. If the Oxyhydroxide Reduction hypothesis proposed by (BGS) is correct and if arsenic was present in an adsorbed form on iron hydroxide for thousands of years and existed in a solution for thousands of years in the aquifer groundwater of the Bengal Basin without being flushed out to sea, how did the people of Bangladesh and West Bengal of India avoid arsenic poisoning when thousands of people drank water from dug wells for thousands of years and from thousands of tube wells for 60 to 70 years, prior to the 1970s?

2. How did millions of people in Bangladesh who had been drinking water from millions of tube wells during the interval between the 1960's and prior to 1975, before the construction of dams/barrages and diversion of surface water by India from the Ganges, Tista, and 28 other common rivers of Bangladesh and India, lack signs of arsenic poisoning?"

The available data refutes the reduction hypothesis as the main cause for the mobilization of arsenic into groundwater for the following reasons:

1. The lag time for the development of arsenic lesion (karatosis, melanosis etc) in Bangladesh and West Bengal of India according to S.K. Shaha varies from 2-5 years; according to DPHE/BGS/DFID 10 years; and according to other investigators around the world the lag time varies from 8-14 years.

2. In Bangladesh and West Bengal, prior to 1975 and 1960 no significant abstraction of groundwater occurred. Extensive use of groundwater in Bangladesh started after 1975, where as in West Bengal it started after 1960.

3. Prior to 1960, both in Bangladesh and West Bengal, millions of infants, as well as young and old people drank water from thousands of dug wells for thousands of years.

4. According to DPHE, in 1948 there were 50,000 tube wells in use in Bangladesh and, according to Dipankar Chakrabortti, 50,000 tube wells in West Bengal, from which millions of people drank water.

5. In order to establish groundwater arsenic poisoning in Bangladesh as a natural disaster, DPHE/BGS/DFID investigators presented pre-1975 hydrological data for three major rivers in Bangladesh but, despite of our repeated requests, did not include post-Farakka/post-1975 data for any rivers in Bangladesh.

6. DPHE/BGS/DFID investigators presented only four-groundwater hydrographs and have failed to present pre-Farakka/pre-1975 data. They mentioned significant draw down of the water table in Bangladesh and presented some hydrographs of groundwater level of Dhaka, Bogra, Jessore, Joydebpur and Kishoreganj (DPHE/BGS/DFID, 1998a, 2001b).

7. DPHE/BGS/DFID investigators conducted a mineralogical study in just three "hot spot" locations, based on only 21 samples of 55,000 sq.miles of Bangladesh and reported the absence of arsenopyrite minerals. They also reported the presence of pyrite minerals in the sediments, but did not, however, investigate the conditions for the presence of arsenic-bearing minerals in the "neo-oxidation zone" that was created after 1975. They did not map the vertical and aerial extent of the "neo-oxidation zone" in Bangladesh and West Bengal of India.

8. In West Bengal, arsencic poisoining was first dectected in the population in 1985, and in Bangladesh in 1994.

9. DPHE/BGS/DFID investigators did not investigate the pre-and post-Farakka surface water and groundwater relationship in Bangladesh and West Bengal of India.

If dug wells had never penetrated below the historic zone of fluctuation, arsenic poisoning might have been avoided or rarely occurred and could explain the absence of any historical records of arsenic problems.

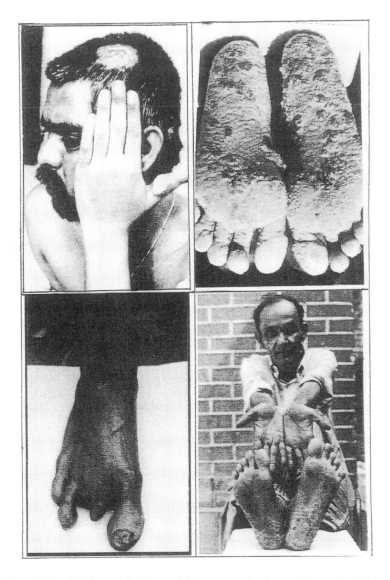

Figure 1. Arsenic lesions on feet (upper right); cancer on head caused by arsenic lesions (upper left); amputed leg due to gangrene caused by arsenic lesions (lower left); arsenic lesions on feet and hands (lower right, photographs: after Wilson, R., 2004).

Prior to the 1990's and 1980's, the people of Bangladesh and West Bengal of India never suffered from arsenic-related diseases. The above data/explanations evidently support the conclusion that the groundwater arsenic poisoning in Bangladesh and West Bengal of India is a recent environmental problem related to the recent abstraction of groundwater and the harvesting of river waters in the upstream region of India, and that the oxidation and subsequent reduction of arsenic minerals is the plausible mechanism for the mobilization of arsenic into groundwater.

(Harvey et.al. 2001) conducted a limited investigation in a 16 sq. km area located in the Munshigonj district of Bangladesh. Like all other proponents of reduction hypothesis, they have also rejected the oxidation hypothesis. In their article they stated that, "The inverse relation of dissolved sulfate with arsenic (Fig. 1C) in the natural groundwater, and the presence of acid volatile sulfide (AVS) in the sediments near the dissolved AS peak, suggest that oxidative dissolution of pyrite has not liberated arsenic. Instead, low dissolved sulfur levels appear to limit the precipitation of arsenic sulfides near the arsenic peak."

The inverse relation of dissolved sulfate with arsenic and the presence of acid volatile sulfide (AVS) in the sediments near the dissolved arsenic peak do not reject the oxidative dissolution of pyrite/arsenic bearing minerals for the liberation of arsenic into groundwater. In Bangladesh, after 1975, the "neo-oxidation zone" beneath the thousands of years old historical "oxidation zone" was created due to increased draw down of the water table caused by the over-pumping of groundwater and diversion of river waters. (Fig. 2)

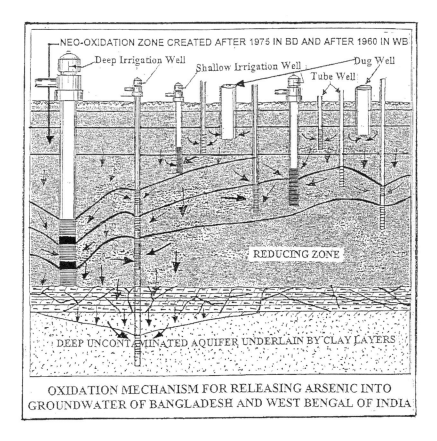

Figure 2. Illustration of oxidation mechanism for the releasing and mobilization of
arsenic into groundwater in Bangladesh and West Bengal of India .

In order to fully refute the "oxidation hypothesis" in the context of
groundwater arsenic poisoning in Bangladesh and West Bengal of India,
(Kinniburgh et.al. 2003) presented the following arguments in the book
entitled, "Arsenic in groundwater: Geochemistry and occurrence", edited by
Alan H. Welch and Kenneth G. Stollenwerk". They stated that:

Pyrite oxidation hypothesis- the hypothesis was strongly advocated by
West Bengal Scientists in 1995 (Chatterjee et.al. 1995; Chodhury et.al.
1997). It is based on the idea that arsenopyrite, or later As-rich pyrite, was
initially present in the sediments and has been at least partially oxidized
as a result of the recent seasonal lowering of water table. This lowering
has been attributed to the use of groundwater for irrigation and, by some,
to the construction of the Farakka barrage (a controversial dam that was
completed in 1975 across the River Ganges in West Bengal close to the

West Bengal-Bangladeh border and which diverts River Ganges water to the Bhagirati-Hoogly River and ultimately to Calcutta). This hypothesis therefore supports the notion that the release of arsenic to the groundwater is a recent phenomenon induced by man's activities. Certainly such a hypothesis is a possibility and needs to be considered. However, proponents of the hypothesis have offered little scientific evidence in support of it other than demonstrating the presence of pyrite in the sediments.

The river water diversion/harvesting during summer months in the upstream territory of India through dams/barrages construction on more than 30 common rivers of Bangladesh and India for the last three decades is directly related to the oxidation mechanism (Husain, 1999, Bridge and Husain, 2000a, 2000b, 2001c).

Water resource systems are dynamic in nature. Surface and groundwater resources are integral parts of the same hydro-geological component. They respond both in quantity and quality to natural changes and human activities such as diversion of surface water and abstraction of groundwater. As a result, the water chemistry changes with time when the water moves through the changing environment. In some areas this change is slow, but in other areas where environmental conditions have changed, the change is rapid. The hydro-chemical parameters are not uniform and constant in a deltaic environment such as Bangladesh.

Human activities create conditions that promote the migration of contaminants. For example, the abstraction of groundwater results in the lowering of water table levels. The diversion of water by Farakka and other dams has resulted in increased use of groundwater and caused a significant lowering of the groundwater table in the Bengal delta. The diversion of surface water, for the last 25 years by India, from about 30 rivers that flow into Bangladesh has produced major environmental changes. In terms of flow, the Ganges River was the eighth largest river in the world. Prior to 1975, the average flow during the dry season at the Hardinge Bridge point in Bangladesh was 2000 Cum/Sec. The present flow at this point due to India's 30 years of unilateral diversion is only 400 Cum/sec. (Fig. 3).

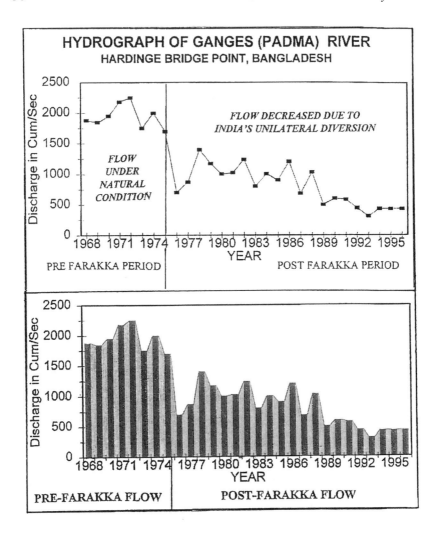

Figure 3. Illustration of the pre and Post Farakka average annual discharge of the Ganges River at Hardinge Bridge Point, Bangladesh (Hebblethwaite, 1997).

Besides the abstraction of groundwater and diversion of surface water from the common rivers of Bangladesh and India, no other human or natural cause is known to have occurred in Bangladesh that could have destabilized the arsenic-bearing minerals present in the sediments and brought about a significant geochemical change in the sediments and groundwater in the Bengal basin. In addition to Farakka, India has constructed the dams/barrages in Feni, Muhuri, Selonia, Gomti, Sonai, Khowai, Dhalai,

Manu, Juri, Sonai Bardal, Kushiyara Rivers in the east, Piyan, Bhogai, Jinjiram, Dharla, Sangil, Tista, Buri Tista, Deonai-Jamuneshwari, Ghoramara, Talma and Karatoa Rivers in the north, and Ichamati-Kalindi, Benta-Kodalia, Bhairab-Kabodak, Khukshi, Atrai, Punabhaba and Mahananda Rivers in western Bangladesh (Fig. 4).

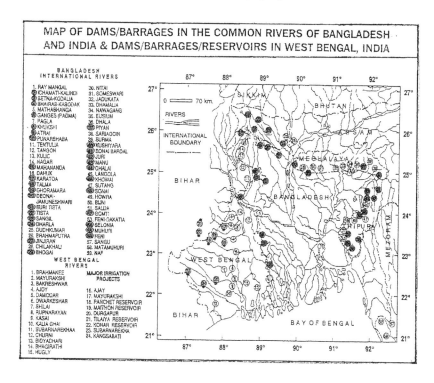

Figure 4. Illustration of dam/barrages in the common rivers of Bangladesh and India (shaded circle) and dams/barrages/reservoirs in West Bengal of India (circle with a dot).

Prior to 1975, most of the following problems occurring in Bangladesh were not present and the severity of some of the problems that were present prior to 1975 has greatly increased:

(1) Arsenic poisoning of groundwater
(2) Severe floods (1988 and 1998, 2004)
(3) Depletion of surface water resources
(4) Depletion of groundwater resources
(5) Desertification
(6) Extinction of aquatic species
(7) Negative impact on fish industry

(8) Drop in organic matter content in the soil

(9) Destruction of agriculture and horticulture

(10) Inland saline water intrusion

(11) Loss of navigable waterways

(12) Riverbank erosion

(13) Climate change

(14) Loss of professions

(15) Outbreak of environmental diseases

(16) Land subsidence (from water table lowering) and

(17) Social instability due to symptoms of arsenic poisoning etc.

The diversion of surface water from the rivers and the over-use of groundwater are the human activities largely responsible for the destabilization and destruction of the eco-system of the Bengal Basin and current environmental problems in Bangladesh. Currently no one has collected hydrological data from the 30+ common rivers of Bangladesh and India. We have presented the hydrograph of Ganges River as evidence that supports the recent oxidation conditions that became present in Bangladesh after 1975.

In southeast Bangladesh (Lakshmipur and Noakhali) a significant lowering of the water table took place due to the diversion of surface water from the Ganges by Farakka and other dams/barrages in other common rivers of Bangladesh and India. During 1977-1997 in Lakshmipur, the observation well NA012 and NA018 showed a water table draw down of 2-3m and in Noakhali, the observation well NA020 showed a water level draw down of 2-3m. In Barisal, the draw down observed in well BA013 was between 3-4m, in Pirojpur well BA015 the draw down ranged from 2-3m and in Madaripur well FA36, the draw down ranged from 1-2 m (fig. 5).

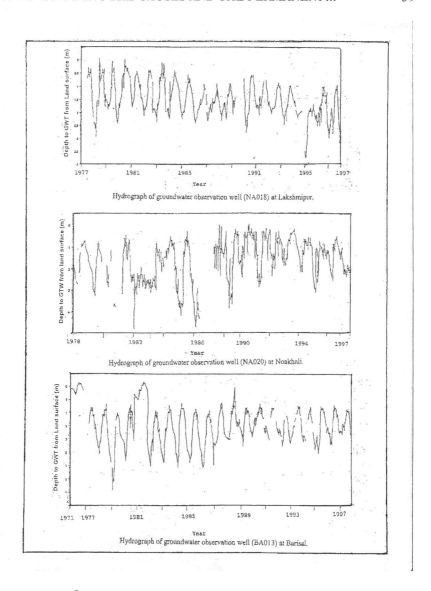

Figure 5. Illustration of significant lowering of the water table after 1975 in the southeast Bangladesh (arsenic hot spot area) caused by over -pumping of groundwater and diversion of river water from the Ganges and other common rivers of Bangladesh and India (after Aggarwal et.al. 2001).

The high concentration of arsenic in southeast Bangladesh is the result of oxidation caused by the lowering of water table and presence of arsenic bearing minerals in the "neo-oxidation zone". In the neo-oxidation zone, one may or may not find arsenic-bearing minerals due to 30 years of increased draw down of the water table. On the other hand, migration of arsenic from the up-gradient and cross-gradient sources/areas can not be ruled out.

Like the Ganges catchments, the Meghna catchment has been severely impacted by the abstraction of groundwater and diversion of river waters for the last three decades. Besides Farakka, the others dams that India constructed in the common rivers of Bangladesh and India are Feni, Muhuri, Selonia, Gomti, Sonai, Khowai, Dhalai, Manu, Juri, Sonai Bardal, Kushiyara in the east, Tista, Piyan, Bhogai, Jinjiram, Dharla, Sangil, Buri Tista, Deonai-Jamuneshwari, Ghoramara, Talma and Karatoa in the north, and Ichamati-Kalindi, Benta-Kodalia, Bhairab-Kabodak, Khukshi, Atrai, Punabhaba, and Mahananda in the west of Bangladesh (Fig. 4).

Kinninburg et.al. are not aware of the impact of the harvesting of water from these common rivers in summer months in the Indian territory and as a result have failed to understand the hydrological, hydrogeological and hydrochemical history of Bengal Basin. The distance of the Farakka barrage [from the affected areas in the Bengal Basin *ed.*] is not very significant for the creation of arsenic poisoning. It is the reduction of water in the Bangladesh part of the Ganges River as a result of the Farakka barrage that has caused the lowering of the water table in a large area including southeast Bangladesh. Prior to 1975, before the construction of Farakka barrage, there was no shortage of water in the Ganges River in Bangladesh. In summer before the construction of the barrages, the Ganges River used to supply large volumes of water in the southeast region (Fig. 3).

(Chakrabortti, 2004) reported that, "so far we have analyzed more than 700 dug wells from the arsenic affected areas of West Bengal and Bangladesh. We have found 90% dug wells are safe with respect to arsenic (<3 - 35 microgram per litre; Average 15 microgram per litre). There are some areas where we have found arsenic contamination above 50 microgram per litre (maximum 330 microgram per litre). One such area in Bangladesh is Samta village where many dug wells are arsenic contaminated and the dug wells are shallow (10- 20 ft), recent and 2-3 ft diameter and waters have a foul smell" (Husain, 2004).

Chakrabortti's data clearly shows that some dug wells are not free of arsenic contamination in arsenic-contaminated areas in Bangladesh and West Bengal. The presence of arsenic in dug wells further suggests that the oxidation of arsenic minerals is the plausible mechanism for the creation of the groundwater arsenic poisoning in Bangladesh and West Bengal.

The high concentrations of phosphate in some samples may be related to arsenic-bearing phosphate nodules, which are common inorganic-rich deposits.

However, reviews of the Nickson et.al. Kinneinberg et.al. (DPHE/BGS/DFID), Harvey et. al., and other works on the arsenic problem in Bangladesh and West Bengal of India have revealed that neither the reduction proponents nor the oxidation proponents have properly investigated the oxidation mechanism thus far. Therefore the reduction mechanism is not a principal cause for the mobilization of arsenic into groundwater in the context of Bangladesh and West Bengal of India.

5.   OXIDATION MECHANISM

Dipankar Das and others conducted a geochemical survey in the six districts of West Bengal bordering the western part of Bangladesh. These districts are Mulda, Murshidabad, Bardhaman, Nadia, North 24-Pargana and South-24 Pargana. Their subsurface investigation and some laboratory analysis, revealed the presence of arsenopyrite minerals in the sediments. They stated that, "the source of arsenic in groundwater and in the soil is from pyrite minerals containing arsenic." However, they did not discuss how arsenic is released in groundwater from arsenopyrite. They cited the oxidation of pyrite process presented in the literature from the U.S. But in their conclusions they state that, "the way that arsenic enters the groundwater in these six districts is not well understood. Our borehole analyses show arsenic-rich FeS_2 in sediment layers. Since iron pyrite (FeS_2) is not soluble in water, the question therefore arises as to how arsenic from pyrites enters the water." (Das et.al., 1999)

Although pyrite is not soluble in water, it decomposes when exposed to air or in aerated water and proceeds rapidly in confined environments without the addition of oxygen from external sources as the acid level reaches a pH of 3.5 or lower.

A USA Government article on acid-mine drainage (AMD) describes the process thus:

> "The formation of acid drainage is a complex geochemical and microbially mediated process. The acid load ultimately generated is primarily a function of the following factors: Microbiological Controls; depositional environment; Acid/base balance of the overburden; Lithology; Mineralogy; and hydrologic conditions."

Chemistry of Pyrite Weathering:

$2\ FcS_2 + 7\ 0_2 + 2\ H_20 \rightarrow 2\ Fc_2{}^+ + 4\ SO_4 + 4\ H^+$ (Equation I)
$4\ Fe^{2+} + 0_2 + 4\ H^+ \rightarrow 4\ Fe^{3+} + 2\ H_20$ (Equation 2)
$4\ Fe^{3+} + 12\ H_20 \rightarrow 4\ Fe(OH)_3\ \{12\ H^+$ (Equation 3)

$$FeS_2 + 14\ Fe^{3+} + 8\ H_2O \rightarrow 15\ Fe^{2+} + 2\ SO_4^{2-} + 16\ H^+ \text{ (Equation 4)}$$

In the initial step, pyrite reacts with oxygen and water to produce ferrous iron, sulfate and acidity. The second step involves the conversion of ferrous iron to ferric iron. This second reaction has been termed the 'rate-determining' step for the overall sequence."

"The third step involves the hydrolysis of ferric iron with water to form the solid ferric hydroxide (ferrihydrite) and the release of additional acidity. This third reaction is pH dependent. Under very acid conditions of less than about pH 3.5, the solid mineral does not form and ferric iron remains in solution. At higher pH values, a precipitate forms, commonly referred to as 'yellow boy'."

"The fourth step involves the oxidation of additional pyrite by ferric iron. The ferric iron is generated by the initial oxidation reactions in steps one and two. This cyclic propagation of acid generation by iron takes place very rapidly and continues until the supply of ferric iron or pyrite is exhausted. Oxygen is not required for the fourth reaction to occur. The overall pyrite reaction series is among the most acid producing of all weathering processes in nature."

The article goes on to state that the raising and lowering of the water table (wetting and drying) in the reacting environment provides optimal conditions for the weathering of pyrite. The changes in the geochemical environment due to high withdrawal of groundwater resulted in the decomposition of pyrites and the release of arsenic.

Similar pyrite reactions have been described by many workers around the world (Schreiber et.al., 2003).

Dipankar Das et al., mineralogical studies by XRD(X-ray diffraction) shows the presence of $FeSO_4$. (Welch et. al., 1988) studied arsenic in the groundwater of the western USA and suggested that the "mobilization of arsenic in sedimentary aquifers may be, in part, a result of changes in the geochemical environment due to agricultural irrigation. In the deeper subsurface, elevated arsenic concentrations are associated with compaction caused by groundwater withdrawal." (Bridge and Husain, 2000a, 2000b, 2001c)

If the time of arsenic contamination is after 1975 in Bangladesh, a probable explanation is that the changes in geochemical environment due to the high withdrawal of groundwater resulted in the decomposition of arsenic-bearing minerals that were stable in a reducing environment. These arsenic oxides if introduced to the reducing conditions below the water table are reduced to poisonous oxide forms.

The DPHE/BGS/DFID report states that, "the greatest arsenic concentrations are mainly found in the fine-grained sediments especially the gray clays. A large number of other elements are also enriched in the clays including iron, phosphorus and sulfur. In Nawabganj, the clays near the surface are not enriched with arsenic to any greater extent than the clays below 150 m; in other words, there is no evidence for the weathering and deposition of a discrete set of arsenic-rich sediments at some particular time in the past. It is not yet clear how important these relatively arsenic-rich sediments are for providing arsenic to the adjacent, more permeable sandy aquifer horizons. There is unlikely to be a simple relationship between the arsenic content of the sediment and that of the water passing through it."

The arsenic is associated with low energy sediments, and organic matter would tend to be associated with the lower energy environments also. Organic matter is present in the sediments below the water table in Bangladesh. Arsenic along with other trace elements, when present in the environment, is enriched in organic-rich sediments. Sulfur from decay of organic matter combines with iron to form sulfides in reducing environments and these sulfides will incorporate arsenic if arsenic is present. When the groundwater table was lowered by increased irrigation during the dry season and the sediments were exposed to the oxygen from the atmosphere in a moist environment, arsenic-rich sulfides associated with organic matter and other reduced-arsenic bearing minerals would oxidize in this moist environment and release arsenic. Bacterial decay of the organic matter would produce H_2CO_3, HCO^{3-} CO^{3--}, the kind of sulfates present are dependent on pH, and hydrogen sulfide, below an Eh of 3 or H_2SO_4 above an Eh of 3. With the appropriate concentrations below an Eh of 3 and between a pH of 3 to 9 these carbonate and sulfur species would react with the ferrous iron in solution to produce siderite and pyrite. (Figure 1 after Robert M Garrels, Solutions, Minerals, and Equilibera).

(Breit, et.al. 2001) discovered the presence of arsenic in iron oxide and sulfides in the sediments above and below the capillary fringe zone at Brahmanbaria area adjacent to the Tripura State of India. They also reported that, "holocene sediments collected above the capillary fringe are yellow-brown and characterized by abundant 0.5 N HCl extractable ferric iron. Sediment below the water table is grey to black and contains 90% of 0.5 N HCl extractable iron as ferrous. Pleistocene sediment recovered from one borehole is also characterized by a yellow-brown color due to grain coating ferric oxides. Total arsenic concentrations range from <1 to 16 ppm in most brown and grey samples. An exception to this range are the high contents of arsenic (200 to 500 ppm) found in orange-brown ferric oxides bands a few centimeters thick that develop in clayey-silt above the capillary fringe." They further stated that, "exposure of some grey Holocene sediment to humid air for one week oxidized 50% of the extractable iron and arsenic emphasizing the sensitivity of these elements to redox environment. A few samples of cuttings collected near Brahmanpara (south of Brahmanbaria) are substantially enriched in sulfur (0.3 wt.%) likely due to deposition in an estuarine environment. XAFS analyses of these sediments are consistent with arsenic contained in a sulfide phases".

The Breit et. al. findings revealed that arsenic bearing minerals are present in the Bengal Delta sediments. DPHE/BGS/DFID and other proponents of the reduction hypothesis have failed to collect adequate mineralogical and geo-chemical data to find the presence of arsenic-bearing minerals in the sediments. Arsenic-bearing minerals of different kinds are present in the sediments. The distribution and the abundance of minerals in the Bengal Delta sediments varies with geologic conditions and the recent abstraction of groundwater and harvesting of river water (Bridge and Husain, 2000). The Breit et. al. findings clearly revealed that the oxidation mechanism has been playing a major role in releasing arsenic in the groundwater of Bangladesh and West Bengal.

6.   HOW OXYGEN ENTERED THE NEO-OXIDATION ZONE

Proponents of the reduction hypothesis have difficulty in understanding how oxygen entered into the "neo-oxidation zone" or into the deep aquifer. Oxygen entered into the deep aquifer in the dewater zone along with the lowering of the water table. (Schreiber et. al. 2003) describe the following processes, observed in the state of Wisconsin, responsible for the source of oxygen, all of which are applicable in the context of Bangladesh and the West Bengal of India:

"(1). Oxygen-rich water infiltrates to the SCH in the recharge area; (2) vertical leakage of oxygen-rich water through the Sinnipee confining unit initiates a regional oxidation of sulfides in the SCH; (3) partial dewatering of the St. Peter aquifer due to extensive groundwater withdrawal exposes the SCH to oxygen; and (4) boreholes provide a direct conduit for atmospheric oxygen to reach the SCH in areas where the static water level is coincident with, or below, the SCH"

The vertical and the areal extent of the neo-oxidation zone during the dry season in Bangladesh has greatly increased in the last three decades due to the abstraction of groundwater caused by 6-11 million hand tube wells, 0.5 million shallow tube wells and 55, 000 deep tube wells, and to the harvesting of river water in the upstream territory of India from more than 30 common rivers in Bangladesh and India. Prior to 1975, the area was under water for thousands of years. After 1975, due to the diversion of river water and abstraction of groundwater, millions of shallow tube wells have become inoperable during dry season due to the draw down of the water table. These wells supply atmospheric oxygen (direct conduit) to the dewater zone (cone of depression) caused by pumping of groundwater by shallow and deep irrigation wells (<100 m) during the dry season, thus oxidizing the arsenic-bearing minerals present in the sediment (Fig. 2).

Breit et. al. (2001) discovered that, "exposure of some grey Holocene sediment to humid air for one week oxidized 50% of the extractable iron and arsenic emphasizing the sensitivity of these elements to redox environment." The Breit et. al. findings further revealed how atmospheric oxygen reacts with arsenic-bearing minerals in the cone of depression/dewater zone/neo-oxidation zone in the subsurface aquifer sediments in the Bengal Delta.

Like DPHE/BGS/DFID, Aggarwal et. al. have failed to understand the situation in Bangladesh, because they have failed to collect pre- and post-1975 hydrological data from the common rivers of Bangladesh and India. They did not even collect adequate pre- and post-1975 groundwater hydrological data. Moreover, they misinterpreted the groundwater level data in their study area and as a result have failed to understand the relationship between surface water and groundwater; three decades of diversion of river water; and the abstraction of groundwater in Bangladesh.

Charles Harvey and his group also did not collect pre- and post-1975 river water discharge data in Bangladesh. They collected groundwater level data between 1988-1997 from 183 hydrographs in their study area. They reported a significant lowering of water table in their study area.

DPHE/BGS/DFID reported a significant lowering of the water table in the Ganges and the Jamuna catchment areas due to the abstraction of groundwater. However, they did not collect adequate pre- and post-Farakka river water discharge data, water table draw down data, or mineralogical data to examine the oxidation mechanism for releasing arsenic in groundwater in Bangladesh and West Bengal.

7.   AGE OF THE ARSENIC POISONING

The age of the groundwater arsenic poisoning in Bangladesh and West Bengal of India is an important indicator in finding the source and cause of the poisoning, and a solution to the arsenic disaster in Bangladesh. Based on their isotopic data, proponents of the reduction hypothesis (DPHE/BGS/DFID, 12001, Agarwall et. al. 1999a, 2003b, and Geen et. al. 2003) believe that groundwater arsenic poisoning in Bangladesh was present prior to 1960, before the massively increased irrigation and diversion of river water from 30+ common rivers in Bangladesh and India that began post 1975.

We do not disagree with the isotopic findings regarding the age of the groundwater propounded by DPHE/BGS/DFID, Aggarwal et. al. and Geen et. al., but we do not agree with them regarding the age of groundwater arsenic poisoning. The age of the water and the time arsenic entered the water are two different things. If arsenic was tied up in minerals that were stable below the water table when the sediments were first deposited and released only when oxidation occurred as the groundwater table was lowered at a later date, then the date of arsenic contamination relates to the groundwater lowering not the age of the water. The fact that some of the wells were below the WHO limits for arsenic when they were first tested and later tests detected an increase in concentration of arsenic above safe limits is an indication that recent local changes in the environment caused the change.

Oxidation of arsenic pyrite and other arsenic-bearing minerals is one possibility for the change. If the water diversion from rivers and the over-pumping of groundwater are continued, this process will contaminate both new and old uncontaminated water whether the water is 25 years old, or thousands/millions of years old (Husain, 2001).

(BGS/DPHE/DFID, 2001) reported that, "irrigation wells (STWs and DTWs) are typically shallow (<100 m) with multiple screens in an unconfined aquifer. The water level is commonly near the surface and within the limit of suction pumps (7 m). The pump intake is set above the screen level, but the screens are set lower (typically 30 m bgl for STW and 100 m bgl for DTW), depending on where the appropriate coarse lithology is encountered. Pumping of this type of well causes vertical gradients to be developed as the well induces flow from the water table to the well screen. This depletion of the water table is replenished during the wet season as long as total abstraction does not exceed the available resources."

DPHE/BGS /DFID also reported that, "the decline in water levels due to abstraction for irrigation during the dry season through the use of shallow and deep sidewalls can be significant, especially in areas of thick near-surface silt and very fine sand layers with with low specific yields. In low-lying areas of increased annual abstraction for irrigation, as in the Jamuna and Ganges delta floodplains, shallow tube well use may be halted due to decline of water levels below the suction level before the end of the dry season. In such areas, crop irrigation has to be completed using water from deep tube wells. Such a regional decline in water level renders many hand-operated suction pumps inoperative towards the end of the dry season."

Prior to 1975, during dry season there was no shortage of river water and no significant draw down of water level below the thousands of years old historical oxidation zone. After 1975, a significant draw down of the water level started due to the river water harvesting from more than 30 common rivers in Bangladesh and India taking place in the upstream territory of India, and to the creation of a "neo-oxidation zone" caused by the over pumping of groundwater.

During the dry season, atmospheric oxygen enters into the "neo oxidation zone" causing the break down of arsenic-bearing minerals. During the wet season, the "neo oxidation zone" is replenished thus releasing arsenic into groundwater. DPHE/BGS/DFID have found a zone of maximum concentration of arsenic that ranges from 15-30 m. On the other hand, Harvey et. al. (2001) reported a zone of dissolved arsenic (>90%) As(III) that ranges from 30-40 m. The thirty years of increased draw down and recharge have caused migration of poison arsenic from the "neo oxidation zone" into deep aquifer (reducing zone) thus contaminating the thousands-of-years-old uncontaminated water.

DPHE/BGS/DFID, Agarwall et.al.,and Geen et.al., and other proponents of the reduction hypothesis have failed to collect relevant hydrological, hydro geological, hydro chemical, historical medical and historical groundwater use data from dug wells and tube wells, and based on their isotopic study have been suggesting that the groundwater arsenic poisoning has been present in Bangladesh and West Bengal of India for thousands of years.

For thousands of years, prior to 1975 and the construction of dams/barrages by India and India's unilateral diversion of surface water from the Ganges, Tista, and 28 other common rivers of Bangladesh and India, the people of Bangladesh drank groundwater from dug wells. During a period of about 60-70 years prior to 1975, some several millions tube wells were installed in Bangladesh. In 1940, about 50,000 tube wells were in use in Bangladesh (former East Pakistan). Millions of people (infants, young and old) drank water from these wells. No occurrences of arsenic diseases were recorded for those people who drank water from these tube wells.

8.   CONCLUSION

1. The groundwater arsenic poisoning in Bangladesh and West Bengal of India appears to be a recent man-made disaster which has been created by the over pumping of groundwater from irrigation wells and the harvesting/diversion of river water from more than 30 common rivers in Bangladesh and India.

2. The presence of arsenic-bearing sulfides above and below the capillary fringe zone of Bengal Delta sediments, the increased lowering of the water

table for the last three decades during the dry season and recharge of the dewater zone during the wet season have created "neo oxidation zone" and the oxidation mechanism, thus releasing arsenic into the groundwater of Bangladesh and West Bengal of India.

The lowering of the water table resulted in the exposure to air in the zone of aeration. This exposure resulted in the oxidation of arsenic minerals previously present below the water table in the Bengal sediments. The arsenic oxides migrated to the groundwater and were reduced to poisonous forms in the reducing environments below the water table.

3. The historical medical data, historical groundwater use data from the dug wells and tube wells, arsenic toxicological data, mineralogical data, hydrological and hydro geological data reject the presence of arsenic poisoning in Bangladesh and West Bengal of India for thousands of years as proposed by the proponents of the reduction hypothesis. These data also reject the hypothesis that the groundwater arsenic poisoning in the Bengal Delta is a natural disaster and that the reduction mechanism is the principle cause for the mobilization of arsenic into groundwater.

4. The presence of arsenic poisoning in the reducing zone is the result of over pumping of groundwater by irrigation wells due to the pulling of arsenic poison water from the "neo oxidation zone" to the reducing zone thus contaminating the deep aquifer.

5. The presence of high concentration of arsenic in the zone of 15-40 m depth and also in some wells >100 m is the result of three decades of pumping of groundwater from shallow and deep irrigation wells, because, "the pump intake is set above the screen level, but the screens are set lower (typically 30 m bgl for STW and 100 m bgl for DTW), depending on where the appropriate coarse litho logy is encountered." (DPHE/BGS/DFID, 2001)

9.   SOLUTION TO THE ARSENIC POISONING IN BANGLADESH

The best solution appears to be the restoration of the thousands-of-years - old natural environment by restoring the river flow and groundwater levels that existed prior to the 1975 commission of Farakka, Tista and other dams/barrages that India constructed in the common rivers in Bangladesh and India (Fig. 4).

The river water should be filtered, treated, continually tested and delivered through a closed system to provide a safe water supply for the nation.

10.   RECOMMENDATIONS

(a). The over pumping of groundwater must be stopped and surface water should be used for drinking and cooking purposes. The deep uncontaminated groundwater should be monitored properly and may be used on a safe yield basis.

(b). The natural groundwater level that existed prior to 1975 should be restored by removing all dams/barrages that India constructed in the common rivers of Bangladesh and India. The removal of dams/barrage and the dredging of rivers will decrease the number of disasters in both Bangladesh and in the upstream region of India.

(c). The flushing of arsenic contaminants may take a long time but the removal of dams/barrages affecting Bangladesh will provide plenty of water during the dry season for drinking, irrigation and industry. The restoration of groundwater levels in Bangladesh that existed prior to 1975 will prevent migration of arsenic contaminants from the adjoining states of India.

(d). In order to fully understand the oxidation mechanism for releasing arsenic into the groundwater in Bangladesh and West Bengal of India, scientists must investigate the following conditions:

(i). Presence of arsenic bearing minerals in the "neo-oxidation zone" and the cone of depression and, if arsenic-bearing minerals are absent, the cause for the absence of arsenic-bearing minerals in that zone.

(ii). Evidence for the lowering of the water table and penetration of atmospheric oxygen intoto the newly created unsaturated zone, "neo-oxidation zone" during dry season.

(iii). Evidence for the recharge of groundwater in the "neo-oxidation zone" during dry and wet seasons.

These three conditions must be examined based on the following data/parameters:

1. Pre-and post-1975 river water discharge data of more than 30 common rivers in Bangladesh and India. This data will present the relationship between pre- and post-1975 river water harvesting and the draw down of groundwater level.

2. Pre- and post-1975 groundwater level data. This data will help to construct pre- and post-1975 groundwater table maps, groundwater flow direction, construction of a thickness map of a "neo-oxidation zone" that was created after 1975.

3. Sediment samples must be collected from the "neo-oxidation zone", cone of depression and below the capillary fringe zone for mineralogical study.

(e). In order to determine the migration of arsenic contaminants from the upgradient/off-site sources, groundwater flow tests along the border of Bangladesh and West Bengal, Bangladesh and Tripura, and Bangladesh and Assam should be conducted.

f). In West Bengal scientists, should also examine these three oxidation conditions based on pre- and post-1960's groundwater level data, river water discharge data and mineralogical data, because the over-pumping of groundwater in West Bengal started after 1960, earlier than in other regions. Scientists should also investigate the impact of Ajay, Mayurakshi, Panchet, Maithon, Durgapur, Tilaiya, Konar, Subarnarekha and Kangsabati dams/barrages/reservoirs on the hydrology, hydrogeology and hydrochemistry in West Bengal of India. (Fig. 4)

(f). India should abandon the "Inter Basin Water Transfer Link project of India", because the implementation of this project may cause arsenic and other environmental disasters similar to those that Bangladesh and India are facing today.

ACKNOWLEDGEMENT

We acknowledge with special thanks the contributions of "Arsenic Discussion Group" members who have shared arsenic-related data of the Bengal Basin with us. The authors are also thankful to several news media, researchers, students, policy makers and environmentally concerned people for their concern and important questions related to the arsenic problem of Bangladesh and West Bengal of India.

REFERENCES

Wilson, R. 2004. Chronic Arsenic Poisoning: History, Study and Remediation, Harvard, U.S.A., (http://phys4.harvard.edu/).

Husain, M.T. 2004. Arsenic Poisoning in Hand dug wells in Bangladesh and West Bengal. Shetubondhon (http:groups.yahoo.com/group/shetu bondhon).

Smedley, P.L. 2003.Arsenic in Groundwater -south and east Asia. In: Arsenic in Groundwater: Geochemistry and Occurrence (Welch, A.H., and Stollenwerk, K.G.). Boston/Dordrecht/London, Kluwer Academic Publishers.

Kinniburg, D.G., Smedley, P.L., Davies, J., Milne, C.J., Gaus, I., Trafford, J.M., Burden, S., Huq, S.M.I., Ahmad, N. and Ahmed K.M. 2003. The scale and causes of the groundwater arsenic problem in Bangladesh. In: Arsenic in Groundwater: Geochemistry and Occurrence (Welch, A.H., and Stollenwerk, K.G.).Boston/Dordrecht/London, Kluwer Academic Publishers.

Schreiber,M.E., Gotkowitz,M.B., Simo, J.A., Freiberg, P.G. 2003. Mechanisms of Arsenic Release to Groundwater from Naturally Occurring Sources, eastern Wi sconsin: In Arsenic in Groundwater: Geochemistry and Occurrence (Welch, A.H., and Stollenwerk, K.G.).Boston/Dordrecht/London, Kluwer Academic Publishers.

Harvey, C.F., Awartz, C.H., Badruzzaman, A.B.M., Blute, N.K., Yu, W., Ali, M.A., Jay , J., Beckie, R., Niedan, V., Brabander, D., Oates, P.M., Ashfaque, K.N., Islam, S., Hemond, H.F. and Ahmed, M.F. 2003. Response to comments on "Arsenic Mobility and Groundwater Extraction in Bangladesh". *Science.* 300.

Geen A.v., Zheng, Y., Stute, M. and A hmed K.M. 2003. Comment on "Arsenic Mobility and Groundwater Extraction in Bangladesh" (II). *Science.* 300.

Aggarwal, P.K., Basu, A.R. and Kulkarni, K.M. 2003. Comment on "Arsenic Mobility and Groundwater Extraction in Bangladesh" (I). *Science.* 300.

Harvey, C.F., Awartz, C.H., Badruzzaman, A.B.M., Blute, N.K., Yu, W., Ali, M.A., Jay, J., Beckie, R., Niedan, V., Brabander, D., Oates, P.M., Ashfaque, K.N., Islam, S., Hemond, H.F. and Ahmed, M.F. 2002. Arsenic Mobility and Groundwater Extraction in Bangladesh, *Science,* 298.

Husain, M.T., 2002. The Age of the Groundwater Arsenic Poisoning in Bangladesh. *The Daily Independent*, Dhaka, Bangladesh.

Breit, G., Foster, A.L., Whitney, J.W., Uddin, N. Md., Yount, J.C., Welch, A.H., Alam, M.Md., Islam, S.Md., Sutton, S. and Newville, M. 2001. Variable Arsenic Residence in sediment from Easter Bangladesh: Clues to Understanding Arsenic Cycling in the Bengal delta. Arsenic in Drinking Water - An International Conference, Columbia University, U.S.A.

Husain, M.T. 2001. The Roots of the Groundwater Arsenic poisoning in Bangladesh. *The Daily Bangladesh Financial Express* , Dhaka, Bangladesh.

DPHE/BGS/DFID. 2001. Arsenic contamination of groundwater in Bangladesh, Vol 2: Final report.

Bridge, T.E., and Husain, M.T. 2001. Arsenic Crisis: A Challenge for Scientists to find the Origin of Arsenic that Caused the Poisoning Disaster in Bangladesh. *The Daily Bangladesh Observer,* News From Bangladesh, Dhaka, Bangladesh.

Aggarwal, P.K., Basu, A.R. and Por eda, R.J. 2000. Isotope hydrology of groundwater in Bangladesh: Implications for characterization and migration of arsenic in groundwater, IAEA-TC project (BGD/8/016). IAEA, Vienna.

Bridge, T.E., and Husain, M.T. 2000. Groundwater Arsenic Poisoning and a solution to the Arsenic Disaster in Bangladesh, *The Daily Star, The Daily Independent,* Dhaka, Bangladesh.

Bridge, T.E., and Husain, M.T. 2000. The Increased Draw Down And Recharge in Groundwater Aquifers And Their Relat ionship to the Arsenic Problem in Bangladesh. *News From Bangladesh*, Shetubondhon (http :groups.yahoo.com/group/shetubondhon).

Husain, M.T. 1999. Oxyhydroxide Reduction and Agrochemical Hypothesis: Myth or Reality? *The Daily Bangladesh Financial Express* , Dhaka, Bangladesh.

DPHE/BGS/DFID. 1999. Groundwater Studies For Arsenic Contamination in Bangladesh. Phase 1: Rapid Investigation Phase, Final Report.

Nickson, R.T., McArthur, J.M., Ravenscroft, P., Burgess, W.G. and Ahmed, K.M. 1999. Mechanism of arsenic release to groundwater, Bangladesh and West Bengal. 1999. *Applied Geochemistry* 15 (2000) 403-413.

Chowdhury, T.R., Basu, G.K., Mandal, K.B., Biwas, B.K., Samanta, G., Chowdhury, U.K., Chanda, C.R., Lodh, D., Roy, S.L., Sa ha, K.C., Roy, S., Kabir, S., Quamruzzaman, Q., Chakraborti, D. 1999. Arsenic Poisoning in the Ganges Delta, *Nature.*

Bridge, T.E., and Husain, M.T. 1999. Clean Healthy Water for Bangladesh -An Emergency Supply is Desperately Needed to Protect the People of Bangladesh from the Arsenic Disaster *News From Bangladesh, The Weekly Bangla Barta, L.A, USA, The Weekly Bangladesh, N.Y, USA.*

Bridge, T.E., and Husain, M.T. 1999. Arsenic Disaster in Bangladesh -An Urgent Call to Save a Nation, Arsenic International Confe rence, N.Y, *The Daily Star, Bangladesh, News From Bangladesh., The Weekly Bangladesh, NY, USA* .

Nickson, R.T., McArthur, J.M., Ravenscroft, P., Burgess, W.G. and Ahmed, K.M. 1998. Arsenic poisoning of Bangladesh Groundwater. *Nature.* Volume 395, page 338.

Das, D., Samanta, G., Mandal, K.B.,Chowdhury, T.R., Chanda, C.R., Chowdhury, P.P., Basu, G.K., and Chakraborti, D. 1996. Arsenic in Groundwater in Six Districts of West Bengal, India. *Environmental geochemistry and Health (* 1996), 18, 5-15 (0269-4042 #1996 Chapman & Hall).

Sullivan, K.A. and Aller, R.C.,1996. Diagenetic Cycling of Arsenic in Amazon Shelf Sediments. , Vol 60, No.9,pp.1465-1477.

CHAPTER 3

EVALUATION OF HEAVY METAL AVAILABILITY IN THE MINING AREAS OF BULGARIA

Penka S. Zaprjanova[1], Violina R. Angelova[2], Krasimir I. Ivanov[2]
[1]*Institute of Tobacco and Tobacco Products, Markovo, 4108, Bulgaria;* [2]*University of Agriculture, Department of Chemistry, Acad. G. Bonchev Str., Block 24, Sofia 1113, Bulgaria*

Abstract: This experiment is a comparative study of the efficiency of 12 different extractants in determining the movable and easily assimilated by plants forms of lead and cadmium in a region polluted by the ore mining industry. The main part of both elements is blocked in slightly soluble compounds, and thus they can barely be reached by plants. Both the total quantity and the movable forms of lead and cadmium depend mainly on the possibility of oxidizing processes occurring in the soil and on the transformation of the major pollutants - PbS and CdS in compounds with higher solubility. This leads to their mechanical removal due to their getting into the soil solution or their adsorption and remaining in an exchangeable form in some of the soil components. The concentration of lead and cadmium in the plant species, typical for the investigated region, show that the risk assessment in using soils, contaminated by heavy metals, for agricultural activities cannot be based solely on the data regarding the total amount of these elements. Particularly important is their form, as well as the percentage of exchangeable forms, assimilable by the plants

Key words: mobile forms, Pb, Cd, defining, acid soils

1.   INTRODUCTION

The fate of various metals, including chromium, nickel, copper, manganese, mercury, cadmium, and lead, and metalloids, including arsenic, antimony, and selenium, in the natural environment is of great concern, particularly near former mine sites, dumps, tailing piles, and impoundment (Adriano, 2001). Soil, sediment, and water in these areas may contain higher-than-average concentrations of these elements, in some cases due to past mining and (or) industrial activity, which may cause the formation of the more bioavailable forms of these elements.

In order to estimate effects and potential risks associated with elevated elemental concentrations that result from natural weathering of mineral deposits or from mining activities, the fraction of total elemental abundances in water, sediment, and soil that are bioavailable must be identified. Bioavailability is the proportion of total metals that are available for incorporation into biota (bioaccumulation).

It is well known that the level of heavy metals assimilation by plants corresponds to the concentration of the exchangeable forms of those elements in soil. It is commonly accepted to call exchangeable the cations, reversibly adsorbed in the soil, which are kept in the soil due to electrostatic forces and which can be substituted by equivalent quantity cations of neutral salts. In contrast to the cations of the alkaline and alkaline-earth metals, the ions of the transition metals are able to form coordinate bonds, because of which the methods of extraction of the exchangeable cations of sodium, potassium, calcium and magnesium are not suitable for extraction of the exchangeable cations of heavy metals. The approaches that different authors use vary. The ability of each extractant to extract only one chemical species of soil lead and Cd is doubtful. Moreover, the kind of extracting solution as well as the conditions of treatments can influence the results. Most important for the correct defining of the exchangeable forms of heavy metals in soil is the right choice of extractant. According to Gorbatov and Zyrin (1987), the cation of the extracting salt has to meet the following requirements: (i) it must have a great adsorption energy; (ii) it should not interact specifically with the soil components and should not form sediments and/or complex compounds with them; (iii) the ion radius should be close to the ion radius of the extracted (substituted) cation and (iv) it should not create obstacles during the analysis. The second and the fourth conditions are also valid for the anion of the extracting salt. The strict fulfillment of all those conditions is very difficult, especially in case of simultaneous extraction and defining of several elements.

On the other hand, the concentration of the salt solutions for extracting the exchangeable ions of heavy metals should be chosen in such a way, that it is sufficient for full extraction of the ions from the soil-absorbing complex, because the ionic strength of the solution is able to exert an influence not only on the exchangeable ions of heavy metals, but also on their other compounds in the soil (Zyrin et al., 1974). Most researchers recommend the concentration of the extracting solution to be from 0.05 to 1 M.

Commonly used as soil extractants for the determination of mobile species and defining plant-available Cd, Pb, Zn and Cu, are unbuffered neutral salts ($NaNO_3$, $CaCl_2$, NH_4NO_3) (Jackson and Alloway, 1991). Gupta and Hani (1981) recommend the use of 1.0 M $NaNO_3$ as an extractant of Cd, while Sauerbeck and Styperek (1984) recommend the use of 0.05 M $CaCl_2$, because the solution of this salt does not extract the organic substance and has little influence on pH of the soil.

According to Gorbatov and Zyrin (1987) 0.5 M or 1 M $Ca(NO_3)_2$ is more appreciable for estimating exchangeable lead (II) content in soils. According to some authors, the use of calcium chloride is not recommended because the chloride ion has the ability to form stable compounds with heavy metals, which shifts the equilibrium to the liquid phase and leads to increased results.

In several publications, exchangeable lead is estimated with the technique commonly used to determine exchangeable alkaline and alkaline-earth cations, i.e. by extraction with 1N NH4Ac at pH 7.0 (Scokart et al., 1983) or pH 4.8 (Zyrin at al., 1974; Anderson, 1976).

Very often researchers use also complex forming extractants (EDTA, DTPA) for the determination of both mobile and potentially mobilizable species. According to Knezek and Ellis (1980) and Silanpaa and Jonsson (1992) EDTA can be used successfully as extractant for defining mobile forms of heavy metals in soils of different types, while DTPA is inapplicable as extractant in acid soils (Dolar and Keeney, 1971; Cattenie et al., 1981; Haq and Miller, 1992).

Norvell (1984) modifies the pH of the soil-extraction mixture from 7.6 to 5.3 by addition of NH_4Ac and the ratio soil-extractant and suggests that this extractant (DTPA pH 5.3) can be used for extraction of available essential and toxic metal from acid and metal-enriched soils.

In the case of simultaneous studying of the exchangeable forms of several elements, results can be derived which are difficult to explain, because of the possible occurrence of parallel reactions of complex-forming and sedimentation of the hardly soluble compounds with the participation of the replacing cations, as well as with the resulting change of the pH of the system.

In spite of the large number of papers on the defining of plant-available heavy metals, the comparison of the obtained results and establishing of common relationships is practically impossible. This gave us a reason to carry out a comparative study which would permit us to evaluate the effectiveness of 12 of the most broadly used extractants under identical conditions.

We took six soil samples with varying levels of pollution from mining regions of the East Rhodopes, Bulgaria. One of the important reasons for using samples from East Rhodopes is that despite the serious lead and cadmium pollution of this area over many years, the flora, which is typical of these types of soil, has not been unduly affected. Finding a connection between the total content of lead and cadmium in the soil, the quantity of movable forms and their real assimilation by plants could contribute to the development of new objective criteria for ecological evaluation and the possibilities of using this region for agricultural activities.

2.   MATERIAL AND METHODS

2.1   Soils

The polluted soils used in this experiment were sampled from the vicinity of the contaminated area of the mining regions of the East Rhodopes, Bulgaria. The ores mined in this region contain mainly sulfides of lead and cadmium. Soils were collected from the surface (0-20 cm depth) of six sites in the region of the town of Madzharovo. The soil samples were air dried, homogenized in an agitate mortar and sieved. A fraction with particle size <1 mm was taken for analysis. For determination of the physical and chemical properties of the soils, the following analytical methods were used: the total amount of humus - by a method of Turin; the soil texture analysis - by the pipette method; described by Atanassov et al. (1979); and pH (H_2O) was determined using soil-water ratio of 1:2.5. Some general characteristics of the soil samples are given in Table 1. The results presented in Table 1 show that the soils are of alluvial-meadow type. Coarse sand (45.0 – 62.8 %) and fine sand (20.0 – 36.0 %) dominate in the soil content. The silt clay and the clay fraction are within the normal limits for this type of soil. The content of humus is low, and the soil reaction in all cases is strongly acid.

Table 1. Some characteristics of the investigated soils

Soil №	pH (H_2O)	Humus, %	Particle size distribution, mm	
			Coarse sand 2 -0.2	Fine sand 0.2-0.02
Biotope 1	3.46	0.52	62.80	26.16
Biotope 3.1	3.77	0.73	59.24	20.76
Biotope 3.2	4.70	0.88	45.74	36.02
Biotope 4	3.88	0.83	45.40	24.56
Biotope 5	3.45	0.48	54.26	20.44
Biotope 6	3.79	0.47	60.06	17.34

Table 1 Cont. Some characteristics of the investigated soils continued.

Soil №	Partical size distribution, mm	FAO (1992)

	Silt 0.02-0.002	Clay <0.002	Silt +clay <0.02	
Biotope 1	5.04	6.00	11.04	Calcaric Fluvisols
Biotope 3.1	7.20	12.80	20.00	Calcaric Fluvisols
Biotope 3.2	7.92	10.32	18.24	Calcaric Fluvisols
Biotope 4	7.55	22.49	30.04	Calcaric Fluvisols
Biotope 5	9.57	15.73	25.30	Calcaric Fluvisols
Biotope 6	9.00	13.60	22.60	Calcaric Fluvisols

For extraction of the accessible for plants heavy metals we used the following extractants:

1. 1M KNO_3

2. 0.1 M $NaNO_3$ (Gupta and Hani, 1981)

3. 1M NH_4NO_3 (Symeonides and McRae, 1977)

4. 0.1M $Ca(NO_3)_2$

5. 2 M KCl

6. 1 M NH_4Cl (Krishnamutri et al, 1995)

7. 0.1M $CaCl_2$

8. 1M $MgCl_2$ (Martens and Chesters, 1967)

9. 1 M NH_4OAc, pH 7.0 (John et al, 1972)

10. 1 M NH_4OAc, pH 4.8 (Anderson and Nilsson, 1974)

11. 0.05 M Na_2EDTA, pH 6.0 (Claytion and Tiller, 1979)

12. 0.005 M DTPA+0.1 M TEA, pH 5.3 (Norvell, 1984)

Soil samples are shaken for 1 hour at 20°C with extractants 1-12. After shaking, the soil-solution system was centrifuged and filtered. For extractants 1-11 the ratio soil-liquid was 1:10, and in the case of extractants 12 – it was 1:5. Total content of heavy metals in soils was determined after decomposed over sand bath heater for 3 h with 21 ml of concentrated HCl + 7 ml of concentrated HNO_3. After cooling, it is transferred into a 50 ml flask and water is added to the mark.

2.2   Plants

Representative samples have been taken from dominant species of herbaceous, shrub and tree plant species (from 4 families: Fabaceae; Poaceae; Crassulacea; Polygonaceae) from 4 biotopes.

The flora in biotope 2 is represented by *Elytrigia repens* (Common Couch), *Trifolium arvense* (Rabbit-foot Clover), *Sedum album* (White Stonecrop).

Biotope 4 is located on the biggest tailings pond (total area of about 1 km^2) in the region, where technical and biological recultivation has been carried out. The biological recultivation has been implemented by using traditional grass mixes of the following species: *Medicago lupulina* (Black Medick), *Lotus corniculatus* (Bird's Foot Trefoil), *Vicia hirsuta* (Hairy Tare), *Lolium perenne* (Perennial Ryegrass), *Trifolium arvense* (Rabbit-foot Clover). Amphitheatrically, shrub and tree species have been planted in tiers: *Rosa canina* (Dog Rose), *Berberis vulgaris* (Common Barberry), *Betula pendula* (European White Birch), *Gleditschia triacanthos* (Honey Flower), *Robinia pseudoacacia* L. (Black Locust), *Populus canescens* (Gray Poplar).

The herbaceous vegetation in biotope 5 is represented by *Rumex acetosella* (Common Sheep Sorrel) and *Sedum album* (White Stonecrop). The flora in biotope 6 is represented mainly by *Elytrigia repens* (Common Couch), and the shrub and tree species are represented by single specimens: *Crataegus monogyna* (Hawthorn), *Prunus spinosa* (Blackthorn), *Rosa canina* (Dog Rose), *Betula pendula* (European White Birch), *Juniperus communis* (Common Juniper).

Biotopes 1, 5 and 6 are entirely denuded of vegetation. Because of that, for the purpose of a better representation of the plant species, vegetation probes have been taken also from biotope 2. The soil taken from this biotope contains low quantities of lead (14.7 mg/kg) and cadmium (3.1 mg/kg).

The contents of Pb and Cd in the plant material were determined after the method of dry mineralization. A 1 g sample was weighed in a quartz crucible and put into a furnace (T = 400°C) until ashing occurred. After cooling to room temperature, 1 ml HNO$_3$ (1:1) was added, evaporated in a sand bath and put again into the furnace (T = 400°C). The procedures were repeated until the ash was white. It was finally dissolved in 2 ml 20 p.cent.HCl, transferred in a graduated 10 ml flask and brought to volume with bidistilled water.

Quantity measurements for content of Pb and Cd in soils and plants samples were performed with AAS 3030B flame variant from the firm "Perkin Elmer".

3.   RESULTS AND DISCUSSION

3.1   Soils

The quantities of heavy metals extracted with the different extractants are shown in Table 2 and Table 3. As can be seen from the Tables 2 and 3, a large variation in the quantities of Pb and Cd extracted from the soils by all 10 extractants exist. This requires the two elements to be observed separately.

Table 2. Quantity of the extracted lead (mg/kg) in using various extractants

№	1 M KNO$_3$	0.1 M NaNO$_3$	1 M NH$_4$NO	0.1 M Ca(NO$_3$)	2 M KCl	1 M NH$_4$Cl	0.1 M CaCl$_2$
1	5.3	0.75	7.0	2.5	7.3	10.5	1.0
3.1	4.5	0.25	3.5	1.0	10.3	7.7	1.0
3.2	1.8	1.25	1.8	0.25	7.7	3.7	2.5
4	0.50	0.75	2.8	1.0	7.8	2.5	1.5
5	4.8	0.75	5.3	5.5	9.3	3.5	2.3
6	1.8	0.25	0.90	0.75	2.0	1.0	1.0

Table 2 Cont. Quantity of the extracted lead (mg/kg) in using various extractants continued.

№	1 M MgCl$_2$	0.05M EDTA	0.005 M DTPA	NH$_4$Ac (pH 4.8)	NH$_4$Ac (pH 7.0)	Total Pb
1	11.0	26.2	0.45	0.25	0.25	1477
3.1	3.5	16.2	0.10	1.5	1.5	735.1
3.2	0.25	17.2	0.50	0.75	0.05	223.4
4	2.3	8.75	0.25	2.5	1.5	243.4
5	3.5	4.0	0.15	5.0	2.2	2633.9
6	1.0	2.6	1.8	0.90	0.60	911.8

Table 3. Quantity of the extracted cadmium (mg/kg) in using various extractants

№	1 M KNO$_3$	0.1 M NaNO$_3$	1 M NH$_4$NO	0.1 M Ca(NO$_3$)	2 M KCl	1 M NH$_4$Cl	0.1 M CaCl$_2$
1	0.125	0.100	0.200	0.175	0.150	0.200	0.100
3.1	0.225	0.150	0.300	0.225	0.350	0.375	0.525
3.2	0.525	0.225	0.65	0.725	1.10	1.175	0.325
4	5.225	4.075	8.175	8.425	9.425	10.5	8.725
5	0.225	0.075	0.175	0.275	0.325	0.125	0.125
6	0.225	0.100	0.175	0.300	0.225	0.125	0.150

Table 3 Cont. Quantity of the extracted cadmium (mg/kg) in using various extractants continued.

№	1 M $MgCl_2$	0.05M EDTA	0.005 M DTPA	NH_4Ac (pH 4.8)	NH_4Ac (pH 7.0)	Total Pb
1	0.025	0.200	0.085	0.100	0.100	3.95
3.1	0.150	0.300	0.180	0.250	0.250	2.90
3.2	1.025	1.125	0.470	0.950	0.375	7.50
4	9.100	5.925	7.40	10.50	7.00	12.2
5	0.100	0.050	0.080	0.500	0.125	2.75
6	0.125	0.025	0.040	0.075	0.100	3.20

3.1.1   Defining of the mobile forms of Pb

The content of Pb in the studied samples varies within a wide range: in all cases it exceeds the accepted maximum allowable concentrations for Bulgaria. It is observed that the lowest level of pollution is determined for sample No 4 (tailings pond), which contradicts expectations. Explanation of this fact should be sought in the location of the site, from where the samples are taken, and the possibilities of the occurrence of the chemical processes of lead. As we already pointed out, the main reason for pollution of the studied region is the mining of lead ores, containing mainly PbS (galenite). It is known that PbS is easily oxidized by the oxygen from the air to $PbSO_4$, which bonds with CO_2 in the presence of moisture to $PbCO_3$:

$$PbS + 2O_2 = PbSO_4$$
$$PbSO_4 + H_2O + CO_2 = PbCO_3 + H_2SO_4$$

The proportion of the quantities of the lead compounds determines the total content of lead in the soil and the exchangeable part of the lead accessible for the plants. The dryer and better aerated soils stimulate the formation of the well-soluble lead sulfate, which easily gets into the soil solution and leads to decrease of the total quantity of lead and increase of its moveable forms. This is confirmed in general by the results presented in Table 2 and explains the low content of lead in the sample from the tailings pond. In case of over-moisturized soils the above-mentioned processes are less probable, which is in synchrony with their higher level of pollution.

The results presented in Table 2 do not allow single-meaning conclusions regarding the efficiency of the different extractants. While the small quantity of exchangeable lead when using $NaNO_3$ is natural and understandable, it is difficult to explain the lower efficiency of the bivalent ions (Ca^{2+} and Mg^{2+}) in comparison to the efficiency of K^+, especially when using the extractant

KCl. In all cases, however, the quantity of moveable lead does not exceed 8 % of its total quantity.

Considerable differences in the quantities of heavy metals extracted with EDTA and DTPA are found, which means that differences in their ability to chelate heavy metals is observed. EDTA solution extracts a greater part of the Pb contained in soil. This is explained on one hand by the affinity between EDTA and these element and the ability of Pb to form stable complex compounds with EDTA, and on the other hand by the fact that this extractant removes a great deal of the metals bound with Mn and Fe oxides (Li and Shuman, 1996). This is also confirmed by the results we have obtained.

The comparison of the results obtained by using NH_4OAc with a different pH deserves particular attention. Acetate ions probably exert solving action on the compounds of the humic acids with heavy metals, which is also manifested by coloring of the acetate solutions with organic (humic) substances. According to Zyrin and Sadovnikova (1985) NH_4OAc extracts the unstably bound compounds of heavy metals with the humus and as a result the solution is entered by 2-3 times more heavy metals than with other extractants. Usually 1 M NH_4OAc (pH 7) is used for extraction of the exchange or easily mobile forms of heavy metals from neutral or weak acid soils, which are considered to be fully available by plants. The results we have obtained show that with 1 M NH_4OAc (pH 4.8) more exchangeable Pb is extracted, than with 1 M NH_4OAc (pH 7). The registered values, however, are too low and contradict our previous experiments where with ammonium acetate there was up to 70 % extraction of the exchangeable Pb in the alkaline carbonate soils. This result can be explained by the low pH and the small quantity of humus in all samples.

3.1.2   Defining of the mobile forms of the Cd

The main reason for the pollution of the studied region by cadmium is CdS, which is present together with galenite in the mined ore. The results presented in Table 2 and Table 3, however, show opposite tendencies in the degree of pollution of the different samples. The most heavily polluted lead samples contain cadmium in quantity, which is close to the maximum allowable concentrations, while its content in the tailings pond and the comparatively less polluted lead sample No 6 considerably exceeds maximum allowable concentrations. This fact could be explained by the chemical processes that occur and by the different solubility of the obtained compounds of lead and cadmium.

Essential difference between the chlorides and nitrates used as extractants were not observed. In general, the chlorides extract a bit more exchangeable

Cd, which, as well as in lead, is best expressed in KCl. This can be explained by the formation of the stable complex $[CdCl_4]^{2-}$, while the nitrate ions do not form a stable complex with cadmium. These results confirm what has been found by Barrow (1987), that greater sorption is reached with chloride solutions compared to those reached with nitrate solutions. Among the nitrates, the greatest quantity is extracted with $Ca(NO_3)_2$. Ca^{2+} is adsorbed by CEC to a greater extent than the single valent ions (Na^+, K^+, NH_4^+) replacing respectively greater quantities of heavy metals.

Considerable differences in the quantities of Cd extracted with EDTA and DTPA are not found, meaning that no difference in their ability to chelate heavy metals is observed. DTPA extracts some more Cd than EDTA in soils from samples 4, 5 and 6. This corresponds to what is found by Alloway and Morgan (1986) and Mellum et al.(1998), who claim that DTPA extracts most Cd, due to the high stability constant of the Cd- DTPA complex (Smith and Martell, 1977).

NH_4OAc with pH 4.8 and NH_4OAc with pH 7.0 extract almost the same quantities exchangeable Cd, because the registered values do not differ considerably from the ones extracted with other extractants.

In cadmium, as well as in lead, the quantity of the exchangeable forms is comparatively small and does not exceed 20 %. The only exception is Sample No 4 (the tailings pond), where the content of cadmium is the highest, and the quantity of its exchangeable forms reaches 86 %. The results we have obtained do not allow a satisfactory explanation of this obvious fact, most probably related to the oxidizing processes that occur at different rates.

3.2   Plants

The results obtained regarding the concentration of heavy metals in the examined dominant herbaceous, shrub and tree species are presented in Figures 1 and 2. The results show that in most cases most of the heavy metals accumulate in the roots. This is explained by the fact that during the penetration into the plasma there occures inactivation and deferring of considerable quantities of heavy metals, probably due to the formation of less available compounds with the organic substance. Interestingly, the examined herbaceous species have different abilities to accumulate lead and cadmium. The higher values of Pb have been registered in the roots of the Common Sheep Sorrel (60.4 mg/kg) from biotope 3.1, and for Cd – in the roots of the White Stonecrop (6.6 mg/kg) from biotope 2.

It is well known that herbaceous species do not tend to accumulate heavy metals, which is confirmed by the present results. Because of a considerable difference in the plant species characteristic for the different biotopes, it is difficult to find correlation dependencies between the results obtained from the investigation and soils and the plant species. Evident, however, is the relation between the total concentration of Pb in the soil and its concentration in the roots of the plants. For example, when the concentration of Pb in the soil is 243.4 mg/kg (biotope 4) the registered values reach up to 3mg/kg, while when the concentration of Pb in the soil is 735.1 mg/kg (biotope 3.1), they reach up to 60.4 mg/kg. The results are different for the Rabbit-foot Clover. There is observed considerable difference in the accumulation of Pb in its roots, despite the registered values of Pb in the soils from biotope 4 (6.1 mg/kg) and biotope 3.2 (17.2 mg/kg). The results obtained might be explained by the exchangeable forms of Pb in the investigated biotopes. In biotope 3.2 about 8% of the total quantity of Pb is located in a form that is assimilable for the plants, while in biotope 4 it is below 4.0 %.

The results for Cd in the White Stonecrop and the Common Sheep Sorrel are highly interesting – the concentration of Cd in the roots is higher than the concentration of Cd in the soil. For example, the White Stonecrop accumulates 6.6 mg/kg in its roots when the concentration in the soil is 3.1 mg/kg (biotope 2), and 3.2 mg/kg when the concentration in the soil is 2.9 mg/kg (biotope 3.1). Similar results have been obtained for the Common Sheep Sorrel. The concentration of Cd in the roots reaches 2.3 mg/kg, when the concentration in the soil is 2.9 mg/kg (biotope 3.1). A possible explanation for these results is the fact that both biotopes characterize with constant (during spring and summer) draining of subterranean waters from the surface of the tailings pond and from the appertaining slopes, forming a micro-water basin, whose waters drain into the Arda River.

In most cases in the overground parts of the herbaceous species, smaller quantities of heavy metals compared to the root system were found, which shows that their moving through the conducting system is impeded. The highest values of Pb have been registered in the Rabbit-foot Clover (6.6 mg/kg) in biotope 3.2, while for Cd in the White Stonecrop – 6.6 mg/kg in biotope 3.2.

Pb

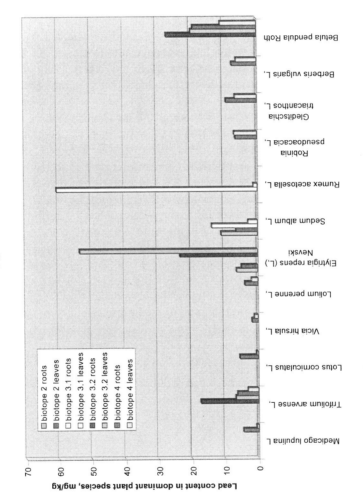

Figure 1. Lead accumulation in dominant plant species in East Rhodopes, Bulgaria

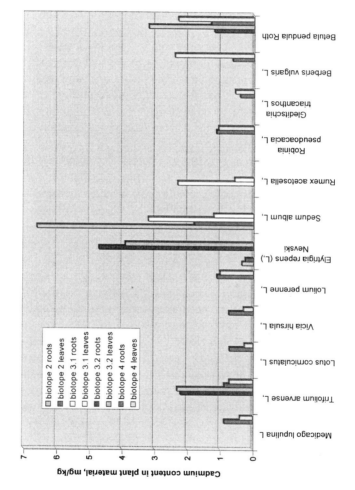

Figure 2. Cadmium accumulation in dominant plant species in East Rhodopes, Bulgaria

Shrub and tree species (*Robinia pseudoacacia*; *Gleditschia triacanthos*; *Berberis vulgaris*; *Betula pendula*), as seen from the results, presented in Figures 1 and 2, have different abilities to accumulate heavy metals. In both in the shrub and the tree species, the main part of lead and cadmium is deferred in the roots, as the concentration of Pb in the roots varies from 6.4 mg/kg to 27.0 mg/kg, and of Cd - from 0.4 mg/kg to 1.3 mg/kg. Considerably higher values have been registered in the stems and leaves. The highest values of Pb have been registered in the overground parts of the Honey Flower – 6.4 mg/kg, while for Cd – in the overground parts of the Common Barberry - 2.4 mg/kg.

Our results show that the concentration of heavy metals in the herbaceous, shrub and tree species is comparatively low and corresponds to the small quantity of mobile forms of lead and cadmium. An exception is observed in biotope 4 (the tailings pond), where the concentration of Cd in the studied species is small, while its concentration in the soil is the highest, and the quantity of its mobile forms reaches 86 %. The explanation of this fact can be sought in the characteristics of biotope 4 – lack of water drainage, strong over-moisturizing of the soil during the rainy seasons, as well as in the antagonistic impact of other elements, contained in the lead and zinc ores – zinc, selenium, etc. Finding a definite answer to this question, however, requires further investigation.

4.   CONCLUSION

A comparative study of the efficiency of 12 different extractants in determining the moveable and plant-assimable forms of lead and cadmium in a region polluted by the ore mining industry has been carried out. It has been found that the main part of both elements is blocked in slightly soluble compounds, and thus they can be hardly reached by plants. The exceptions and contradictory results that have been observed for some of the samples (mainly Sample No 4) do not allow us to draw just one single conclusion. It is obvious, however, that both the total quantity and the moveable forms of lead and cadmium depend mainly on the possibility of oxidizing processes occurring in the soil and on the transformation of the major pollutants - PbS and CdS in compounds with higher solubility. This leads to their mechanical removal due to their getting into the soil solution or their adsorption and remaining in an exchangeable form in some of the soil components. The complete clarification of those issues requires fractioning of the soil and determination of the content of lead and cadmium in any of the forms, as well as searching for a correlation with their concentration in plants.

The results for the concentration of lead and cadmium in the plant species, typical for the investigated region, show that the risk assessment in using soils contaminated by heavy metals for agricultural activities cannot be based solely on the data regarding the total amount of these elements. Of particularly important is their form, as well as the percentage of exchangeable forms, assimilable by the plants.

REFERENCES

Adriano, D.C. 2001. Trace elements in terrestrial environments: biogeochemosrty, bioavailability and risks of metals . Springer-Verlag, New York

Alloway, B. J., Morgan, H.1986. The behaviour and availability of Cd, Ni and Pb in polluted soils. *In* Contaminated soil, Assink, J.W. and W. J. van den Brink (Eds). pp. 101 -113, Martinus Nifhoff Publishers, Dordrecht, The Ne therlands.

Anderson, A., Nilsson, K. O. 1974. Influence of lime and soil pH on cadmium availability to plants. *Ambio.* 3, 198-200.

Anderson, A. 1976. On the determination of ecologically significant fractions of some heavy metals in soils, *Swed. J. Agric. Res*. 6, 1

Atanassov, I., Totev, T., Stefanov S. and Dimitrov, G. 1979. Handbook of soil science practicals. Zemizdat, Sofia

Barrow, N.J., 1987. Reactions with variable charge soils. pp. 18 -32, Martinus Nijhoff, Dordrecht, The Netherlands

Cattenie, A., Kang, G., Kiekens, L. and Sajjapougse, A. 1981. Micronutrient status. *In* Characterization of soils in relation to their classification and managment for crop production: Examples from same Areas of the Humid Tropics, Greenland, D .J. (Ed). pp.149-163, Clarendon Press, Oxford

Clayton, P. M. and Tiller, K. G. 1979. A chemical method for the determination of the heavy metal content of soils in environmental studies. CSIRO Aust. Tech. Pap.no 41: 1 -17.

Dolar, S.G. and Keeney, D. R. 1971 . Availability of Cu, Zn and Mn in soils. I. Influence of pH, organic matter and extractable phosphorus. *J. Sci.Fd. Agric*. 22, 273-278.

FAO, 1992. Guidelins for soil profile description. Rome, p.72.

Gorbatov, V. S. and Zyrin, N. G. 1987. Comparison of extr acting solutions applied to displace exchangeable cations of heavy metals from soils, Vestnik Moscovskovo Universiteta, Ser. 7 Soil Science. 2,22

Gupta, S. and Hani, H. 1981. Einflub von leicht extrahierbaren Boden -Cd auf die Reaktion verschiedener Testpflanzen (unter spezieller Beruckisichtigung der Wurzelaustauschkapazitat) und einige microbiolodische Parameter. Korresp. Abwass. 4, 211-213.

Haq, A. U. and Miller, M. H. 1992. Prediction of available soil Zn, Cu and Mn using chemical extract ants. *Agron. J.* 74, 779-782.

Jackson, A. P. and Alloway, B. J. 1991. The bioavailability of cadmium to lettuce and cabbage in soils previously treated with sewage sludge. *Plant Soil.* 132, 179-186

John, M.K., Van Laerhoben, C. J. and Chuah, H.H. 1972. Factors affecting plant uptake and phytotoxicity of cadmium added to soils. *Environ. Sci. Tech.*. 6, 1005-1009.

Knezek, D.D. and Ellis, H. P. W. 1980. Essential micronutrients IV. Copper , iron, manganese and zinc. In Applied soil trace elements , Davies, B. E. (Ed). pp. 259-268, John Wiley, Chichester

Krishnamutri, G.S.R., Huang, P. M., Van Rees, P. M., Kozak, L.M. and Rostad, H. P. W. 1995. A new soil test method for the determination of plant available cadmium in soils. *Commun. Soil Sci. Plant Anal.*. 26, 2857-2867.

Li, Z. and Shuman, L. M. 1996. Redistribution of forms of zinc , cadmium and nickel in soils treated with EDTA. *Sci. Total Environ.*. 19, 95-100.

Martens, D.C. and Chesters, B. R. 1967. Comparison of chemical tests for estimation of the availability of soil zinc. *J. Sci. Food Agric.* 18, 187-193.

Mellum, H. K., Arnesen, A. K. M. and Singh, B. R 1998. Extractability and plant uptake of heavy metals in alum shale soils. *Commun. Soil Sci. Plant Anal*. 29, 1183-1190.

Norvell, W. A., 1984. Comparison of chelating agents as extractants for metals in diverse soil materials. *Soil Sci. Soc. Amer*. J. 48, 1285-1292.

Sauerbeck, D. and Styperek, P. 1984. Predicting the cadmium availability from different soils by CaCl$_2$ extractionIn: Prosscessing and use of sewage sludge, Hermite, P. L. and H. Ott (Eds). pp. 431-435, D. Reidel Publishing Co.

Scocart, P.O., Meeus-Verdinne, K. and De Borger, R. 1983. Mobility of heavy metals in polluted soils near zinc smelters. *Water, Air Soil Pollution*. 20, 451

Silanpaa, M and Jansson, H. 1992. Stat us of cadmium, lead, cobalt and selenium in soils and plants of thirty countries. FAO Soils Bulletin 65. Food and Agricultural organization of the United Nations, Rome, Italy

Smith, R. M. and Martell, A. 1977. Critical stability constants. Volume 3. Other organic ligands. Plenum Press, New York, NY

Symeonides, C. and Mc Rae, S. G. 1977. The assessment of plant available cadmium in soils. *J. Environ. Gual.* 6, 120-123.

Zyrin, N.G., Obukhov, A.I. and Motuzova, G. V. 1974. Forms of trace elements in soils and methods to studt them. In Proc. X International congress of soil science, Moscow, 2.

Zyrin, H. G., and Sadovnikova, L. K. 1985. Chemistry of the heavy metals , As and Mo in soils. Moscow, p. 208.

CHAPTER 4

AVERAGE PARTICLE SIZE RATIOS AND CHEMICAL SPECIATION OF COPPER AND ZINC IN ROAD-DUST SAMPLES

Adnan M. Massadeh[1] and Qasem M. Jaradat[2]

[1]*Department of Medicinal Chemistry and Pharmac ognosy, Faculty of Pharmacy, Jordan University of Science and Technology, P.O. Box 3030, Irbid 22110, Jordan, email: massadeh@just.edu.jo, Fax: 00962 -2-7095019;* [2]*Chemistry Department, Faculty of Science, Mutah University, Al-Karak, Jordan*

Abstract: Road-dust samples were taken from areas of high traffic flows in Irbid City; Jordan. A series of <90, 90-<106, 106-<125, 125-<250, 250-<1000 and 1000- < 2000 μm of road-dust particle size fractions were investigated throughout to evaluate the ratio between metal content in each particle size fraction and total metal content. Atomic absorption spectrometry technique was used throughout. Sequential procedure was used for chemical speciation in road dust samples of <90 μm. This procedure permits a reproducible evaluation of the partitioning of Cu and Zn among the various chemical forms in which they are present in street dust. Two reference standard materials BCR -CRM 142R and NIST-SRM 2709 were tested to validate the propo sed method. Results show that there was no significant difference between the measured values for Cu and Zn and their certified values with RSD of less than 5%.

Key words: copper, zinc, chemical speciation, road dust, atomic absorption spectrometry.

1. INTRODUCTION

Copper and zinc are essential trace elements for humans, for example Cu is required for normal biological activity of many enzymes and for haemoglobin formation (Underwood, 1977). Zn is also a trace element needed as a building block for many enzymes and by cells of the immune system. The presence of Cu and Zn at high concentrations is considered a potential health hazard to both humans and their environment. Chronic exposure to Cu can cause liver and kidney damage, anemia, chills, malaise and shortness of breath, and stomach and intestinal irritation. People with Wilson's disease are at greater risk for health effects from overexposure to Cu. Exceeding or staying below normal concentrations of Zn the in blood will cause various medical disorders. Thus, it is important to study these elements in different environmental samples including road dusts.

The variation in particle size may be due to the various types of pollutant sources. For example, wear releases rubber particles with an average of 20 μm. The amount of heavy metals contamination increases as traffic density increases (Massadeh and Snook, 2002). Rubber is the main source for Zn, plus metal plating on car parts. Moreover, Zn is also, associated with tyre wear and oil spills on road surfaces (Ellis and Revitt, 1982). The largest vehicle-emitted particles of 300-3000μm diameter are subject to rapid gravitational settling whilst the 5-50 μm fraction also settles to a close source due to turbulent deposition upon surfaces (Al-Chlabi and Hawker, 1997). Cited authors recommended using different particles of less than 100 μm and 750 μm with different acid extraction methods or smaller for heavy metal analysis in road dust and roadside soils (Massadeh and Snook, 2002; de Miguel et al., 1997; Bjelkas and Lind mark, 1994). It is believed that roadside dust smaller than 125 μm originated from airborne particulates, while larger dusts, for instance 500 μm, might originate from roadside soils.

In this study, Cu and Zn levels of different particle sizes of road-dust samples taken from two high-traffic roads in Irbid City in the northern part of Jordan were evaluated to find the ratio between each particle size's metal content and the total metal content; to show their distributions in each particle size under test and to deduce the relationship between the deposited road-dust particle size and heavy metal contents; and to demonstrate the chemical speciation of Cu and Zn in the five steps using the sequential extraction procedure recommended by Tessier et al., 1979.

2.   EXPERIMENTAL

2.1   Reagents and Materials

All reagents and chemicals used were analytical reagent grade. Copper nitrate, Zinc nitrate (Scharlau, Spain) were used to prepare standards; 70% HNO3, 35-38% HCl (Scharlau, Spain) and doubly de-ionised water were used. All glassware, plasticware, sample collection bottles and other containers were thoroughly cleaned by soaking in 10% (v/v) HNO_3 for 48 hours being rinsed thoroughly with de-ionised water prior to use. Standard solutions of the elements under test were prepared to produce 1000 ppm of copper nitrate and zinc nitrate. Serial dilutions of the stock solution with doubly de-deionised water were used to generate working solutions.

2.2   Area of Study and Sampling Procedure

In this study, samples of road dust were collected from two major roads in Irbid city, Jordan, during the period between October 2001 and July 2002. Samples were collected by gently sweeping the road surface with a plastic brush into dustpan, taking care to collect smaller particles that might be lost by suspension into the atmosphere (Wang et al., 1998). At first the samples were stored at 4 °C in a refrigerator before the drying step. The sampling points are designated as road dust from three different sites along the street with subscripts (THRD1), (THRD2), (THRD3) for Thirty Street and (HARD1), (HARD2), (HARD3) for Al-Hashmi Street.

Pre-cleaned polyethylene containers were used for collection of road-dust samples. Road dust sampling tools comprised brushes and plastic spatulas. Road-dust samples were collected and retained in polyethylene bottles before being brought back to the laboratory for subsequent analysis.

2.3   Instrumentation and Procedures

A Varian model (Spectra AA-10) flame atomic absorption spectrometer with computer system was employed throughout the experiment for determination of Cu and Zn. Measurements were performed at 324.8 and 213.9 nm for Cu asnd Zn, respectively. A slit width of 0.7 nm was used. The flow rate of acetylene used was 1.5 ml/min. Air–Acetylene flame was used throughout. Sartorious analytical balance (model A 120 S), shaker with water bath (model D-3162 Kottermann labortechnik, type 3047, West Germany model), PTFE vessel and oven (model D4C Genlab, Widnes, England) were also employed throughout this project. A series of stainless-steel sieves with apertures of 90, 106, 125, 250, 1000 and < 2000 μm were used to sieve road dust into these size ranges.

2.4   Acid Digestion Method

The procedure used by Massadeh and Snook, 2002 with some modifications was applied in this work. Road-dust samples were dried in the oven for 24 hours at 105 °C. After cooling, each sample was passed through a 2 mm mesh to remove large particles such as stones, glass and other debris and then samples were passed through a series of stainless steel sieves with apertures of 90, 106, 125, 250 and 1000 and < 2000 μm, respectively. After that, a 0.5 g of each fraction size of road-dust samples was digested in 10 ml of mixture solutions of $HClO_4$, HCl, HNO_3 with a ratio of 2:3:5 v/v in a pre-cleaned polyethylene flask. This suspension was shaken for 24 hours, diluted with 10 ml of deionized water, and then filtered using Whatmann filter paper (No. 41) into a 25 ml volumetric flask. The extract solution for each road-dust sample was completed with deionized water to the mark and kept in the refrigerator at 4 C° until analysis. Chemical analysis for Cu and Zn concentrations in road-dust samples were carried out for the different particle sizes using atomic absorption spectrometry (AAS).

2.5   Sequential Extraction Method

Sequential extraction techniques have been widely used by geochemists to study the distribution of heavy metals in various environmental samples such as soil, sediment and dust (Tessier et al., 1977).

The chemical speciation of metals is widely used to show the average relative abundance (%) of heavy metals in each phase in the analysed road-dust samples of the study area. The sequential procedure was carried out as follows in the sequence steps:

1. Exchangeable fraction: 0.5 g dry weight dust sample (<90 μm) was extracted for 1 hour with 8 ml of 1M $MgCl_2$ (Scharlau, Spain) at pH 7 with continuous agitation.
2. Bound to carbonate: the residue from step (i) was leached with 8 ml of 1M CH_3COONa ((Scharlau, Spain) adjusted to pH 5 with CH_3COOH (Scharlau, Spain) for 6 hours.
3. Bound to Fe-Mn oxides: the residue from step (ii) was extracted with 0.04 M hydroxyl amine hydrochloride ($NH_2OH.HCl$) in 25% (v/v) CH_3COOH for 12 hours with continuous agitation.
4. Bound to organic matter: the residue from step (iii) were added 3 ml 0.02 M HNO_3 and 5 ml of 30% H_2O_2 adjusted to pH 2 with HNO_3 in 20% (v:v) HNO_3 was added and the sample was diluted to 20 ml and agitated continuously for 30 minutes.
5. Residual fraction: the residual from step (iv) was digested with a ratio of 5:1 of HF: $HClO_4$ mixture.

3.   RESULTS AND DISCUSSION

3.1   Analysis of Standard Reference Materials

Two reference standard materials BCR-CRM 142R and NIST-SRM 2709 were analysed to assess the accuracy and the reliability of the proposed method. Table 1 shows the results obtained.

Table 1. Copper and Zinc levels (μg g^{-1}) in reference standard materials BCR -CRM 142R and NIST-SRM 2709.

	Certified value (μg g^{-1})	Measured value (μg g^{-1})
BCR-CRM 240R		
Cu	69.7± 1.4	68.2 ± 4.2
Zn	93.3 ± 2.8	91.5 ± 5.1
NIST-SRM 2709		
Cu	34.6 ± 0.69	33.9± 1.6
Zn	106± 3.18	104± 4.9

These results are derived from five replicate injections and the errors quoted represent 95% confidence limits. No significant difference was observed between the certified reference values and the results obtained by the acid digestion method. Also, the quality control (QC) samples, by analyzing known concentrations of standard solutions of heavy metals during the analysis of samples, give accurate and precise results.

3.2   The Average Particle Ratios

To achieve better comparisons in particle size classification, it is important to use the average of particle ratios rather than the absolute particle concentrations (Al-Rajhi et al., 1996). This may lead to finding the correlation between groups fractionated into particle sizes the average of particle size ratios which use the ratio between the metal concentration per gram of a certain size to the sum of that metal concentration for all different particle sizes (Al-Rajhi et al., 1996).

Results displayed in Table 2 show the ratio of Cu concentrations for each fraction to the total Cu concentrations in all particle sizes. Ratio ranged from 0.033-0.281 and 0.027-0.237 in THRD and HARD Streets, respectively. This means ratios in the two main streets are different, and the highest ratios were found in the particle sizes of <106-125 and 125-<250μm in THRD and HARD Streets, respectively.

Table 2. Percentage distribution of each fraction size in road -dust samples and the ratio of Cu content in each fraction to total copper concentration.

	Particle size(μm)	% Distribution	Cu (μgg⁻¹)	*Ratio
Thirty Street				
	< 90	6.2	3.5	0.090
	90- <106	14.8	10.2	0.261
	106-<125	16.2	11.0	0.281
	125-<250	19.7	8.5	0.217
	250-<1000	25.3	4.6	0.118
	1000-< 2000	17.8	1.3	0.033
Al-Hashmi Street				
	<90	5.5	4.0	0.045
	90-<106	13.6	11.4	0.128
	106-<125	14.2	17.5	0.197
	125-<250	20.1	21.1	0.237
	250-<1000	27.4	12.6	0.142
	1000- < 2000	19.2	2.4	0.027

* It represents the Cu content in each fraction to total Cu content.

For Zn, ratios ranged from 0.073-0.240 and 0.0818-0.3866 for THRD and HARD Streets, respectively (Table 3). The highest ratios were found in the particle sizes of 125-<250 and 250-<1000μm in THRD and HARD streets, respectively.

Table 3. Percentage distribution of each fraction size in road -dust samples and the ratio of Zn content in each fraction to to tal Zn concentration.

	Particle size(μm)	% Distribution	Zn (μgg⁻¹)	*Ratio
Thirty Street				
	< 90	6.2	21.1	0.079
	90- <106	14.8	52.1	0.194
	106-<125	16.2	52.7	0.196
	125-<250	19.7	64.4	0.240
	250-<1000	25.3	58.4	0.218
	1000-< 2000	17.8	19.6	0.073
Al-Hashmi Street				
	<90	5.5	19.3	0.082
	90-<106	13.6	49.0	0.208
	106-<125	14.2	53.4	0.226
	125-<250	20.1	59.1	0.250
	250-<1000	27.4	91.2	0.387
	1000- < 2000	19.2	23.0	0.098

* It represents the Zn content in each fraction to total Zn cont ent.

The total Cu (Table 2) and Zn (Table 3) contents in dust from Al-Hasmi Street are higher than those from Thirty Street.

3.3 Chemical Speciation of Copper and Zinc

In this study road-dust samples of < 90 μm from each road were examined for chemical forms using the sequential extraction procedure mentioned before. The fine-grade fractions of <90 μm were used due to their higher surface area, which have a positive effect on the mount of heavy metals that can be bounded at their surfaces. The average relative abundance (%) of Cu in different geochemical phases in the analyzed road-dust samples in Thirty Street (Figure 1) are arranged in the following order: organic matter (28.2%) > residual (23.3%) > Carbonate (18.9 %) > exchangeable (17.9%)> Fe-Mn (11.7%), while, the average relative abundance (%) of Cu in different geochemical phases in the analyzed road-dust samples in Al-Hashmi Street (Figure 2) are arranged in the following order: organic matter (25.3 %) > residual (24.8%)> carbonate (20.2%) > Exchangeable (15.9%) > Fe-Mn (13.8%).

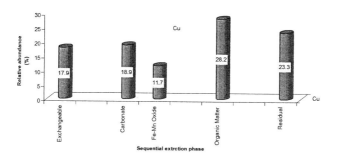

Figure 1. The average relative abundance (%) of Cu in different geochemical phases in the analyzed road-dust samples collected from Thirty Str eet.

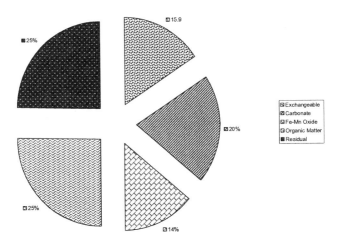

Figure 2. The average relative abundance (%) of Cu in different geochemical phases in the analyzed road - dust samples collected from Al -Hashmi Street.

For Zn, the average relative abundance (%) in different geochemical phases in the analyzed road-dust samples in Thirty Street (Figure 3) are arranged in the following order: carbonate (30.7%) > exchangeable (22.8%) > Fe-Mn(18%) > residual (16.2%) > organic matter (12.3%), while, the average relative abundance (%) of Zn in different geochemical phases in the analyzed road-dust samples in Al-Hashmi Street (Figure 4) are arranged in the following order: carbonate(27.7%)> exchangeable (21.2%) > Fe-Mn (19.3%) > residual (19.0%) > organic matter (12.8%).

This means that low yields of Zn in road-dust samples appear to be largely concentrated in the insoluble fraction.

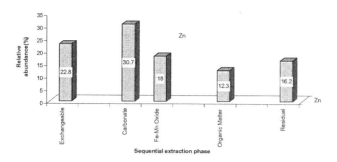

Figure 3. The average relative abundance (%) of Zn in different geochemical phases in the analyzed road-dust samples collected from Thirty Street

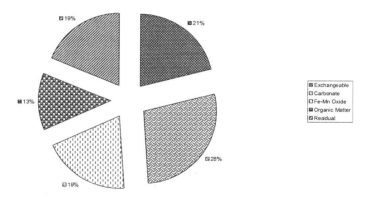

Figure 4. The average relative abundance (%) of Zn in different geochemical phases in the analyzed road-dust samples collected from Al-Hashmi Street.

4.   CONCLUSIONS

In the work presented here, ratios of average particle size concentration for Cu and Zn to the total particle sizes concentrations were evaluated (Tables 2 and 3). The ratio of Cu and Zn in each examined particle size to average total heavy-metal level did change over a period of several months. This means that metal content in each particle size is different from each particle size to another particle size. The chemical speciation forms for Cu and Zn in road dust samples were dominant in organic and residual steps for Cu, and in carbonate steps for Zn. The average relative abundance (%) of Cu in different geochemical phases in the analyzed road-dust samples in the two streets were arranged in the following order: organic matter > residual > Carbonate > exchangeable > Fe-Mn, and the average relative abundance (%) of Zn were arranged in the following order: carbonate> exchangeable > Fe-Mn > residual > organic matter.

The sequential extraction step for organic matter produced low yields for Zn, which means that Zn in road-dust samples appears to be largely concentrated in this insoluble organic fraction. Thus it is reasonable to conclude that the two likely sources for this insoluble organic matter are vehicle exhaust emissions and asphalt paving material (Kleeman et al., 2000).

ACKNOWLEDGMENTS

The authors would like to thank Hokmia Al-Khateeb for her technical assistance.

REFERENCES

Abollino M, Aceto M, Malandrino, Mentasti E, Sarzanini C, Barberis R, Environ. *Pollut.*, 2002, 19, 177-193.

Al-Chlabi AS, Hawker D, 1997, *Sci. Tot. Environ.*, 206, 195-202.

Al Rajhi MA, Al-Shayeb SM, Seaward MRD and Edwards HGM, 1996, *At. Environ.*, 30 (1), 145-153.

Al Rajhi MA, Seaward MRD and AL-Aamer AS, 1996, Metal levels indoor and outdoor dust in Riyadh, Saudi Arabia, *Environ. Inter.*, 22 (3), 315-324.

Beckwith P, Ellis JB and Revitt DM, 1985 Heavy metals in the Environment, 1, 174-176, CEP, Edinburgh.

Bjelkas J, Lindmark P, 1994, Rep. VARIA 420, Swedish Geotechnical Inst.,"Pollution of ground and highway runoff."

de Miguel E, Llamas JF, Chacon E, Berg T, Larssen S, R Øyset O, Vadset M,1997, *Atm. Environ.*, 31 (17), 2733-2740.

Ellis JB, Revitt DM. Incidence of heavy met als in street surface sediments : solubility and grain size studies. *Water, Air and Soil Pollution* 1982; 17:87-100.

Kleeman MJ, Schauer JJ, and Cass GR, 2000, Size and composition distribution of fine particulate matter em itted from motor vehicles, *Environ. Sci. and Techn.* 34, 1132-1142.

Massadeh AM and Snook RD, 2002, J. Envirn. Monit., 4, 567 -572.

Ogunsola O, Ouluwole A, Asbiojo O, Olaniyi H, Akeredolu F, Akanle O, Spyrou N, Ward N and Ruck W, 1994 Sci. *Total Environ.,* 146 (147), 175-184.

Stone M, Marsalek J,1996, *Water Air Soil Pollution*. 87(1/4), 149-168.

Tessier A, Campbell P, Bisson M, 1979, *Analyt. Chem.,* 51 (7), 844-851.

Underwood EJ, E. J, 1977, Trace elements in human and animal nutrition, 4th edi tion, Academic Press, Inc., New York.

Viklander M, 1998, *J. of Environ. Eng.* , August, 761-766.

Wang WH, Wong MH, Leharne S, Fisher , Fractionation and Biotoxicity of Heavy Metals in Urban Dust Collected from Hong Kong and London, Environmental Geochemis try and Health (1998), 20, 185 -198.

PART III: MODELING

CHAPTER 5

CONTAMINANT FATE AND TRANSPORT IN THE COURTROOM

Charles M. Denton, Esq.[1] and Michael G. Sklash, Ph.D., P. Eng.[2]

[1]*Varnum, Riddering, Schmidt & Howlett LLP, Bridgewater Place, P.O. Box 352, Grand Rapids, MI 49501-0352;* [2]*The Dragun Corporation, 30445 Northwestern Highway, Suite 260, Farmington Hills, MI 48334*

Abstract: Modeling of contaminant fate and transport in soils, groundwater, air and other environmental media can be a critical component of permitting, remedial action planning and design, site characterization, and source identification. Modeling of emissions and contamination can result in si gnificant cost-savings as compared with additional sampling and analysis, and is frequently utilized by potentially responsible parties as well as governmental agencies for air permitting, remedial investigations, corrective measures studies, and engineering design. The science of modeling has been challenged in federal and state courtrooms across the country. Questions have been raised as to the reliability, predictability and specificity of contaminant fate and transport modeling by various adversaries, including former operators, neighboring owners, and liability insurers. This paper will discuss the scientific rigors of environmental modeling, its many uses and acceptability within the scientific community, as well as its limitations. The paper will also review and discuss recent court rulings and evidentiary issues regarding contaminant fate and transport modeling in litigation and related expert witness testimony.

Key words: Groundwater models, *Daubert*, expert witness, rules of evidence, CERCLA, modeling process, limitations of modeling, fate and transport, MODFLOW, case studies.

1.   INTRODUCTION

The main objective of this paper is to discuss expert witness testimony utilizing computer models to simulate the fate and transport of chemicals in the environment. A "model" provides an approximation of a field situation. Models are commonly used to simulate the movement of groundwater, surface water, air, and sediments, as well as to simulate the distribution of chemicals in these media. Models can be used to explain current conditions, predict future conditions, and recreate past conditions.

There are two conditions that distinguish modeling for the courtroom from modeling used for typical consulting or research. First, because of the financial stakes typically associated with court cases, modeling for the courtroom often attracts more intense scrutiny than models created for other purposes. Second, although the modeling process for the courtroom must be fully documented with great detail and all of the attendant technical support, what is actually presented to a judge, jury, or panel in the courtroom must be skillfully distilled and simply explained so that the non-technical adjudicators can more readily understand the modeling process, the modeling results, and the modeling implications relative to the dispute.

In this paper, the focus is on modeling contaminant fate and transport in groundwater, although the scientific and legal principles are generally applicable to movement of chemicals in other media as well. Groundwater contamination modeling has been more widely conducted and is the subject of more litigation than other media. The paper reviews the main steps in the modeling process, highlighting the scientific rigors, limitations and uncertainties associated with modeling. The procedural and evidentiary rules that apply to expert witness testimony in federal court are summarized, with analysis of recent court opinions in this area. Finally, selected case studies of contaminant fate and transport for judicial disputes are provided.

2.   WHY MODEL CONTAMINANT FATE AND TRANSPORT

The reasons for creating a model of chemical movement are varied. In the environmental field, these can be grouped under three headings: prospective, retrospective, and cost-savings models.

1.   Prospective modeling is used to predict the outcome of some engineering modification of the field situation. This could include simulating the effectiveness of active remediation (*e.g.,* determining the

capture zone of a pump and treat groundwater remediation system), pollution control (*e.g.,* defining a wellhead protection zone), or permitting (*e.g.,* predicting the outcome of intrinsic remediation or air emissions pollution control technologies).

2. Retrospective modeling is used to look back at the cause of an environmental impairment. For example, retrospective modeling could be used to determine which industrial site is responsible for a groundwater plume when operations on many adjacent properties used the same chemicals, or to determine which operation(s) was (were) responsible at a known release source when there have been multiple facility operators over time.

3. Cost-savings modeling is used to minimize monitoring and sampling costs. Once the model for a site is operational and the processes that are occurring are well understood, regulators are more likely to allow cost-saving measures, such as a reduced monitoring frequency and/or a lesser number of monitoring wells. This may be useful, for example, in determining a mixing zone for direct or indirect discharges to surface water. Also, the model could be used to pinpoint and limit areas where more data are required, reducing the need for a "wall-of-information" approach.

Of these three classifications of modeling, prospective and retrospective modeling are more likely to be used by expert witnesses in court cases.

3. CHALLENGES IN MODELING MOVEMENT OF CONTAMINANTS IN THE ENVIRONMENT

In some cases, a simple equation is used in expert testimony to approximate the field situation. For example, for a simple hydrogeologic situation, Darcy's Law can be used to determine how far groundwater has moved during a given period of time, and the Theis equations can be used to determine how much drawdown should be expected at a given time and distance from a pumping well. The data used in these equations, and how well the assumptions are met, are open to scrutiny in court proceedings. Where the field situation is more complex, simple equations are best suited for rough approximations and as independent checks of more sophisticated computer models (Mandle, 2002).

When a field situation is too complex to approximate with simple analytical equations, "computer models" are used. With the advent of powerful and affordable desktop computers, and sophisticated, commercially available, state-of-the-art model codes, "computer modeling" has become a common tool for environmental consultants. Therefore, it is not surprising that computer modeling frequently occurs as part of expert witness testimony in complicated environmental cases, especially with the visual presentation capabilities.

"Groundwater flow models" simulate the distribution of groundwater elevations (called "hydraulic head" or "head") in a field situation. The output from these models can be used to determine groundwater flow directions, groundwater velocity, and temporal changes in head due to groundwater withdrawals or additions. For example:

1. In *Michigan Citizens for Water Conservation v. Nestle Waters North America, Inc.*, experts for both the plaintiff and the defendant used groundwater flow models to predict the effects of proposed commercial groundwater extraction on the water levels in wetlands and surface waters (example of simulating changes of head with time).

2. In *Aero-Motive Co. v. Beckers,* the plaintiff's expert used a groundwater flow model to determine the historical timing of releases of Trichloroethylene based on the extent of the contaminant plume (example of using groundwater velocity).

3. In *United States Aviation, et al. v. Tuscon Airport Authority, et al.*, the plaintiff's expert criticized the defense expert's groundwater flow model to demonstrate that the model was conceptually faulty. Since no groundwater contamination was detected at a monitoring well location that was encompassed by the simulated plume, either the groundwater flow model (flow direction) was faulty or the contaminant transport model was faulty.

"Solute transport models" are also commonly used in expert witness testimony for environmental contamination lawsuits. These models are used to predict or recreate chemical concentration distributions in groundwater. These models are more complex than flow models because they rely on a groundwater flow model *plus* characterization of the chemical source and chemical fate and transport factors (advection and dispersion, retardation, and chemical reactions).

If the purpose of the contaminant fate and transport model is only to assess whether a groundwater plume has impacted or will affect a location, it is preferable to use a simpler advective transport model. These models simulate chemical transport using only advection and retardation. These "particle tracking" codes simulate groundwater flow paths and estimate groundwater velocity by combining a groundwater flow model with a particle tracking code. More complex full transport models with many variables are sometimes necessary, but are subject to more scientific and legal challenges as to their accuracy and acceptability.

3.1   The Scientific Rigors of Modeling

The literature is in general agreement that there are at least six (6) steps involved in the groundwater modeling process (*e.g.,* American Society for Testing and Materials [ASTM], 1993; Anderson and Woessner, 1992; Spitz and Moreno, 1996; Ohio EPA, 1995; MDEQ, 2001; Alaska Department of Environmental Conservation, 1998; and others). These steps consist of: (1) defining the model objective; (2) developing the conceptual model; (3) selection of the modeling code; (4) model design; (5) calibration, verification, and sensitivity analysis; and (6) simulation (Figure 1).

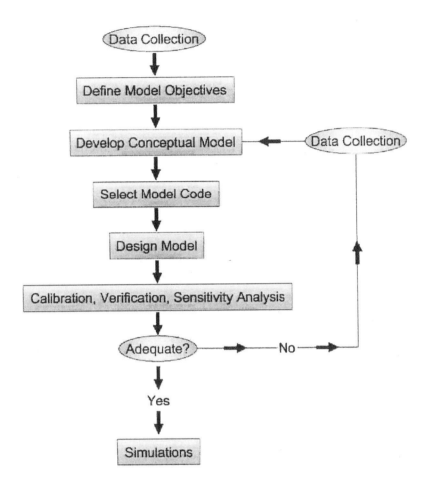

Figure 1. The Steps in the Modeling Process

Decisions made by the modeler during each of these steps profoundly affect the conclusions of the model, and therefore each step is potentially critical to the outcome of the litigation.

ASTM (1993, 1994a, 1994b, 1994c, 1995, 1996a, and 1996b) and others provide guidelines on these modeling steps. Texts such as Anderson and Woessner (1992), Spitz and Moreno (1996), and others, together with treatises such as Reilly and Harbaugh (2004), Mandle (2002), and others, offer useful suggestions to ensure that the modeling effort is reliable, defensible, and useful.

The first step in modeling is to define the modeling objective(s). The modeling objectives should be specific and measurable. Typical modeling objectives found in expert testimony for environmental contamination disputes include: (1) identification of the source of the contamination (which site is responsible for the release?); (2) determination of the timing of the release (which operator is responsible for the release?); and (3) evaluation of the effects of competitive uses of groundwater (does groundwater pumping affect others' use of groundwater or impair surface water features such as wetlands and streams?). Agreement between legal counsel and the expert on the objective(s) for the modeling at the outset is critical to a satisfactory outcome, as is identification of the information needs for a reliable model so that timely field work and/or litigation discovery can be conducted accordingly.

The second step in the modeling process is to develop a conceptual model for the field situation. The conceptual model is a simplified, systematic summary of the relevant hydrogeologic information. Developing the conceptual model is a critical step in the modeling process for two reasons. First, since the computer model simulates the conceptual model, not the actual field conditions, the more closely the conceptual model mimics the actual field conditions, the more closely the model results will portray the field situation. Second, the conceptual model is important because it helps to determine which type of model and code would be best suited for the field conditions.

The conceptual model should be based on site, local and regional hydrogeologic information, as well as professional judgment. Although data presented in map form are ultimately required for input into the model, the conceptual model can be presented in the courtroom using maps, cross-sections, or block diagrams. Block diagrams are useful in the courtroom because they are more easily understood by non-technical people (Figure 2).

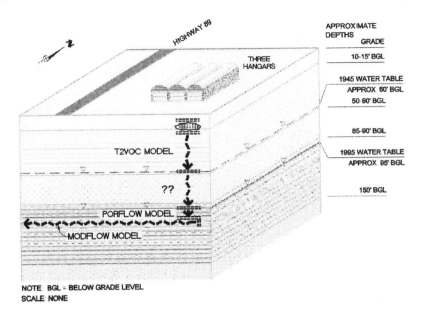

Figure 2. Block Diagram and Model Domains for Tuscon Ai rport Model

For groundwater flow models, the conceptual model should consider relevant hydrogeological information, including: (1) thicknesses and areal extents of the various soil or rock strata; (2) distribution and anisotropy of hydraulic conductivity; (3) significance of flow in fractures, if any; (4) distribution of groundwater recharge; (5) distribution of porosity; (6) the observed heads in the various strata; (7) observed groundwater flow directions; (8) interconnections between aquifers and continuity of aquitards; (9) relationships between groundwater and surface water features; (10) boundary conditions; and (11) pumping rates and changes in rates.

For transport models, the conceptual model should also consider whether an advective transport or full transport model is required, the nature and timing of the chemical release(s), and the behavior of the chemical(s) in the subsurface.   This paper will focus only on flow models and advective transport models.   Advective transport models go beyond simpler flow models and assume that the chemical transport is controlled by the groundwater velocity and retardation.   Retardation refers to the reduced velocity of certain chemicals that travel slower than the groundwater due to adsorption on materials in the aquifer.

The conceptual model must also consider model dimensionality in time and space. Time dimensionality considers whether inputs change with time (*e.g.,* pumping rates, recharge, etc.), and whether the distribution of heads needs be known at various times. For example, if the pumping rate in a well is increased in five years and then again in ten years, how will that affect the flow in a nearby stream? Computer models use time steps and stress periods to denote time dimensions.

Spatial dimensionality in a conceptual model refers to whether a one-, two-, or three-dimensional model is required. Two- and three-dimensional models are typically required for groundwater contamination projects. Two-dimensional models, either vertical or horizontal, are appropriate where the hydrogeologic and flow conditions vary in only two directions (*e.g.,* groundwater flow in a thin, confined aquifer, or groundwater flow from a hillside toward an adjacent stream). Three-dimensional models are required where the geology and flow conditions vary in three dimensions (*e.g,.* regional groundwater flow through interbedded aquifers and aquitards). The dimensionality of the conceptual model should be highlighted in exhibits used in the courtroom.

The third step in the modeling process is the selection of the modeling code or program. The code should be capable of handling all of the processes involved in the conceptual model. For example, in order to examine the interactions of groundwater and surface water, the model code must have that capability. The model code should also be able to withstand scrutiny by an opposing expert, regulator and judge. Preferably, the selected model code is in the public domain and is well-documented in reputable refereed journals. This minimizes the likelihood of challenges to the code in court, in addition to defending selection and use of the model.

MODFLOW, a groundwater flow model developed by the USGS in the 1980s, is an excellent example of a code that has been verified (compared successfully to other solutions), is well-documented in the literature, and has been used successfully hundreds, if not thousands, of times. Domenico and Schwartz (1998) describe MODFLOW as ". . . the *de facto* standard code for aquifer simulation." MODPATH, another USGS model (Pollock, 1989), is a particle tracking code used for modeling advective transport in association with the MODFLOW model. Like MODFLOW, the MODPATH program is a widely-used and generally acceptable code that would not likely be challenged in court proceedings.

The fourth step in the modeling process is model design. This step involves the input of hydrogeologic data into the computer code to represent the conceptual model. In models such as MODFLOW, the space that is modeled (the model domain) is discretized, that is, divided into sub-areas or sub-volumes (Figure 3) called "cells" (in two-dimensional models) or "elements" (in three-dimensional models).

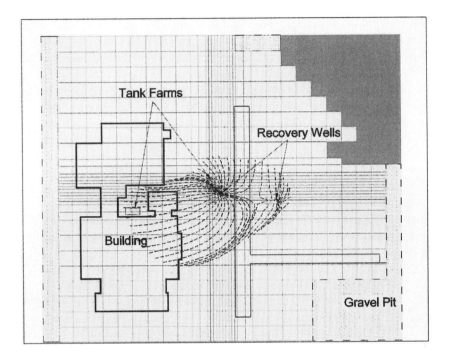

Figure 3. Discretized Model Domain

Each cell or element is assigned hydrogeologic input data consistent with the conceptual model parameters (*e.g.,* hydraulic conductivity, thickness, porosity, recharge, etc.). A range of cell sizes can be selected for the model domain, with smaller cells used to reflect heterogeneity dictated by the conceptual model and to better simulate areas where lateral changes in head are expected to be significant (*e.g.,* near pumping wells).

Model design also includes discretized time and the selection of boundary conditions. The selection of time and space discretization and boundary conditions are critical because they affect the numerical results of the model. Ideally, time and space discretization are as small as possible. Boundary conditions for head and groundwater flow conditions at the edges of the model are selected to approximate the natural groundwater flow conditions. Model boundaries should be placed along natural groundwater flow boundaries (such as groundwater divides and lakes) if possible, or far enough away from pumping wells to be unaffected by the effects of pumping.

The fifth step in the modeling process is generally referred to as "calibration". This calibration step usually includes calibration, verification (history matching), and sensitivity analysis. While steps one through four of the modeling process largely determine the reliability, predictability, and specificity of the model, this fifth step of the modeling process quantifies these factors.

Calibration is typically a comparison between modeled heads and observed heads to evaluate how well the modeled system replicates the field situation. The calibration target or objective is defined prior to conducting the calibration. Although there are no absolute standards for the calibration target, the difference between modeled and observed heads, called the "residual," should be minimized. Mandle (2002) and Spitz and Moreno (1996) suggest that a calibration is acceptable if the residual is less than 10% of the range of the observed heads. For example, if there is a 30 foot difference in heads across the model domain, modeled heads should be within +/- 3 feet of the corresponding observed heads to be considered acceptable.

The calibration results should be presented in maps, tables, or graphs (*e.g.*, Figure 4). Residual maps and graphs comparing observed and modeled heads are useful for identifying spatial bias in the residuals. Spatial bias should be minimized or explained.

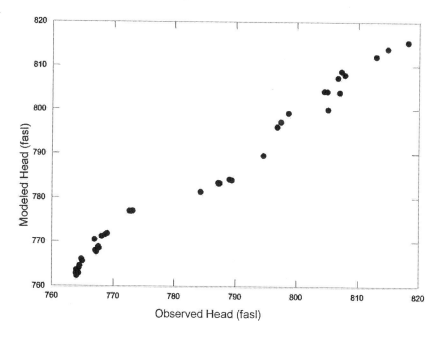

Figure 4. Calibration Plot – Expert's Aero-Motive Model (5 layer)

If the head residuals of a model run are not acceptable, selected parameters, such as hydraulic conductivity and recharge, are adjusted within a reasonable range, the model is re-run, and the residuals are recalculated. These parameter adjustments, which are accomplished manually by trial-and-error or automatically, are repeated and the model re-run until there is an acceptably small residual. This adjustment process is justified considering that regionalized model input is based on point field data, data for the model are estimated between borehole observations, and measurement techniques for parameters such as hydraulic conductivity are not exact. What is not acceptable is to make parameter adjustments that cannot be supported by the actual hydrogeologic conditions (Reilly and Harbaugh, 2004).

Since the distribution of modeled heads depends largely on the recharge to hydraulic conductivity *ratio* (not on their individual magnitudes), many different combinations of recharge and hydraulic conductivity can produce the same distribution of modeled heads. Therefore, a good calibration does not mean that a unique solution has been determined. There are two steps involved in further constraining the model.

First, "model verification" is a test of the calibrated model's ability to simulate observed heads that occurred under a different set of stresses. This may involve other historic head data or it could involve simulation of an aquifer test (where groundwater is removed via a pumping well and the resultant head changes in surrounding wells are documented with time). If the model calibration was reasonably correct, the head distributions resulting from the simulated aquifer test will be similar to the observed aquifer test data.

Second, other types of calibration data can be used to constrain the model. These comparisons include groundwater flow directions (*i.e.,* compare the orientation of groundwater contours), hydraulic gradients (*i.e.,* compare the spacing of head contours, both vertical and horizontal hydraulic gradients), water balance, and consistency with natural isotopic concentrations (such as tritium) or chemical distributions. For transport models, comparisons of modeled and observed data may include chemical migration rates, chemical migration directions, and chemical concentrations. These factors may be used to determine the most representative hydraulic conductivity values. For example, if the groundwater flow model and particle tracking indicate that a monitoring well is located within a plume from a landfill, the major ion chemistry should be indicative of landfill leachate impact. Similarly, one could compare groundwater ages based on tritium or oxygen-18 data to travel times determined from a model.

Once the calibration is acceptable, "sensitivity analysis" is performed to evaluate how much each parameter affects the model outcome, that is, model uncertainty. Sensitivity analysis involves iterative runs of the calibrated model adjusting only one parameter by arbitrary amounts but within the expected range of variation for that parameter, to evaluate how that parameter affects the model outcome. The results of sensitivity testing are plots of some cumulative residual against the particular value of the parameter in the sensitivity run (Figure 5). Sensitivity analysis is most important for parameters for which little field data is available. Augmenting sensitive data can improve the reliability of the model.

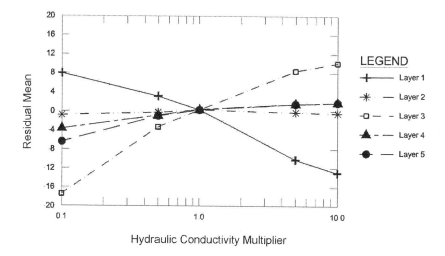

Figure 5. Sensitivity Analysis for Expert's Aero -Motive Model

The sixth and final step is referred to as "simulation," which means to run the calibrated model and evaluate the results in light of the objective(s) stated in the first step of the modeling process.

3.2 Limitations of Models

It is important to remember that groundwater models do not precisely and absolutely predict the behavior of an actual groundwater flow system or the actual movement of chemicals in the environment. However, well-prepared models can reasonably forecast contaminant fate and transport. There are many reasons for this degree of uncertainty. For example, in all modeling cases, the field hydrogeologic conditions are not known everywhere. In fact, only a small percentage of the actual field situation is observed and measured, and the remainder is inferred. Furthermore, except for measurements of hydraulic head and thicknesses of geologic units, field parameter measurements, such as hydraulic conductivity, are not exact.

A groundwater model is not reality, it is a reasonable approximation. The uncertainty in a groundwater model should be reflected in the information that is presented with the model. Figure 6 shows two ways in which uncertainty from the sensitivity analysis can be presented.

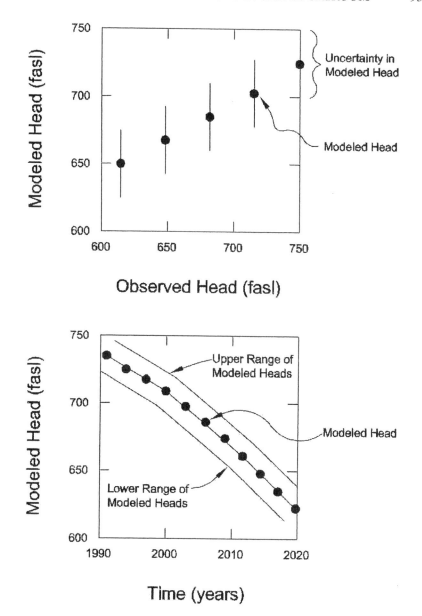

Figure 6. Uncertainty Presented in Calibration Plot (top), Hydrograph (bottom)

The upper plot shows a calibration plot, where the uncertainty in modeled head is expressed as a range of values. Similarly, the lower plot shows a simulated hydrograph where the modeled head range is provided.

3.3   The Best Use of Models

For complex hydrogeologic situations, groundwater models are the only means to synthesize all of the heterogeneity, transient conditions, and the various processes that affect a field situation.   Despite the degree of uncertainty that accompanies groundwater modeling, due largely to a less than perfect characterization of a field situation that affects every case, the amount of uncertainty can be limited by competent hydrogeologic characterization, developing a realistic conceptual model, and using good modeling practices.

As noted previously, the uncertainty in a model can be further constrained by calibrating the model to other types of field data and observations. For example, critique of a groundwater flow model of a site in the Eastern U.S. where the modeler had chosen the end of the model domain at a cranberry bog, found that the modeler assumed that all of the groundwater discharged into the bog.   However, the target chemical was observed in monitoring wells downgradient of the bog, and that model boundary therefore needed to be moved.   In another modeling case reviewed from the Midwest U.S., the regulator's model contained numerous breaches in the aquitard that overlaid the first confined aquifer. Based on review of the geology, hydraulic head, and chemical data, the aquitard appeared to be continuous.   Ultimately, field work was performed with a drill rig and a shovel to test for the presence or absence of holes in the aquitard.   The regulator selected six locations where their model simulated breaches, but the aquitard was encountered at all six borings thus undermining the model's reliability.

These examples illustrate the road to the best use of models. Follow the modeling steps summarized in this paper and in the relevant literature. Use other types of field data to further constrain the model. Finally, ask a simple question of the model and its components: "Does this make sense in light of all of the observed conditions at the site?"

4.   EXPERT ENVIRONMENTAL TESTIMONY

In virtually every environmental case, expert testimony will be required, whether it is used to prosecute or defend against a claim. This section details the procedural and evidentiary rules applicable to expert witnesses in federal court, and analyzes recent court opinions applying those rules in the specific context of environmental litigation.   (State court rules and requirements can vary, but are often based on the federal precedent discussed below; nevertheless, local practice in the relevant jurisdiction should be consulted.)

4.1   Evidentiary Rules Governing Experts

4.1.1   Federal Rule of Evidence 702/*Daubert*

Federal Rule of Evidence 702 governs the ultimate admissibility of an expert's proposed testimony at trial. That Rule provides:

> If scientific, technical, or other specialized knowledge will assist the trier of fact to understand the evidence or to determine a fact in issue, a witness qualified as an expert by knowledge, skill, experience, training, or education, may testify thereto in the form of an opinion or otherwise, if (1) the testimony is based upon sufficient facts or data, (2) the testimony is the product of reliable principles and methods, and (3) the witness has applied the principles and methods reliably to the facts of the case.

Fed. R. Evid. 702.

Under the seminal United States Supreme Court case of *Daubert v. Merrell Dow Pharmaceuticals, Inc.*, 509 U.S. 579, 113 S. Ct. 2786 (1993), a court is to perform a "gate-keeping" function to ensure that expert testimony is "not only relevant, but reliable." *Id.* at 589. The Supreme Court subsequently held that this gate-keeping function applies not only to scientific or technical knowledge, but to all types of expert testimony. *See Kumho Tire Co., Ltd. v. Carmichael*, 526 U.S. 137, 149, 119 S. Ct. 1167 (1999).

The *Daubert* court expressly declined to set forth a bright line test for the admission of expert testimony. Rather, the Supreme Court identified a number of factors to be considered by a court when making this determination:

- Whether the theory or technique employed by the expert is generally accepted in the scientific community;
- Whether the theory or technique has been subjected to peer review and publication;
- Whether the theory or technique can be or has been tested;
- Whether there are standards controlling the technique's operation; and
- Whether the known or potential rate of error is acceptable.

The Supreme Court has repeatedly emphasized the flexibility available to judges in determining the reliability of proffered expert testimony: "[T]he trial judge must have considerable leeway in deciding in a particular case how to go about determining whether particular expert testimony is reliable." *Kumho Tire*, 526 U.S. at 152.

The focus of this flexible inquiry, however, must always be the *methods* employed by the expert rather than the *results* of the expert's work: "The inquiry envisioned by Rule 702 is, we emphasize, a flexible one. Its overarching subject is the scientific validity and thus the evidentiary relevance and reliability of the principles that underlie a proposed submission. The focus, of course, must be solely on principles and methodology, not on the conclusions that they generate." *Daubert*, 508 U.S. at 594-95. Any attack on the expert's conclusions will almost always be for the jury to decide—a judge may only examine the principles and methods employed by the expert to determine their scientific validity and admissibility. If those principles and methods pass such scrutiny, then the expert's testimony is admissible and available for the jury to evaluate.

4.1.2   Federal Rule of Evidence 703

The other principal evidentiary issue to be analyzed with respect to expert testimony is what exactly an expert is allowed to use as the basis for his or her opinion. According to Federal Rule of Evidence 703:

The facts or data in the particular case upon which an expert bases an opinion or inference may be those perceived by or made known to the expert at or before the hearing. If of a type reasonably relied upon by experts in the particular field in forming opinions or inferences upon the subject, the facts or data need not be admissible in evidence in order for the opinion or inference to be admitted. Facts or data that are otherwise inadmissible shall not be disclosed to the jury by the proponent of the opinion or inference unless the court determines that their probative value in assisting the jury to evaluate the expert's opinion substantially outweighs their prejudicial effect.

Fed. R. Evid. 703.

Consequently, an expert may base his or her opinion on a wide range of sources, including otherwise inadmissible hearsay, if others in his or her field reasonably rely on such materials. This Rule gives an expert witness a good deal of leeway in forming his or her expert opinion, including, for example, reliance on another consultant's field sampling if found to be reliable. Note, however, that some states have adopted evidentiary rules that differ from Rule 703, and which require that *all* information on which an expert's opinion is based be in evidence. *See, e.g.,* Michigan Rule of Evidence 703.

4.2   Procedural Rules Governing Expert Witnesses

In federal court, the disclosure of expert witnesses is governed by Federal Rule of Civil Procedure 26. More specifically, Rule 26 requires each expert witness who has been retained by a party to the lawsuit to give testimony to provide a signed, written report to the other party. (Non-testifying consultants are not subject to this requirement.) The Rule also sets forth a laundry list of items that must be included in an expert's written report, including the following:

- A complete statement of all opinions to be expressed by the expert and the bases and reasons therefore;
- A list of all data or other information considered by the expert in forming his or her opinion;
- Any exhibits to be used as a summary of or support for the expert's opinion;
- A list of all publications authored by the witness in the preceding ten years;
- The compensation to be paid to the expert; and
- A list of any other cases in which the expert has testified as an expert witness, either at trial or in deposition, in the preceding four years.

Fed. R. Civ. P. 26(a)(2). The deadline for disclosure of expert witnesses will typically be established by the Court in a case management order or other pre-trial scheduling order governing the conduct of the case.

Counsel should note that expert witness reports are subject to the continuing duty of supplementation imposed by Federal Rule of Civil Procedure 26(e). That Rule contains a specific provision regarding the supplementation of expert witness information:

"With respect to testimony of an expert from whom a report is required under subdivision (a)(2)(B) the duty extends both to information contained in the report and to information provided through a deposition of the expert, and any additions or other changes to this information shall be disclosed by the time the party's disclosures under Rule 26(a)(3) are due."

Fed. R. Civ. P. 26(e)(1).

This disclosure deadline is 30 days prior to trial, unless the Court orders otherwise.

4.3   Admission of Expert Testimony in Environmental Litigation

As noted above, the key inquiry when determining the admissibility of a proposed expert's testimony is whether the methodology employed is adequate under Rule 702 and *Daubert*—the expert's ultimate conclusions typically should not bear on the threshold issue of admissibility. The following cases offer excellent illustrations of the application of these rules in the context of environmental litigation, an area of the law in which expert testimony is often a focal point to explain relative responsibilities for historical contamination.

4.3.1   *Freeport-McMoran Resource Partners Ltd. Partnership v. B-B Paint Corp.*, 56 F. Supp. 2d 823 (E.D. Mich. 1999)

The court's decision in *Freeport-McMoran Resource Partners Ltd. Partnership v. B-B Paint Corp.*, 56 F. Supp. 2d 823 (E.D. Mich. 1999), demonstrates the danger of failing to comply with Federal Rule of Evidence 702 and the Supreme Court's 1993 holding in *Daubert*. *Freeport-McMoran* involved a CERCLA claim for contribution regarding certain environmental cleanup costs incurred by the plaintiff in connection with a former co-disposal landfill site that received both non-hazardous and hazardous wastes. *See id.* at 829.

The plaintiff was a potentially responsible party ("PRP") who sought contribution from other alleged PRPs toward the costs incurred by the plaintiff while engaging in response activities at the waste site in question. The plaintiff brought suit against numerous defendants, claiming that each of them was responsible for some portion of the contamination found at the disposal site. *See id.* According to the plaintiff, each defendant shipped drums of waste to a certain incineration facility, which disposed of the liquid wastes contained in the drums and then sent the drums on to the final waste disposal landfill site. *See id.* at 830. The plaintiff alleged that, after the liquids were drained at the incineration facility, the drums still contained sludges, solids and/or residues, which constituted hazardous wastes previously belonging to the defendants and which caused contamination at the landfill site. *See id.* This "trans-shipper" PRP liability was the crux of the dispute.

The plaintiff had no direct evidence to substantiate its claim that any of the defendants' wastes were hazardous and actually disposed at the landfill site. *See id.* at 831. The plaintiff therefore attempted to rely on testimony from an expert witness to establish that "defendants' waste was transshipped after processing at the [shipper's] site for ultimate disposal at the [final disposal] Site." *Id.* at 831. The plaintiff's expert based his proffered opinions "on [his] review of records and information, including deposition testimony, relating to the drummed wastes of each Defendant...as well as [his] personal knowledge and expertise relating to hazardous substances." *Id.*

Certain of the defendants challenged the plaintiff's expert's testimony and filed a motion to exclude that witness's testimony. *See id.* In support of their motion, the defendants cited the expert's deposition testimony, which indicated that he had "no personal knowledge" about any of the defendants' alleged waste shipments, and that "he conducted no studies, experiments, or literature searches upon which to rest his opinions." *Id.* at 833. Most significantly, the plaintiff's expert admitted that he "did not follow any published professional standards in reaching his opinions, that there are no peer reviewed standards for the work he performed, and ... he did not perform any scientific tests to confirm his conclusions and that he did not evaluate any margin of error." *Id.*

After reciting the standard for admissibility of expert testimony under Federal Rule of Evidence 702 and *Daubert*, the court engaged in a specific analysis of the proffered expert testimony, and determined that it was inadmissible for failure to meet the *Daubert* threshold:

This Court is required, pursuant to *Daubert* and its progeny, to inquire into whether [the expert's] proposed expert testimony reflects valid scientific knowledge and whether the proffered expert testimony is relevant to the case at hand. [citation omitted] Applying the *Daubert* factors, the Court finds that [the expert's] proposed testimony does *not* comply with the *Daubert* standards. The proposed expert's testimony does not reflect a "theory or technique" that can and has been tested. . . . [The expert] can point to no peer review standards for the work he performed. He has not computed a "known or potential rate of error," nor has he identified "standards controlling the technique's operation." Most importantly, plaintiff has not demonstrated that [the expert's] theory or technique enjoys "general acceptance" within a "relevant scientific community." The record is utterly lacking in *any* indicia that would establish *any* of the *Daubert* factors.

Id. at 833.

The court then summarized its holding with regard to the proffered expert testimony, stating:

[T]his Court finds that proposed expert [witness's] testimony does not meet the *Daubert* requirement of scientific reliability. [The expert] is unable to substantiate his conclusions with any source other than his own "experience." Plaintiff has failed to meet its burden of establishing that [the expert's] conclusions are "based on sound science" by the "objective, independent validation of the expert's methodology." [citation omitted] [The expert] has utterly failed to point to any "objective source—a learned treatise, the policy statement of a professional association, a published article in a reputable scientific journal or the like -- to show that [he] has followed the scientific method. . . ."

Id. at 834.

It is important to note that, after deciding to exclude the proffered testimony of the plaintiff's proposed expert, the court went on to grant many of the defendants' motions for summary judgment. In doing so, the court noted each time that the plaintiff had been left with no evidence to contest the defendants' motions as to the lack of proof that they had arranged for disposal of hazardous wastes at the site, having lost the testimony of its proffered expert. *See id.* at 843-51.

4.3.2   *United States v. Dico, Inc.*, 266 F.3d 864 (8th Cir. 2001)

In contrast to the *Freeport-McMoran* case, *Dico, Inc.* is an example of establishing the proper foundation for expert testimony and thereby overcoming challenges to admissibility. *Dico, Inc.* was an action brought by the government under CERCLA to recover response costs incurred in connection with the cleanup of contaminated groundwater allegedly impacting the Des Moines, Iowa public water supply. *See Dico, Inc.*, 266 F.3d at 866. The EPA claimed that the defendant was responsible for the release of certain contaminants from its property, while the defendant attempted to show that the contamination had come from other sources. *See id.* at 868.

At trial, the defendant moved to exclude the testimony of the government's expert witness, citing *Daubert* and claiming that "his methodology was unreliable." *Id.* at 869. The court addressed three specific challenges to admission of the expert's testimony, and found them all to be without merit.

The court first found that the defendant's objection to the expert's proffered testimony on the grounds that the expert had ignored other possible sources of contamination was not supported by the record. *See id.* at 870. In so holding, the court found that the computer model used by the expert, MODFLOW, was "sanctioned by the EPA and is considered a standard model that is acceptable and commonly used by hydrologists." *Id.* at 870. This finding is significant precedent for the admissibility of groundwater modeling using MODFLOW. The court also concluded that the expert had properly considered all of the evidence that the defendant had accused him of ignoring. *See id.*

The defendant's second objection, a challenge to the expert's methodology on the grounds that the expert's "continuous line theory" used to test soil samples was unreliable, also failed. *See id.* at 871. Not only did the court find that the expert had not relied on the results of this test to reach his conclusions, but, more importantly, the court held that "the sufficiency of the factual basis of [the expert's] continuous line theory was open to any challenge [the defendant] desired to mount on cross-examination, but that sufficiency was not a basis for excluding [the expert's] testimony altogether." *Id.*

The final challenge the defendant offered to the expert's testimony, that it was unreliable because it was based on insufficient data, was also rejected by the court. *See id.* at 872. The court held that, "because 'the factual basis of an expert opinion goes to the credibility of the testimony, not the admissibility, and it is up to the opposing party to examine the factual basis for the opinion in cross-examination,' the District Court properly refused [the defendant's] invitation to exclude [the expert's] testimony on this ground." *Id.* In other words, this objection (like the previous one) would be for the fact-finder to evaluate in weighing the significance or credibility of the testimony, but would not exclude the evidence.

Dico, Inc. is an excellent reminder of the basic principle underlying the admission of expert testimony in environmental litigation -- if the expert's methodology is sound, it is highly unlikely that a court will preclude that witness from testifying. Conversely, if the expert's methodology is suspect, as it was in *Freeport-McMoran*, a party runs the risk of being left without any expert testimony whatsoever on critical issues.

4.3.3 *Dura Automotive Systems of Indiana, Inc. v. CTS Corp.*, 285 F.3d 609 (7th Cir. 2002)

The City of Elkhart, Indiana's water supply became contaminated by Trichloroethylene (TCE). The U.S. Environmental Protection Agency (EPA), after remediating the contamination, sued several entities including Dura Automotive Systems ("Dura") to recover the costs of the cleanup. Dura, in turn, sought contribution from CTS Corporation ("CTS"), claiming that CTS was responsible for some of the pollution and should therefore be required to share Dura's cleanup liability. *Id.* At a *Daubert* hearing, the District Judge disqualified Dura's sole expert witness. *Id.* The District Court then found that the remaining evidence was insufficient to create a genuine issue of material fact and granted summary judgment against Dura. *Id.*

At issue was whether some of the groundwater contamination beneath CTS's plastic plant had seeped into the City well field in the late 1970s or early 1980s, contributing to the pollution. *Id.* This could only be the case, however, if CTS's plant was within the well field's "capture zone," the area within which groundwater, if present, could be expected to flow to the well field. *Id.* The size of the capture zone would depend upon such things as the porosity of the soil and the rate at which the well field pumps water. *Id.* The more water the well field pumps, the larger the capture zone, because the removable groundwater beneath the field causes groundwater to be drawn in by gravity from other areas. *Id.* An environmental consulting firm retained by the EPA in the original lawsuit placed CTS's plant outside the well field capture zone. *Id.* Thus, in its contribution action against CTS, Dura employed an expert witness who was to contradict EPA's conclusion and testify that CTS's plant was within the well field's capture zone. *Id.*

Dura designated a hydrologist as its one and only expert witness. *Id.* At his deposition, however, Dura's expert witness admitted that he was not an expert in mathematical models of groundwater flow and that the modeling on which he relied for his conclusion that CTS's plastics plant was indeed within the well field's capture zone had been done by other employees of his consulting firm, using two models, QuickFlow (a 2-dimensional steady-state and transient groundwater flow interactive analytical element model) and SLAEM (single-layer analytic element model). *Id.* at 612. When CTS moved that Dura's expert be barred from testifying, Dura responded by producing affidavits from four employees of Dura's consulting firm who had worked on the Dura project. *Id.* These four employees, professional groundwater-flow modelers, attested that the models they had used, QuickFlow and SLAEM, were reliable and were appropriate for determining the well field's capture zone in the late 1970s. *Id.* CTS moved to strike these affidavits pursuant to Federal Rule of Civil Procedure 37 on the ground that Dura's disclosure of additional expert witnesses was untimely, since the deadline for filing expert reports had expired six months earlier. *Id.* The District Judge granted the motion to strike the affidavits, and held that, without the affidavits, there was insufficient evidence of the reliability of the models such that Dura's expert could not testify. *Id.* On appeal, Dura argued that the affidavits were not experts' reports that Dura had failed to disclose. *Id.* at 612. Rather, Dura argued that the reports were merely attestations showing that its expert was competent to report the results of the modeling exercises undertaken by employees of his consulting firm. *Id.*

In deciding this issue, the court noted that an expert witness is permitted to use assistants in formulating his expert opinion, and normally the assistants need not themselves testify. *Id.* This is consistent with Federal Rule of Evidence 703, as discussed above, which allows an expert to rely on facts or data made known to him or her before the hearing, if reasonably relied upon to base an opinion or inference. However, this analysis becomes more complicated if the assistants are not merely "gophers" or "data gathers" but exercise professional judgment that is beyond the testifying expert's expertise. *Id.* at 613. The court noted that, under the *Daubert* test, a scientist, however well credentialed he may be, is not permitted to be the mouthpiece of a scientist in a different specialty. *Id.* at 614. Separate expert opinions are different than underlying facts or data for purposes of FRE 703. After reviewing the affidavits, the court stated:

> 'It is apparent from these affidavits that [the expert's] assistants did not merely collect data for him to massage or apply concededly appropriate techniques in a concededly appropriate manner, or otherwise perform routine procedures, and that he himself lacks the necessary expertise to determine whether the techniques were appropriately chosen and applied.

Id. at 615.

The court pointed out that there were two crucial issues in the case, the map of the capture zone and whether, if CTS's plant was within it, how much if any of the contamination of the well field was due to the groundwater from beneath that plant. (These disputes were of course highlighted by EPA's conclusion apparently contrary to that of Dura's expert.) The court held that the expert was not competent to testify on the first issue and, without the expert, Dura could not get to the second issue and could not prevail in the case. *Id.* The court noted that, had Dura merely wanted to use QuickFlow and SLAEM to determine the *current* capture zone of the well field, the analysis might be different. *Id.* However,

> Dura wanted to use these models to determine the capture zone 20 years ago. The affidavits make clear that adapting the models to that use required a host of discretionary expert judgments for the affiants, not [the expert] to make.

Id.

Therefore, the Appellate Court affirmed the District Judge's decision to exclude the affidavits. *Id.* at 616. Further, the Appellate Court affirmed the trial court's holding that, without the affidavits, there was insufficient evidence of the reliability of the models, such that the expert could not testify to their validity. *Id.* The third-party contribution complaint of Dura was therefore dismissed because the only supporting evidence of CTS's liability was excluded.

5.   CASE STUDIES

5.1   *Michigan Citizens for Water Conservation v. Nestle Waters North America, Inc.*, No. 01-14563-CE (Mecosta Co. Cir. Ct., Feb. 13, 2004)

The *Nestle* case involved an action by local citizens to enjoin the defendant's pumping operations for a water bottling plant. Nestle Waters North America, Inc. ("Nestle") developed a well field known as "Sanctuary Springs" as a source for spring water. Slip Op. (Feb. 13, 2004) at 5. Nestle was initially approved by the State DEQ for four wells at the site and began commercial pumping.

The plaintiff citizens' group contended that Nestle's use of water from the aquifer was prohibited under both common law theories and Michigan environmental statutes, including the Michigan Environmental Protection Act. *Id.* The central issues in the case were the projected effect of the pumping operations on the water levels in the aquifer and surface waters, as well as the projected impact of those reduced levels on the environment and natural resources. Both sides presented expert witnesses who used computer modeling to project the effects of the pumping operations. The state court relied heavily on this testimony in reaching its ultimate conclusions.

The aquifer in question was a shallow, unconfined aquifer, meaning that it is near the surface and is actually exposed in places of open surface water. *Id.* at 11. Thus, the streams and lakes, as well as certain wetlands and surface springs, were actually features of the subject aquifer. *Id.* As stated, the central dispute in the *Nestle* case related to the extent of the effects and impacts on the surface-water bodies connected with the aquifer.

The court noted that a major problem in analyzing the effects of Nestle's pumping operation was that the pumping operation was largely underground and thus hidden from view. *Id.* at 12. (This, of course, is a challenge in most if not all groundwater contamination cases, as well.) Therefore, the effects of Nestle's pumping had to be determined largely by observing changes in the surface waters over time. *Id.* The analysis was further complicated by several factors. First, Nestle started its commercial pumping at a rate that was much lower than allowed by its permits. *Id.* Second, Nestle had been pumping water for less than two full annual cycles. *Id.* Finally, the environment involved was complicated, consisting of a stream, two lakes, and four wetlands systems. *Id.* Because of the limited data and complicated environment, computer modeling was used by both parties' experts to show the projected effect of the defendant's pumping operations.

As a preliminary matter, relying on state court rules, the court found that both experts were qualified in the field of hydrology and related computer modeling. *Id.* Further, the court found the experts' computer modeling methodologies were generally reliable. *Id.* However, the court noted the experts varied significantly in their use of computer modeling. *Id.* For example, the court noted approvingly that the plaintiff's expert never undertook to model the entire ecosystem hydrologically, but only used modeling to analyze components of the system. Thus, the court found that the plaintiff's model was not designed or used to "balance" the entire system. In contrast, the court found that the defendant's expert attempted to model the hydrologic environment to "balance" multiple factors, an effort that the court found caused him to draw unreasonable conclusions regarding the variables in the system. (The lesson here may be that more complicated modeling creates more opportunities for confusion and challenges.)

The experts also disagreed about the sufficiency of the available data. The defendant's expert argued that there was insufficient data available to accurately predict the effects of the defendant's pumping operation, and that five to ten years of data would be necessary to make a reliable assessment. *Id.* at 10. Therefore, the defendant's expert argued that his approach via computer modeling was the best and only way to predict the effects. *Id.* In contrast, the plaintiff's expert argued that there was sufficient data both to measure the present effects and to project future effects. *Id.* The plaintiff's expert did acknowledge that there was insufficient data for certain components of the system, and that computer modeling was necessary for these specific components. *Id.* (Perhaps, to the extent actual field data can be utilized rather than modeled results, fact-finders will be more receptive.)

After listening to the experts' testimony, and considering both the factual evidence relied upon by the experts and their respective methodologies, the judge as fact-finder determined that the plaintiff's expert testimony and opinion were more credible than the defendant's expert. *Id.* at 10. In reaching this conclusion, the court noted how each of the experts used computer modeling to analyze the complex hydrologic environment involved in the case. *Id.* The judge held that data and observation should carry more weight than modeling when sufficient evidence is available. *Id.* at 15. This is important guidance for other modeling projects.

Because the judge found the plaintiff's expert testimony to be more credible and supportable, the court based its findings regarding the projected impact of Nestle's pumping operations largely on the plaintiff's expert testimony. *Id.* at 24. These findings as to the projected impact of the effects of pumping on the hydrologic environment led to the court's ruling that Nestle's pumping operation should be permanently enjoined. *Id.* at 67. The court's injunction is being appealed in the Michigan Court of Appeals on various legal and factual grounds.

5.2   *Aero-Motive Co. v. Beckers*, 2001 WL 1699191 (W.D. Mich. Oct. 2, 2001)

In *Aero-Motive*, the plaintiff brought numerous legal claims against the defendants, including a claim under CERCLA, to recover response costs involved in the cleanup of historical soil and groundwater contaminants at a manufacturing facility near Kalamazoo, Michigan (the "Site"). *Aero-Motive v. Beckers*, 2001 WL 1699191 (W.D. Mich. Oct. 2, 2001). The plaintiff was the purchaser of defendants' business in 1972. *Id.* at *1. A few years after the purchase, the plaintiff added a warehouse to the original building and installed an underground storage tank ("UST") used to hold liquid wastes. *Id.* When the plaintiff removed the UST several years later, it discovered that contaminants had leaked from the tank into a limited area around it. *Id.* As a result, the State DEQ required the plaintiff to take corrective action to cleanup the release. *Id.*

While the plaintiff was remediating the UST release, it discovered that soils and groundwater underneath the factory and elsewhere at the manufacturing Site were also contaminated with Trichloroethylene ("TCE") and other solvents. *Id.* Subsequent hydrogeological investigations conducted by the plaintiff revealed that the TCE contamination had affected an area one mile downgradient from the Site. *Id.* As a result, the plaintiff removed contaminated soils from the Site, undertook other remedial actions (including groundwater pump-and-treat), and incurred various costs related to these response activities. *Id.*

The plaintiff claimed that the defendants were responsible for most of the on-site and all of the off-site TCE contamination. *Id.* at *2. Specifically, plaintiff claimed that the defendants, while operating the manufacturing facility, used a large pit on the property for disposal of paint and other wastes. *Id.* In addition, the plaintiff contended that the plant degreaser was periodically cleaned by pumping TCE from the degreaser into 55-gallon drums, which were then dumped into the disposal pit. *Id.* The plaintiff also claimed that TCE would sometimes spill onto the floor and be swept into a nearby drain which discharged to soils. *Id.* In contrast, the defendants contended that TCE spills from the plaintiff's degreaser and underground storage tank were the primary sources of the contamination at and from the Site. *Id.*

Thus, the key issues in the case involved disputes over the source(s) of the TCE contamination and when the contamination began (*i.e.*, before or after the business sale). Both the plaintiff and defendants used expert testimony to support their claims on these issues. Plaintiff's expert witness used computer modeling to demonstrate when and where the contamination began, whereas defendants' experts offered rebuttal critiques without any independent modeling.

Two groundwater flow models were prepared for the Aero-Motive Site. The plaintiff first retained an engineering consultant who used a FLOPATH model as part of the design process for a pump-and-treat remediation system. The plaintiff subsequently retained a hydrogeologist to act as an expert witness in the litigation. As part of the expert report for the court proceedings, the plaintiff's expert developed a second groundwater flow model of the area to determine time of travel, using MODFLOW/MODPATH modeling. The hydraulic conductivity distribution in the engineering consultant's model indicated a significantly faster groundwater velocity than that determined using the expert's model and inputs. Because time of travel was a central question in the *Aero-Motive Co. v. Beckers* lawsuit for pre-1972 liability, the existence of two plaintiff models giving significantly different groundwater velocities was problematic.

The Aero-Motive Manufacturing Company acquired the Site in 1964 and began operations during 1965. During 1991, an investigation associated with a waste oil underground storage tank ("UST") removal revealed the presence of TCE and other chlorinated hydrocarbons in the soil and groundwater beneath and around the building. Historically, TCE had been used at the plant to clean metal parts prior to painting. Subsequent investigations indicated widespread soil impact beneath the building, and a 2,000 foot wide TCE groundwater plume that extended approximately 4,400 feet downgradient from the Site. Potential sources of TCE in the soil and groundwater included a stormwater sewer under the building, a disposal pit, two former loading docks, degreasers, a paint room, and the waste oil UST.

The Site and the plume are underlain by unconsolidated glacial deposits. These soils progressively decrease in thickness from about 250 feet at the Site to about 100 feet near the end of the plume. Much of the immediate Site is underlain by clayey soil above the uppermost aquifer. The uppermost aquifer downgradient of the Site is unconfined. It consists of medium to very coarse sand and gravel with localized clayey silt, varying in thickness beneath the Site from less than 10 feet to more than 30 feet. The uppermost aquifer progressively thickens to about 90 feet near the end of the plume. Beneath the uppermost aquifer is a hard, low permeability, clay till confining unit.

The groundwater flow is generally northwestward near the Site, with localized variations, and northward further downgradient. The horizontal hydraulic gradient varies from about 0.015 near the Site to 0.009 further downgradient from the Site. Changes in the hydraulic gradient are consistent with the regional geologic conditions and topography. The depth to the water table ranges from about 38 feet at the Site to less than two feet near the end of the plume. The USGS report for the area indicates that groundwater recharge in the area averages 9.3 inches/year. Aquifer tests conducted for recovery well design and grain size tests provided hydraulic conductivity data for the aquifer.

Using observed groundwater flow directions and chemical signatures, the plaintiff's expert witness concluded in his expert report that four releases were responsible for the impacted soil and groundwater. The main TCE plume originated from impacted soil beneath a warehouse addition that was built on the east side of the plant in 1974. A disposal pit on the west side of the plant (about 500 feet west of the warehouse) was another source of chemicals. A third source of chemicals was discovered beneath a 1967 factory addition on the south side of the plant. The fourth source of chemicals was a small, chemically distinct plume originating from the waste oil UST located on the east side of the plant.

The plant had changed ownership in 1972, and one of the main points of contention in *Aero-Motive Co. v. Beckers* was which owner/operator was responsible for the release. Groundwater velocities were determined both by the plaintiff's expert using a MODFLOW/MODPATH model and from the consultant's FLOPATH model. The groundwater velocity determined from the engineering consultant's model happened to be significantly greater than the groundwater velocity determined from the expert's model. The differences stemmed from a faulty conceptual model and questionable model design by the consultant, although it may have been adequate for its remedial objective. As discussed previously, successfully calibrating a model does not guarantee a unique solution. Review of the engineering consultant's model identified three major errors in their conceptual model and model design.

First, the plaintiff's consultant had inadvertently used the total reported height of rainfall and snowfall, 110 inches, without adjusting for the rain equivalent of the snow. The engineering consultant used this annual precipitation as a basis for assigning a groundwater recharge of 10 to 40 inches per year (Figure 7).

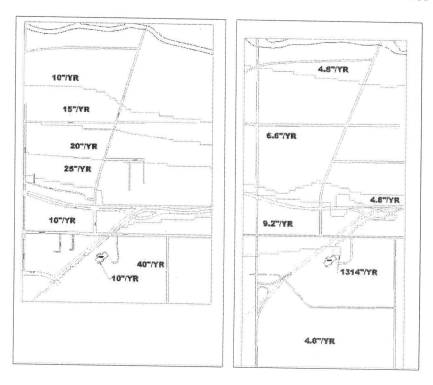

Figure 7. Comparison of Consultant's (left) and Expert's (right) Recharge Distribution, Aero -
Motive Model

This recharge is significantly greater than that reported by the USGS (an
average precipitation of 35 inches per year and groundwater recharge
ranging from 5.9 to 10.7 inches per year, averaging 9.3 inches per year).
Furthermore, the engineering consultant's model assigned the highest
recharge values (40 inches per year) to the area with the lowest hydraulic
conductivity values and the deepest water table. In comparison, the expert's
model input assumed natural recharge ranging from 4.8 to 9.2 inches per
year, plus an artificial recharge area where the cooling water, parking lot
runoff, and remediation water drained (Figure 7).

Second, in order to move the inflated amount of recharge water through
the aquifer and maintain model calibration, the engineering consultant had to
inflate hydraulic conductivities beyond those observed. The engineering
consultant's model used hydraulic conductivity values ranging from 70 to
250 feet/day (Figure 8). Furthermore, the plaintiff's consultant had used 16
hydraulic conductivity zones, far more than are justified given the site data
distribution, but probably useful for improving calibration of the model.

The expert's model used hydraulic conductivity values ranging between 60 to 170 feet/day, with the majority of the model domain at less than 110 feet/day (Figure 8).

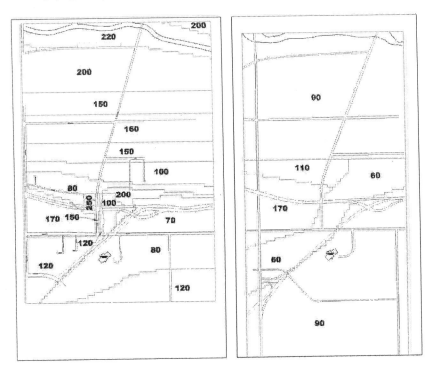

Figure 8. Comparison of Consultant's (left) and Expert's (right) Hydraulic Conductivity Distribution (ft/day), Aero-Motive Model

These values are consistent with field data derived from aquifer tests and grain size data. The expert's model had five hydraulic conductivity zones across the Site, which resulted in a good calibration for heads, hydraulic gradients, and groundwater flow directions.

Finally, the engineering consultant assigned high hydraulic conductivity along a south-north trending road to simulate a preferred pathway along the road bed. Although the water table approaches the surface in parts of the modeled area, the road bed would affect only the uppermost part of the aquifer, if at all.

Defendants filed a motion at the conclusion of discovery to exclude plaintiff's expert witness on the grounds that his expert testimony was unreliable. The court held a *Daubert* evidentiary hearing to examine plaintiff's expert qualifications and methodologies, and review his proposed testimony. *Aero-Motive v. Beckers*, 2001 WL 1698998 (W.D. Mich. Dec. 6, 2001). At the *Daubert* hearing, plaintiff's expert witness reviewed the methodology he employed in reaching his conclusions regarding the sources and timing of the hazardous waste disposal at the Site. The plaintiff's expert explained that he examined historical information on the soil and groundwater around the plant in order to develop a conceptual model of the Site. *Id.* at *2.

The expert used the computer programs MODFLOW and MODPATH to create a model of groundwater flow and to calculate the rate at which chemicals in the plume would move through the groundwater. *Id.* After creating this model, the plaintiff's expert witness compared the results with the actual field data, and found that the model water levels and flow patterns were consistent with actual water levels and groundwater flow at the Site. *Id.* In addition, the plaintiff's expert found that the rate of advance of the modeled plume was consistent with the observed rate of advance of the plume. Based on this modeling, the expert concluded that the primary source of TCE contamination was the area beneath the warehouse addition to the plant and the former location of the disposal pit. *Id.* The expert concluded in his report that these releases occurred prior to the 1972 sale of the property. *Id.*

In considering the factors set out in *Daubert*, the court found that the plaintiff's proposed expert testimony met the standards for expert testimony set out in Federal Rule of Evidence 702. *Id.* The court noted that the computer programs used by the expert are widely-used and well-tested groundwater flow and particle-tracking models that have been subjected to peer review and are commonly accepted in the hydrogeologic community. *Id.* Further, the court found that the data relied on by the expert, including geologic history, maps of the area, and surveys of the water and soil qualities at the Site, were the types of information regularly and reasonably relied on by scientists in the field and deemed generally reliable. *Id.*

In the end, the expert's groundwater model calibrated well and was consistent with the observed hydrogeologic data as well as other background facts known from discovery in the lawsuit. Furthermore, the groundwater velocity determined from the expert's model was consistent with the observed advance of the plume. Although the engineering consultant's model calibrated well, it contained serious flaws arising from a faulty conceptual model and a faulty model design. As noted by Reilly and Harbaugh (2004): "Put in terms of logic, a good match between calculated and observed heads and flow is a necessary condition for a reasonable model, but it is not sufficient. The conceptual model and the mathematical representation of all of the important processes must also be appropriate for the model to accurately represent the system under investigation." As a "reality check" the engineering consultant's precipitation input would have made this region comparable to a rain forest, which should have been a "red flag" of an error, together with the extensive variations necessary to achieve calibration.

Following the *Daubert* hearing, the parties filed cross-motions for summary judgment on their claims. *Aero-Motive v. Beckers*, 2001 WL 1699191 (W.D. Mich. Oct. 2, 2001). The court essentially found that sufficient facts and evidence existed to support contamination arising during Defendants' ownership and operation prior to 1972 that summary judgment was not appropriate for either party, and the case would proceed to trial. *Id.* *9; *Aero-Motive v. Beckers*, 2001 WL 1699194 (W. D. Mich. Dec. 6, 2001) (reconsideration as to pre-1968 contamination during individual ownership). Following the summary judgment rulings, a month before trial the parties reached a settlement of their claims pursuant to an Alternative Dispute Resolution ("ADR") process.

5.3   *Tucson Airport Authority* (Ariz.)

The *Tucson Airport Authority* litigation involved an example of a complex groundwater contamination modeling effort that was plagued by simple modeling errors.

In *Tucson Airport Authority*, several Tucson residents sued the Airport Authority for damages from injuries caused by groundwater contamination. *Tucson Airport Authority v. Smith*, 882 P.2d 1291; 180 Ariz. 165 (App. 1994). Various lessees of the Tucson Airport Authority used Trichloroethylene ("TCE") as an industrial solvent in their operations at the leased premises. *Id.* at 166. Disposal of TCE on the airport property caused groundwater contamination that allegedly impacted many southside Tucson residents. *Id.* The injured residents brought suit against the Tucson Airport Authority and the City of Tucson. *Id.* This suit was settled for $35 million, with an agreement between the parties providing that recovery would be sought only against the defendants' insurers. *Id.*

According to the U.S. EPA (2004), the Tucson International Airport Area ("TIAA") "Superfund" Site covers an area of approximately ten square miles, about fifteen miles south of downtown Tucson, Arizona. Prior to 1981, some 47,000 residents within the TIAA used the groundwater. Beginning in 1942, at least 20 separate facilities in the TIAA used and disposed of waste containing metals, chlorinated solvents, and other hazardous materials. By the early 1950s, one of the City water supply wells near the airport had become impacted by chromium and there were complaints of odor in nearby residential wells. The U.S. EPA and the City of Tucson began to investigate the area in 1981, and the TIAA was declared a "Superfund" Site in 1982. The TCE plume in the upper regional aquifer had spread from the Tucson Airport Authority property to about one-half mile wide and five miles long.

This case study focuses on "Zone E" of the TIAA groundwater plume ("the Site"), and was part of the lawsuit styled *United States Aviation, et al. v. Tucson Airport Authority, et al.* (Superior Court of the State of Arizona in and for the County of Pima No. 298875). Defendant's consultant developed a series of models for the Site. The conceptual model for the Site was complex because of a complex geology, a falling water table, and multiphase flow (Figure 2). Beneath Zone E, the top of the upper regional aquifer is at 150 feet below ground level ("fbgl"). There are four main hydrostratigraphic units above the upper regional aquifer that consist of interbedded silts; sand; gravels; and clays that are in some places cemented (caliche) and are characterized by generally low vertical hydraulic conductivity (10^{-5} to 10^{-7} cm/s). The water table in the upper regional aquifer declined by 50 to 60 feet between the 1940s and mid-1990s, draining much of the overlying units (called the shallow groundwater zone). Because of its low hydraulic conductivity, the groundwater in the shallow groundwater zone drained slowly. Where the flow directions in both the upper regional aquifer and the shallow groundwater zone aquifers were formerly to the northwest, the shallow groundwater zone flow gradually shifted to the west under the influence of the topography of the low hydraulic conductivity material in the lower part of the shallow groundwater zone.

The defendant's expert used three (3) computer models to recreate the history of TCE movement and the contaminant distribution in Zone E soil and groundwater (Figure 2). First, the defendant's expert used the T2VOC code to determine the time required for TCE to migrate vertically from a near surface source to reach the 1940s/1950s water table. T2VOC is a three-dimensional, multi-phase flow model selected to model free-phase TCE movement. Next, the defendant's expert used the PORFLOW code to determine the time required for dissolved TCE to migrate vertically from the top of the clayey soils in the shallow groundwater zone to the top of the gravel sub-unit (located within the lower part of the shallow groundwater zone). PORFLOW can simulate the movement of dissolved TCE with a falling water table scenario. Finally, the defendant's expert used MODFLOW/PATH3D (PATH3D is a particle-tracking code) to simulate horizontal migration of dissolved TCE in the gravel sub-unit of the shallow groundwater zone to determine when TCE reached the area where the gravel subunit discharged into the upper regional aquifer.

Presumably, the defendant's expert used these three (3) separate models because no single model was available that could simulate multi-phase flow in a falling water table scenario. However, the models were faulty in many ways, including the conceptual models, model design regarding data input, model discretization, and model calibration.

The T2VOC model failed to address uncertainties related to assumptions regarding release date, rate, area, and depth. The model did not address the effect of grid size selection on the model results. Furthermore, the model assumed no residual saturation (*i.e.*, the amount of TCE that would be bound up in the soil porosity following gravity drainage), which would typically be 5 to 20 percent. All of these factors would affect the arrival time and amount of TCE at the water table.

The PORFLOW model created by the defendant's expert also had numerous flaws. For example, model calibration was a problem. The modeled water table was 10 feet higher than the observed water table, and the modeled heads in the gravel sub-unit were 20 feet lower than the observed heads. Also, the simulated TCE concentrations were 200 times greater than the observed concentrations. The PORFLOW model was also flawed conceptually. The model increased vertical hydraulic conductivity by a factor of 2.5 to decrease vertical travel time, and a retardation factor was assigned but not activated. All of these factors would affect the arrival time of TCE at the gravel sub-unit.

The MODFLOW/PATH3D modeling was also flawed. Defendant's expert used a computer-generated geology that was biased toward higher groundwater velocity, and selected an effective porosity to achieve the travel time the expert wanted. These factors would affect the arrival time of TCE in groundwater in the gravel sub-unit at the main plume.

In addition to the problems with the individual models, there were also problems associated with the integration of the models. For example, different retardation factors, porosities, and anisotropy ratios were used in the models, even when they were supposed to represent the same volume of soil. Also, significant thicknesses of soil were not included between the three models.

In summary, Tucson International Airport, Zone E Model was an attempted integration of three complex flow and transport models. This modeling effort, however, was plagued by problems in the conceptual model, model design, and calibration. With so many problems and unexplained deviations from field data, it was difficult to place much weight on the results of these models.

6.   CONCLUSION

Modeling of environmental impacts can be a tremendously useful and efficient tool for compliance, remediation, permitting, and other predictive purposes. It seems safe to conclude that environmental modeling, in general, has been accepted by the regulators and relevant scientific community. This methodology, however, relies on various assumptions and interpretations, which account for its efficiencies and potential cost-savings, but also may generate objections when used in courtroom proceedings. Litigation has also challenged modeling used retrospectively to recreate or explain historical occurrences.

The relevant court rules and case precedent do not so much support challenges to modeling *per se* (assuming accepted computer software is properly utilized), but rather are based on the reliability of the inputs, data limitations, and other assumptions made in the modeling process for the particular project. Challenges to the qualifications of the expert witness to conduct modeling can also be submitted. These challenges to the application of modeling methodologies are highly project-specific, and cannot be generalized.

REFERENCES

Aero-Motive v. Beckers, 2001 WL 1699191 (W.D. Mich., Oct. 2, 2001)

Aero-Motive v. Beckers, 2001 WL 1698998 (W.D. Mich., Dec. 6, 2001)

Alaska Department of Environmental Conservation. 1998. Guidance for Fate and Transport Modeling. 9 p.

American Society for Testing and Materials (ASTM). 1993. Application of a Groundwater Flow Model to a Site-Specific Problem. ASTM D 5447-93.

American Society for Testing and Materials (ASTM). 1994a. Standard Guide for Conducting a Sensitivity Analysis for a Groundwater Flow Model Application. ASTM D 5611 -94.

American Society for Testing and Materials (ASTM). 1994b. Standard Guide for Comparing Groundwater Flow Model Simulations to Site Specific Information. ASTM D 5490 -93

American Society for Testing and Materials (ASTM). 1994c. Standard Guide for Defining Boundary Conditions in Groundwater Flow Modeling. ASTM D 5609 -94.

American Society for Testing and Materials (ASTM). 1995. Standard Guide for Documenting a Groundwater Flow Model Application. ASTM D 5718 -95.

American Society for Testing and Materials (ASTM). 1996a. Standard Guide for Conceptualization and Characterization of Groundwater Systems. ASTM D 59 79-96.

American Society for Testing and Materials (ASTM). 1996b. Standard Guide for Calibrating a Groundwater Flow Model. ASTM D 5981 -96.

Anderson, M.P., and Woessner, W.W. 1992. Applied Groundwater Modeling. Academic Press, Inc. 381 p.

Domenico, P.A. and Schwartz, F.W. 1998. Physical and Chemical Hydrogeology (Second Edition). John Wiley and Sons, Inc. New York. 506 p.

Mandle, R.J. 2002. Groundwater Modeling Guidance. Groundwater Modeling Program, Michigan Department of Environmental Quality, 54 p.

Michigan Citizens for Water Conservation v Nestle Waters North America, Inc. , No. 01-14563-CE (Mecosta Co. Cir. Ct., Feb. 13, 2004)

Ohio EPA. 1995. Groundwater Modeling. Chapter 14. Technical Guidance Manual for Hydrogeologic Investigations and Groundwater Monitoring.

Pollock, D.W. 1989. Documentation Of Computer Programs To Compute And Display Pathlines Using The Results From The U.S. Geological Survey Modular Three - Dimensional Finite Difference Groundwater Flow M odel: U.S. Geological Survey Open File Report 89-381, 188 p.

Reilly T.E. and Harbaugh, A.W. 2004. Guidelines for Evaluating Groundwater Flow Models. U.S. Geological Survey, Scientific Investigations Report 2004 -5038.

Spitz, K. and Moreno, J. 1996. A P ractical Guide to Groundwater and Solute Transport Modeling. John Wiley and Sons. Inc., New York. 461 p.

Tucson Airport Authority v. Smith, 882 P.2d 1291; 180 Ariz. 165 (App. 1994).

United States Aviation, et al. v. Tucson Airport Authority, et al. (Superior Court of the State of Arizona, In and For the County of Pima, No. 298875).

CHAPTER 6

A NEW METHOD OF DELINEATING THREE-DIMENSIONAL CAPTURE ZONES WITH MODELS

John P. Glass[1], Scott DeHainaut[2], and Rose Forbes[3]

[1]CH2M HILL, Inc, 13921 Park Center Road, Suite 600, Herndon, VA 20171; [2]CH2M HILL, 318 E. Inner Road, Otis ANG Base, MA 02542; [3]Air Force Center for Environmental Excellence, 322 E. Inner Road, Otis ANG Base, MA 02542

Abstract: The Air Force Center for Environmental Exc ellence operates several groundwater remediation systems at the Massachusetts Military Reservation (MMR). These systems involve multiple e xtraction and injection wells designed to provide complete or partial hydraulic containment of contaminant plumes. It is important in the design and optimization of these systems to be able to accurately delineate their hydraulic capture zones, which are three-dimensional and irregularly shaped. Traditional delineation techniques are based on visual identification of the envelope of pathlines leading to the extraction wells, the pathlines being generated by either backward or forward particle tracking in the simulated flow field. A new technique being used at the MMR involves forward tracking of particles from a dense three -dimensional array of starting locations without actually plotting the pathlines. Instead, the particle-tracking outcome is used to d efine a grid-based three-dimensional continuum of capture parameter, which can then be contoured in two-dimensional projections or otherwise rendered visible by three -dimensional visualization software. The resulting capture parameter is a three -dimensional scalar field that can be considered a quantitative spatial property of the flow regime. It can be combined with other scalar fields defined in the same grid, such as the concentration field, for visualization or for volumetric calculations. Arithmetic comparisons of capture parameter arrays generated at different pumping rates are also useful in sensitivity analysis.

1. INTRODUCTION

Massachusetts Military Reservation (MMR) is located on upper Cape Cod and is underlain by a sand, silt, and gravel aquifer 150 to 300 feet thick.

Past releases of fuels and chemicals to the aquifer have resulted in several plumes of groundwater contamination that are now being remediated under the direction of the Air Force Center for Environmental Excellence (AFCEE). AFCEE currently operates six groundwater extraction, treatment, and reinjection systems at MMR to hydraulically contain and remove contaminated groundwater. At each of these sites, the contaminant plume occupies less than the full thickness of the aquifer. The remediation systems typically consist of multiple extraction and injection wells that partially penetrate the aquifer to focus remedial effort on the contaminated depth intervals. To evaluate the containment performance of these systems and to optimize them hydraulically, it is important to have the capability to accurately delineate their hydraulic capture zones in three dimensions.

A hydraulic capture zone is defined as that portion of the aquifer within which groundwater will flow to an extraction well and be removed. Each extraction well has its own capture zone, and the capture zones of multiple wells may overlap to provide either complete or partial hydraulic containment of a larger aquifer volume. The boundaries of a capture zone separate captured from uncaptured portions of the aquifer. Boundaries of the composite capture zones produced in multiple well systems can be unexpectedly complex. This paper describes a method of capture zone delineation and depiction that has been developed at MMR for such cases.

2.   METHODS

Groundwater flow and contaminant migration at the MMR remediation sites are analyzed using three-dimensional numerical models. The groundwater modeling is implemented using the MODFLOW-SURFACT© (HydroGeoLogic, 1996) code, which is an enhanced proprietary version of the finite-difference modeling code, MODFLOW (McDonald and Harbaugh, 1988), developed by the U. S. Geological Survey (USGS). The flow models applied at the various MMR sites are carefully calibrated to produce accurate simulations of the three-dimensional flow fields and potentiometric head distributions under varying rates of pumping and re-injection.

The output of the groundwater flow model includes files that define the simulated potentiometric head and the simulated vector components of groundwater flow. These results are defined numerically in the context of a three-dimensional finite-difference grid. The potentiometric head part of the solution consists of an array of head values defined at the centers of all the grid cells.   This is the discretized numerical equivalent of a three-dimensional scalar field. The flow part of the solution consists of similar arrays of vector components defined at the boundaries between the finite-difference cells. They constitute the numerical equivalent of a three-dimensional vector field.

The output of the flow model defines the simulated flow field, but further processing is needed to determine the hydraulic capture zones. This is done by a post-processing program that integrates the vector component equations to trace out three-dimensional pathlines starting from any set of points in the simulated flow field selected by the program user. Programs of this type are commonly called particle tracking programs, based on the analogy of an imaginary fluid particle migrating along the selected pathline.

Particle tracking programs are usually used to produce graphical depictions of pathlines as a means of flow field visualization. Their utility for capture zone delineation is obvious since they can show whether a particle that starts at any given place within the simulated aquifer will flow to a specific well or not. However, there is more than one way of using a particle tracking program for this purpose.

2.1 Backward Particle Tracking

A particle tracking program can integrate the velocity component equations either forward in the direction of flow or backward from sinks to sources. One common way of visualizing the hydraulic capture zone of a well is to start particles in a cloud around the well and to track them backwards to show where they came from. An example of this method is shown in Figure 1, which shows the capture zones of the extraction wells at the Landfill-1 remediation site at MMR. This remediation system consists of five extraction wells, a groundwater treatment system, and an infiltration gallery located between Well 4 and Well 5, through which the treated water is returned to the aquifer. The treatment capacity of this system is approximately 700 gallons per minute.

The primary weakness of the backward tracking method is that it only shows the trajectories of particles that are successfully captured. Therefore, the exact location of the capture zone boundary remains unknown, because it is determined by distinguishing between captured and uncaptured particles. In an attempt to overcome this shortcoming, a very large number of particles can be used under the assumption that they will be adequate to show most of the aquifer volume from which captured particles could have come.

A further weakness of the method is that it is purely graphical. Although the flow field and the pathlines may be three-dimensional, their graphical depiction is two-dimensional. In a plan view plot, pathlines that are higher and lower in the flow field are projected vertically so that the three-dimensional shape of the capture zone is not clear.

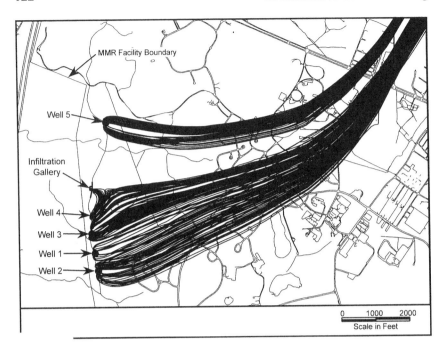

Figure 1. Capture Zones of the Landfill -1 Extraction System Delineated by Backward
Tracking of 334 Particles

2.2   Forward Particle Tracking

The other common procedure for delineating capture zones is to release a
regularly spaced array of particles and track them forward in the simulated
flow field until they exit the model domain through sinks. Pathlines that
terminate at the well of interest can be distinguished from those that
terminate elsewhere. This distinction is the basis for delineating the capture
zone boundaries.

Figure 2 illustrates some of the weaknesses of traditional forward particle
tracking. This figure shows the trajectories of particles started in a two-
dimensional rectangular array in layer 15 of the 33-layer groundwater flow
model. Release of a three-dimensional array of particles would produce a
very confusing figure because the particle tracks in the shallow part of the
aquifer have different trajectories than the deeper particles. The intersection
and crossing of particle tracks projected vertically would make it practically
impossible to distinguish the three-dimensional capture zones. Therefore,
the use of forward-tracked pathlines is not a suitable method for delineating
capture zones in three dimensions.

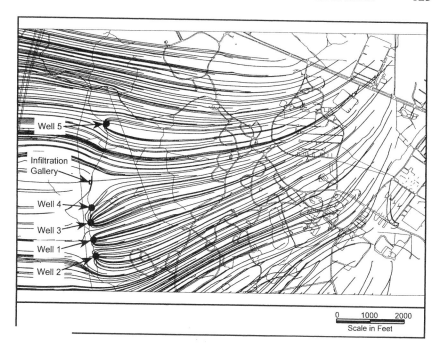

Well 5

Infiltration
Gallery

Well 4

Well 3

Well 1

Well 2

0   1000   2000

Scale in Feet

Figure 2. Capture Zones of the Landfill-1 Extraction System Delineated by Forward Tracking of 330 Particles Released in Model Layer 15

2.3 MMR Capture Zone Procedure

It is significant to note that the hydraulic capture zone can be determined by identifying all of the points within the aquifer from which a particle could be started and still end up in the extraction well. If a correspondence between those starting points and the well can be made, it is unnecessary to draw the pathlines along which the captured particles would flow.

The particle tracking program used at MMR is the USGS code MODPATH (Pollock, 1994). This particle tracker was developed by the USGS for use with the MODFLOW groundwater flow model. The main output of MODPATH is given in two parts: a pathline file that contains the vertices (x,y,z coordinates) of the pathlines, and an endpoint file, which only deals with the starting and ending points of the particles. The endpoint file contains only one data record for each particle. The record lists the coordinates of the particle starting location, the coordinates of its endpoint, the total travel time, and the number of the aquifer zone in which the particle path terminated. The aquifer zones are delineated by the model user and assigned integer zone numbers for identification.

For capture zone analysis, it is convenient to assign a zone number, such as 2, to the model cells containing the extraction well screen. All other cells are assigned some other number, such as 1. After the particle tracker has completed its run, the endpoint file will identify all of the captured particles by zone number 2, and will list the coordinates of their starting points. The uncaptured particles and their starting points will similarly be identified by zone number 1. This can be interpreted as the assignment of a capture parameter to each point in the aquifer from which a particle was started. If the particles were started from a regularly-spaced array of starting points (not necessarily corresponding to the model grid), this results in the numerical equivalent of a three-dimensional scalar field of capture parameter. If the capture parameter array is conceptually extended to a continuum, the capture parameter can be considered to be a three-dimensional step function that exactly distinguishes between the captured and uncaptured portions of the aquifer. In practice, only the gridded numerical definition of the capture parameter is available, so the precision of capture zone definition depends on the density of the particle starting grid.

3.   RESULTS

3.1   Capture Zone Visualization

Visualization of the capture parameter field is done at MMR using the Tecplot® (Amtec, 2003) graphical software package. Tecplot® provides several ways of displaying the capture parameter field. One way, as shown in Figure 3, is to produce filled 2-dimensional contour plots on horizontal layers of the grid and on vertical cross sections. If several layers and cross sections are used, this display can give a sense of the three-dimensional nature of the capture zones while showing their exact boundary locations on a few selected planes. The capture zone in Figure 3 is delineated by the 1.5 capture parameter contour. The capture parameter field was defined by tracking a total of 572,319 particles started on a grid of 123 rows, 141 columns, and 33 layers.

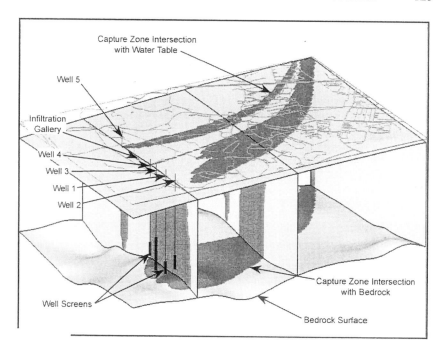

Figure 3. Three-Dimensional Capture Zone of the Landfill-1 Extraction System Visualized by
Contouring the Capture Parameter on Multiple Intersecting Surfaces

Another form of visualization is the three-dimensional iso-surface, as
shown in Figure 4. If the iso-surface value of 1.5 is plotted, the graphics
package interpolates between grid nodes where the capture parameter is 1
and nodes where it is 2, to produce a surface that meets the definition of the
three-dimensional capture zone boundary. The capture zone can then be
displayed either in a shaded two-dimensional view or as an animation, in
which rotational motion provides the viewer with an enhanced sense of the
capture zone's three-dimensional shape.

Visualization software also has the capability of combining the capture
zone iso-surface with iso-surfaces depicting other scalar fields, such as the
potentiometric head field, the hydraulic conductivity field, or the
contaminant concentration field. Figure 5 shows an example of the Landfill-
1 capture zone together with a groundwater contaminant plume
(tetrachloroethylene), which is bounded by the maximum contaminant level
(MCL) of the contaminant. Such a figure shows clearly whether the capture
zone is achieving complete hydraulic containment or not. (Complete
containment is not the remedial objective at Landfill 1.)

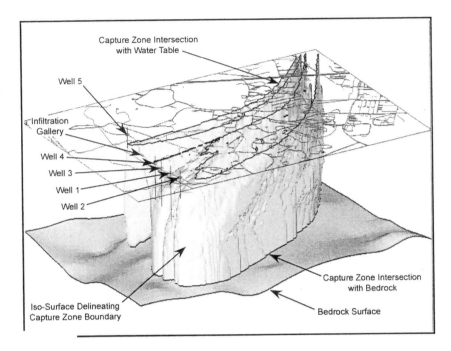

Figure 4. Three-Dimensional Capture Zones of the Landfill -1 Extraction System Visualized
by Plotting the Capture Parameter 1.5 Iso -Surface

3.2   Capture Zone Calculations

Because the scalar field of capture zone parameter is a numerical
approximation of a mathematical entity, it is possible to use it in other ways
besides graphical display. The volume of aquifer contained inside the
capture zone can be calculated simply by adding the volumes of all grid cells
having a capture parameter of 2. This volume then serves as a quantitative
measure of the capture zone that can be compared with the volumes for other
pumping and re-injection rates to assist in optimization of the remediation
system. The Landfill-1 capture zone illustrated in Figure 4 encompasses
3.59×10^9 cubic feet of aquifer volume.

The capture parameter can also serve as a three-dimensional mask to
evaluate the volume or mass of the contaminant plume that is inside or
outside the capture zone. The contaminant plume illustrated in Figure 5
contains approximately 122 kilograms of tetrachloroethylene, of which 65
kilograms is captured and 57 kilograms is not.

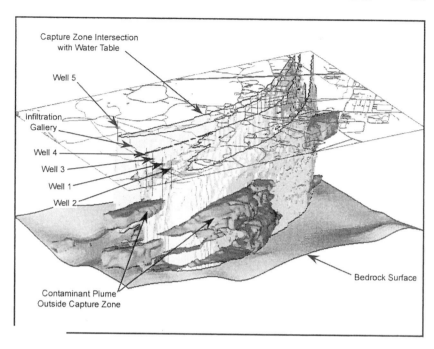

Figure 5. Landfill-1 Capture Zone and the Uncaptured Portion of the Tetrachloroethylene Plume

In a further extension of the capture parameter as a mathematical entity, it is possible to subtract one capture zone from another. The differences between capture zones are then geometric shapes representing the portions of the aquifer that are captured under one pumping scheme but not by another. This, together with the ability to calculate capture zone volumes, facilitates quantitative comparison of different extraction system designs, which is very helpful in optimization of such systems. The same technique could be used to show the changes in the capture zone that would be produced by seasonal variations in the aquifer recharge rate or by any other outside influence that can be simulated with the groundwater flow model.

4.   DISCUSSION AND CONCLUSIONS

Application of the new capture zone delineation procedure uses software for groundwater modeling, particle tracking, and spatial visualization that has been in wide-spread use for a long time. The essential new feature of the method is a computer program that reads the output of the particle tracker and reformulates it as a gridded array of capture parameter relating the particle termination zones to the particle starting points. The capture parameter array defines a scalar field that is a property of the simulated flow field and separates it into captured and uncaptured parts. The capture parameter array is determined by the distribution of particle starting points, but need not be directly related to the flow model grid as long as it is contained within the model domain.

While the example applications shown above deal with steady-state flow simulations and capture zones having no travel time constraints, these are not inherent limitations of the method. Transient three-dimensional flow simulations often produce pathline plots that are very complex and difficult to interpret. The endpoint file produced by the particle tracking program, however, still relates the starting points and termination zones of the particles in the same way as for steady-state flow. Therefore, the capture parameter field can still be assembled and the capture zones displayed just as easily for transient flows as in the steady-state case.

If the analyst wants to define a time-limited capture zone as the volume of the aquifer in which groundwater will be captured within a certain period of time, this can be done using the total travel time recorded for each particle in the MODPATH endpoint file. Particles that take longer than the specified travel time to reach the extraction well can be excluded from the time-limited capture zone.

The new capture zone delineation method also has potential utility in automated hydraulic optimization of hydraulic containment systems. In this regard, the valuable feature of the method is the quantitative characterization of the capture zone that it provides. The simulated capture zone can be quantified in terms of its volume, the volume of contaminant plume that it contains, or by its accurate numerical location in space. Each of these measures could potentially be used as either an objective function or a constraint in hydraulic optimization modeling.

REFERENCES

Amtec. 2003. Tecplot® User's Manual, Version 10. Amtec Engineering, Inc., Bellevue, Washington

HydroGeoLogic. 1996. MODFLOW-SURFACT, Version 2.2. HydroGeoLogic, Inc, Herndon, Virginia

McDonald, M. G. and A. W. Harbaugh. 1988. A Modular Three-Dimensional Finite-Difference Groundwater Flow Model. Techniques of Water-Resources Investigations of the United States Geological Survey, Chapter A1, Book 6. United States Government Printing Office, Washington, D. C.

Pollock, D. W. 1994. User's Guide for MODPATH/MODPATH -PLOT, Version 3: A Particle-Tracking Post-Processing Package for MODFLOW, the U. S. Geological Survey Finite-Difference Groundwater Flow Model. U. S. Geological Survey Open -File Report 94-464. Reston, Virginia

CHAPTER 7

COMPARING AIR MEASUREMENTS AND AIR MODELING AT A RESIDENTIAL SITE OVERLYING A TCE GROUNDWATER PLUME

Dennis Goldman, Ph.D. RG[1], Ron Marnicio, Ph.D. PE[2], Wilson Doctor[3], and Larry Dudus' RG[4],

[1]*Tetra Tech FW, Inc, 1230 Columbia Street, Suite 500, San Diego, CA, 619 -471-3543, dgoldman@ttfwi.com;* [2]*Tetra Tech FW, Inc., 133 Federal Street, 6*[th] *Floor, Boston, MA, 617-457-8262, rmarnicio@ttfwi.com;* [3]*Department of the Navy, 1230 Columbia Street, Suite 1100, San Diego, CA, 619-532-0928, wilson.doctor@navy.mil;* [4]*Tetra Tech FW, Inc., 1230 Columbia Street, Suite 500, San Diego, CA, 619 -471-3509, ldudus@ttfwi.com*

1.   INTRODUCTION

Naval Facilities Engineering Command Southwest Division and Tetra Tech FW, Inc. (TtFW), investigated potential vapor intrusion from volatile organic compounds (VOCs), primarily trichloroethelyene (TCE), in the groundwater underlying a federal housing area at the former Naval Air Station (NAS) Moffett Field (Moffett), California. The investigation was coordinated with the U.S. Environmental Protection Agency (EPA), Region 9 and the California Regional Water Quality Control Board, San Francisco Bay Region (RWQCB).

The Moffett Community Housing (MCH) property is located just south of San Francisco Bay. There are four prominent TCE groundwater plumes within a radius of less than 2 miles of the MCH property. The Middlefield-Ellis-Whisman (MEW) regional TCE plume (a plume that is about 0.5 miles wide and 1.5 miles long) is currently estimated to be located less than a quarter mile to the east of the MCH property. The TCE plume underlying the MCH property appears to originate off-site to the south of the MCH property.

A series of investigations have been completed to characterize the extent of TCE contamination in groundwater and to evaluate the potential for vapors from contaminated groundwater to migrate into the housing units.

2.   MCH BACKGROUND

Prior to federal housing development, the MCH property was used for agriculture. The MCH property covers approximately 110 acres. The property was transferred from the Navy to the Air Force as part of the Base Realignment and Closure (BRAC) program in 1994. In 2000, the housing property was transferred to the Army. MCH includes the Orion Park Housing Area (OPHA) and the Wescoat Housing Area (WHA). OPHA is approximately 72 acres in size. Housing was constructed here during the years 1941 through 1982. OPHA consists of housing units and support facilities along with associated streets, parking areas, and green space. Multi-family residential structures currently occupy most of OPHA. WHA is approximately 40 acres in size. WHA was vacant land prior to development as residential property. The first housing at WHA was built in 1933. The remaining housing units were constructed during the years 1968 through 1982. All housing units, except the homes built in 1933, were demolished in the summer of 2004 to make way for new Army housing construction.

Housing at OPHA and WHA is constructed either slab-on-grade (housing unit with numbers in the 600s and 700s) or over a crawl space (unit numbers in the 800 series). The slab-on-grade buildings were constructed primarily in 1968 and consist of multiple 2-story wood frame units on slabs. The slabs are constructed of 4-inch-thick steel reinforced concrete, placed on 2 inches of sand, a waterproof membrane, and 4 inches of gravel (as described in the original construction drawings). The remaining OPHA units, constructed primarily in the early 1980s, have wood floors over an approximate 18-inch crawl space. The crawl space is accessible through small trap doors located in bedroom or hallway closets. The crawl space is vented with several screened openings located around the base of the units, a few inches above ground surface.

3.   HYDROGEOLOGIC SETTING

Moffett is located at the northern end of the Santa Clara Valley Basin, approximately 1 mile south of the San Francisco Bay. Moffett is relatively flat, ranging from 2 feet below to 36 feet above mean sea level. The Santa Clara Valley Basin contains varying combinations of clay, silt, sand, and gravel that represent the interfingering and layering of estuarine and alluvial depositional environments during the late Pleistocene and Holocene epochs. Buried sand and gravel channel deposits are incised in floodplain and tidal deposits. There appear to be two relative scales of the sand and gravel channel deposits. The larger features have been interpreted as distributary stream deposits. Distributary channels are branches of a main channel that extend out onto the floodplain that form an anastomosing network of relatively permeable material. Smaller sand and gravel features are interpreted as splay and overbank deposits. Splay deposits form when a stream breaks through a levee and deposits its material onto the surrounding floodplain. These deposits generally are thin sheets that have only limited connection to the main channel.

In this depositional environment, streams may cut down through existing floodplain or channel deposits at one geologic time period and at other times bury the earlier sediments. During other geologic episodes, a combination of land subsidence and rises in sea level result in the deposition of fine-grained tidal and shallow marine sediments. The overall result is a complex network of coarser-grained sand and gravel surrounded by fine-grained floodplain and marine silt and clay. Continuity of individual sand and gravel units in this fluvial-dominated depositional environmental setting is limited.

The hydrostratigraphy at Moffett has been defined only in a general sense due to its complexity. Clays and silts are present from ground surface to approximately 12 feet below ground surface (bgs) at most locations. First-observed groundwater is encountered confined below this relatively impermeable layer. Groundwater is observed in the coarser grained soils, ranging from sandy silt to gravely sand. The permeable zones are separated by less permeable zones consisting of silt and silty clays. The thickness and the lateral extent of these zones vary greatly throughout the site. To a depth of about 65 feet bgs, these discontinuous channel deposits are called the A aquifer. The A aquifer has been divided into an upper aquifer zone (A1 aquifer zone) and a lower aquifer zone (A2 aquifer zone). The A1 aquifer zone is made up of 3 to 5 of these discontinuous channel deposits to a depth of about 25 feet bgs.

The depth to groundwater adjacent to the individual housing units modeled ranged from 9.5 to 13.8 feet bgs. The soil types encountered were sandy clay, silty clay, and sand. Groundwater samples were obtained within each of the permeable lenses/layers of the A1 and A2 aquifer zones underlying MCH. There are TCE concentration differences laterally and vertically between these permeable layers. There are places where the first permeable layer had no detectable TCE concentration (less than the method detection level [MDL] of 0.25 [micrograms per liter (µg/L)], while deeper layers had relatively high TCE concentrations (greater than 300 µg/L). There are also places where the first layer had higher TCE concentrations than lower layers. The locations beneath MCH where the first-observed groundwater had concentrations of TCE greater than the laboratory reporting limit of 5.0 µg/L are shown on Figure 1. The maximum TCE concentration measured in any layer in the A1 aquifer zone was 350 µg/L.

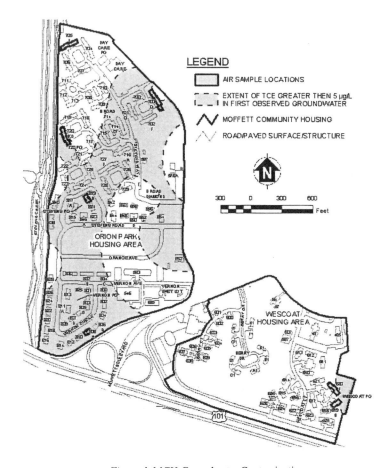

Figure 1. MCH Groundwater Contamination

4.   SITE SAMPLING

TCE at a concentration of approximately 300 µg/L was detected in shallow groundwater upgradient and downgradient of OPHA in 1999 and 2000, respectively. An investigation was conducted in 2002 to evaluate the extent of TCE contamination in shallow groundwater; select housing units for vapor intrusion modeling; and to provide housing unit-specific soil, groundwater, and construction data for subsequent modeling. Continuous soil cores were collected from the ground surface to first-observed groundwater at 62 locations. A sample of first-observed groundwater (typically about 15 bgs) was collected for VOC analysis at each location. Cone penetrometer test (CPT) borings also were advanced to investigate subsurface geology to 55 feet bgs. Field duplicates were collected at a frequency of 10 percent from all sampling locations. Additional field quality control samples included trip blanks.

Six housing units located above TCE-impacted first-observed groundwater and one housing unit located above unimpacted first-observed groundwater were selected for air sampling. The selected housing units were unoccupied and located on the ground floor. Continuous soil cores from ground surface to first-observed groundwater and first-observed groundwater samples were collected adjacent to each selected housing unit. Groundwater TCE concentrations for these units are listed on Table 1. Soil samples from up to three soil strata in the unsaturated zone were collected at each location. Unit-specific soil, groundwater, and building data needed to specify the Johnson and Ettinger (J&E) modeling inputs were collected.

Air samples were collected at each housing unit to evaluate the modeling results. Air samples included an indoor air sample collected over a continuous 24-hour period from a ground floor living area and a corresponding outdoor air sample collected adjacent to the housing unit. Air samples were analyzed for VOCs using EPA Method Toxic Organics (TO)-15 Selective Ion Mode (SIM) (reporting limit for TCE approximately 0.18 micrograms per cubic meter [$\mu g/m^3$]). Field duplicates were collected at 10 percent of the locations.

Additional 24-hour air sampling was performed in November 2003 and May 2004 in these and other units to support additional evaluation of the vapor migration pathway and to verify prior air sample concentrations. Indoor air samples were collected from 25 units (about 40 percent of the housing units located over the TCE plume), including the seven original units sampled in 2002. Samples were collected during two events (November 2003 and May 2004) to evaluate possible seasonal differences, explore potential preferential pathways, and to determine if there may be structural influences on indoor air concentrations of TCE. Indoor air samples were collected from the breathing zone of a ground floor room and near a potential preferential pathway entry point, such as where water pipes or drain lines penetrated the slab in a ground floor bathroom. Ten outdoor air samples were collected concurrently near housing units where indoor samples were collected. Field duplicates were collected at 10 percent of the locations. Additional quality control samples included trip blanks. Air samples were analyzed for TCE using EPA Method TO-15 SIM, with a low reporting limit for TCE of approximately 0.017 $\mu g/m^3$. Average indoor air TCE concentrations from the samples collected for the seven original units are listed on Table 1.

Because the housing units had an undocumented history of use during the nearly 30 years of tenant occupancy prior to being vacated and cleaned, conditions were normalized to the extent possible by implementing a set of procedures during the November 2003 and May 2004 sampling events. The housing units were first ventilated for a period of 24 hours prior to each sampling event. All windows and doors were opened, faucets were turned on and off, and toilets and drains were flushed. All heating and air circulation equipment was turned off and the chimney flues were closed. Following this 24-hour period, the units were closed, windows and doors were shut, and openings of vents such as dryer vents were covered with aluminum foil. The units were then left closed for a period of 48 hours prior to the start of 24-hour indoor air sampling.

Table 1. Measured and Modeled TCE Concentrations and Attenuation Coefficients

	TCE Concentrations			Attenuation Coefficients (Unitless)				
Location	Groundwater (µg/L) [a]	Source Vapor (µg/m³) [b]	Average Measured Indoor Air (µg/m³) [c]	Measured [d]	Original J&E Model Run	Run A [f] [New Defaults]	Run B [g] [Min Location Specific Info]	2002 Vapor Intrusion Guidance (Slab-on-Grade)
619B	≤ 0.25	82	0.14	1.71E-3	3.45E-7	1.24E-6	1.88E-5	1.16E-5
620F	260	85,000	2.03	2.38E-5	3.02E-7	1.09E-6	1.88E-5	1.16E-5
703G	80	28,720	0.11	3.83E-6	3.45E-7	1.24E-6	1.88E-5	1.16E-5
705F[e]	≤ 0.25	86	0.12	1.39E-3	5.87E-7	2.11E-6	1.89E-5	1.16E-5
720D	≤ 0.25	86	0.14	1.63E-3	2.80E-5	1.01E-4	1.88E-5	1.16E-5
808A	≤ 0.25	85	0.15	1.76E-3	1.08E-5	6.49E-6	1.97E-5	Crawl Space
819D	250	78,750	0.11	1.40E-6	3.61E-3	1.94E-5	7.67E-4	Crawl Space

Notes:

(a) Groundwater sample collected in August 2002, assumed steady-state through duration of study. The laboratory MDL is shown with a ≤ symbol when there was no reported concentration.

(b) Estimated soil gas concentration just above the water table was calculated using groundwater concentration multiplied by 1,000 and by the temperature-dependent Henry's Law Constant. If the TCE concentration was ≤ 0.25 µg/L, to be conservative, the MDL of 0.25 µg/L was used for the groundwater concentration.

(c) Average of Navy-collected indoor air samples on 9/3/02, 9/5/02, 11/18/03, 11/25/03, 5/7/04, and 5/14/04.

(d) Average indoor air concentration divided by source vapor concentration.

(e) Off-plume location.

(f) Run A reflects all the location-specific inputs of the original J&E Model runs and the updated 2002 default parameters, where default parameters were used in the corresponding Original Runs.

(g) Run B reflects the 2002 default building parameters and default soil properties linked to the location-specific soil types and the measured depth to groundwater.

5.   ORIGINAL VAPOR MIGRATION AND INTRUSION MODELING

Vapor migration and building intrusion modeling were performed for the seven housing units using the GW-ADV Module of the J&E Model, Version 2.3 (J&E, 2001). This modeling was performed originally in September and October of 2002 using this version of the J&E model with certain of its recommended default parameters from the 1997 *Users' Guide for the Johnson and Ettinger (1991) Model for Subsurface Vapor Intrusion into Buildings* (Environmental Quality Management, Inc. [EQMI], 2000). A substantial amount of location-specific data were also used as input to the original model runs. Table 2 lists the site-specific parameters that were measured at each of the housing units and used in the original runs of the model. Location-specific measurements were made to capture, for example, the fact that the local unsaturated soils tended to be moist due to fine-grained particle size. The original modeling runs used the Environmental Quality Management, Inc. (2000) recommended default parameters for the physical- and chemical-specific properties linked to soil type, the soil-building pressure differential, the floor-wall seam crack width, and the indoor air exchange rate. The combination of location-specific information and default parameters were used to calculate the Source Vapor Concentration (just above the groundwater table), the Infinite Source Indoor Attenuation Coefficient (AC), and the Infinite Source Building Concentration. The subsurface Source Vapor Concentration and calculated ACs from the original modeling runs are shown in Table 1. The original J&E modeling runs represent a Tier III assessment, as described in the *OSWER Draft Guidance for Evaluating the Vapor Intrusion to Indoor Air Pathway from Groundwater and Soils* (2002 Draft Subsurface Vapor Intrusion Guidance) (EPA, 2002).

Table 2. Location-Specific Input Parameters

GROUNDWATER	TCE Concentration
	Soil/Groundwater Temperature
	Depth to Water Table
SOIL	Type Classification
	Layer Thickness
	Dry Bulk Density
	Total Porosity
	Water-Filled Porosity
	Percent Organic Carbon

ENCLOSED SPACE	Floor/Foundation Thickness
	Floor Length
	Floor Width
	Height

6.   ANALYSIS OF MEASURED AND MODELED ATTENUATION COEFFICIENTS

In order to compare the results from the original J&E model runs with the 2002 Draft Subsurface Vapor Intrusion Guidance (for slab-on-grade construction), the differences in model inputs were considered. The inclusion of considerable location-specific information for the original modeling runs (in addition to the location-specific depth to the water table and soil type) and the subsequent changes in a few of the default input parameters (between 1997 and late 2002) make a direct comparison of the results of the original J&E modeling runs to the 2002 Draft Subsurface Vapor Intrusion Guidance AC results difficult. To disaggregate the influence of the location-specific information from the influence of the changed default input parameters on these results, three additional modeling runs were performed for each housing unit (shown in Table 1):

1. The original J&E modeling runs were redone with the 2002 defaults substituted for the 1997 defaults and the location-specific model inputs were retained. The resulting ACs are shown in Table 1 as Run A.
2. The Run A modeling inputs were then further adjusted to reflect only the location-specific depth to the water table and the observed soil type(s). The remaining 2002 soil and building recommended default parameters were applied (i.e., the default soil properties associated with each soil type were used and not the location-specific measured soil properties). The resulting ACs are shown in Table 1 as Run B.
3. Modeling runs were then performed to mimic Figure 3b of the 2002 Draft Subsurface Vapor Intrusion Guidance, but for slab-on-grade construction. Version 3.0 of the J&E Model was used with only the location-specific inputs used in Run B, but with defaults appropriate to a slab-on-grade construction. The AC results are shown as the last column in Table 1.

Run A modeling results show ACs about an order of magnitude larger than was obtained in the original J&E model runs (suggesting more TCE migration and intrusion), but about an order of magnitude smaller (suggesting less TCE migration and intrusion) than was estimated using the

2002 Draft Subsurface Vapor Intrusion Guidance. The significant changes to the defaults presented in the 2000 J&E Model User's Guide (EQMI, 2000) relative to those in the new 2002 Draft Subsurface Vapor Intrusion Guidance include a change in the building default air exchange rate from 0.45 per hour $(hr)^{-1}$ to 0.25 hr^{-1} and in the building crack width/crack area ratio (expressed as Q_{soil}). These parameters are influential in affecting the AC and the resulting quantitative calculation of the indoor air concentration. Run A tends to show the effect of a greater assumed soil TCE flux and greater TCE concentration buildup in the indoor air resulting from less dilution by clean air exchange as reflected in the updated defaults.

Run B modeling results show ACs about two orders of magnitude larger than was obtained in the original J&E model runs (suggesting more TCE migration and intrusion), about an order of magnitude larger than from Run A, and roughly similar results as estimated using the 2002 Draft Subsurface Vapor Intrusion Guidance. Run B most closely mimics the modeling performed for the 2002 Draft Subsurface Vapor Intrusion Guidance. Run B tends to highlight the apparent overly conservative nature of the new building and soil defaults relative to the measured conditions at MCH. These results illustrate the potential importance of using representative location-specific site data whenever possible in this type of modeling.

TCE was detected in all air samples (above-plume and off-plume, breathing zone and preferential pathway, indoor and outdoor). TCE concentrations for the above-plume preferential pathway samples and for the above-plume breathing zone samples were essentially equal in all but 3 of 25 locations. This comparison suggests that, in general, there is no significant vapor migration into indoor air from contaminated groundwater through preferential pathways (which indicates that modeling of vapor migration at these sites should be appropriate). With the exception of three slab-on-grade units), the average TCE concentration at above-plume locations (breathing zone, preferred pathway, and outdoors) were about the same as the average TCE concentration for the off-plume locations. These similarities suggest that vapor migration into indoor air from contaminated groundwater is not accumulating in the housing units and may be related to an elevated regional ambient air TCE condition.

The average measured indoor air breathing zone TCE concentration was divided by the calculated Source Vapor Concentration to yield the Measured AC values listed in Table 1. The Source Vapor Concentration (the TCE vapor concentration at the soil/groundwater interface) was calculated using unit-specific groundwater TCE concentrations and location-specific groundwater temperature measurements with the default groundwater-soil gas equilibrium relationship.

TCE was not detected above laboratory MDLs in groundwater samples collected adjacent to housing units 619B, 705F, 720D, and 808A. It is possible that the measured ACs for these units are related to the apparent ubiquitous nature of TCE in the ambient and indoor air throughout the housing area, and are not representative of vapor migration from underlying groundwater contamination. Nonetheless, the groundwater sample MDL for TCE was used to calculate the unit-specific Source Vapor Concentration and the measured AC for these four units.

7. UNCERTAINTY

The November 2003 and May 2004 air measurements (breathing zone and preferential pathway samples approximately equal) suggest there were no preferential pathways from groundwater to indoor air at the sampled units. Vapor migration modeling should thus be applicable to this site. However, the November 2003 and May 2004 results also suggest that vapor migration into indoor air from contaminated groundwater is minimal and not accumulating in the housing units. Therefore, comparisons of the measured indoor air concentrations and ACs to that modeled using the 2002 Draft Subsurface Vapor Intrusion Guidance may not be applicable.

The TCE concentrations detected in indoor and outdoor air samples located over the TCE plume and off the TCE plume are similar. These TCE concentrations may be considered a localized background concentration. However, whether the localized background concentrations are due to an urban society, the regional TCE plumes, use of home products containing TCE, and/or even construction materials has not been defined. Since there has not been any correction for background TCE concentrations, the significance of the measured ACs and modeled ACs is not clear.

Another factor confounding a comparison of the modeled ACs and the measured ACs for these residential units was the presence of a moisture barrier within the floor/foundation construction. As the condition of this moisture barrier could not be verified, and it was uncertain as to whether it also was an effective vapor barrier for TCE, this barrier was not considered in this modeling. A competent moisture barrier also would typically pose a significant resistance or barrier to vapor transport in the unit.

8. CONCLUSIONS

A detailed analysis of the potential migration and intrusion of TCE from groundwater into indoor air was performed for a number of residential units at MCH. The field and analytical portion of this analysis was designed and implemented to include the direct sampling of indoor and ambient air and the direct measurement of many location-specific soil and building parameters to support vapor intrusion modeling. Location-specific modeling was then performed for a set of slab-on-grade and crawl space housing units, and the collected air concentration measurements were critically evaluated in a series of pair-wise comparisons.

The measured indoor air concentrations of TCE were seen to be very similar to the outdoors ambient concentrations for the housing units. Air sampling at potential preferential pathway entry points to the indoor air also provided no evidence that the subsurface source of TCE was contributing to the indoor concentrations. Consequently, the vapor migration to indoor air pathway was not demonstrated to be dominant or possibly even significant at the MCH site. Given this finding, many of the possible comparisons of measured and modeled metrics lack meaning. However, the original detailed modeling performed for these units was re-examined and explored relative to the recommended modeling procedures, defaults, and recommended assessment values associated with the 2002 Draft Subsurface Vapor Intrusion Guidance. A step-wise location-specific modeling analysis was performed to demonstrate the relative contribution and value of collecting location-specific measurements to support the modeling and to illustrate the sensitivity of the modeling to some of the key default parameters that are difficult to define or measure on a site-specific basis. The exercise illustrated the role of both measurements and modeling in drawing conclusions about the importance of the vapor intrusion pathway at a site, and the importance of gathering location-specific data to support modeling when modeling is appropriate and site conditions differ materially from generic defaults.

REFERENCES

Environmental Quality Management, Inc. (EQMI). 2000. *User's Guide for The Johnson and Ettinger (1991) Model for Subsurface Vapor Intrusion into Buildings* . December.

Environmental Quality Management, Inc. (EQMI). 2003. *User's Guide for Evaluating Subsurface Vapor Intrusion into Buildings* . June 19.

Johnson and Ettinger (J&E), 2001. Model for Subsurface Vapor Int rusion into Buildings, Version 2.3. March.

U.S. Environmental Protection Agency (EPA). 2002. *OSWER Draft Guidance for Evaluating the Vapor Intrusion to Indoor Air Pathway from Groun dwater and Soils (Subsurface Vapor Intrusion Guidance)* . November 29 Publication in Federal register.

PRIOR PUBLISHED DOCUMENTS

Tetra Tech FW, Inc. (TtFW). 2003. *Final Site Characterization and Baseline Human Health Risk Assessment Report for Orion Park and Wescoat Housing Areas* . December 19.

Tetra Tech FW, Inc. (TtFW). 2004. Air Sampling Results for Orion Park and Wescoat Housing Areas. September 24.

CHAPTER 8

DEVELOPMENT AND APPLICATION OF A MULTIMEDIA MODEL TO ASSESS FATE AND TRANSPORT OF ORGANIC CHEMICALS IN A SOUTH TEXAS LAKE

Venkatesh Uddameri and Dhanuja Kumar
Department of Environmental and Civil Engineering, Texas A&M University-Kingsville, MSC 213, Kingsville, TX 78363

Abstract: The coastal bend region of South Texas has undergone significant changes in recent times. Increased agricultural activities, industrialization and population growth have enhanced the releases of organic chemicals in South Texas environments. Several classes of organic contaminants have been detected in air, water and sediment samples collected in this region. There is a growing concern that this increased usage may lead to unacceptable exposures and pollute surface-water bodies used for water supply purposes.

1.   INTRODUCTION

The coastal bend region of South Texas (Figure 1) is currently experiencing significant growth. Cities like Corpus Christi and Victoria have experienced over 10% increase in population within the last few years (Census, 2000). In addition, industrial activities and agricultural land use have grown in the last five years (USDA, 2002). The coastal bend of South Texas is a semi-arid environment with an erratic climate, prone to severe drought (Norwine, 1995). While the coastal bend region is larger than several states of the union, only 2% of its land area is covered by surface-water bodies. Groundwater quality is fairly poor and generally unfit for consumption in this region. As a result, roughly 80% of the region's water demands are met through the use of surface-water resources.

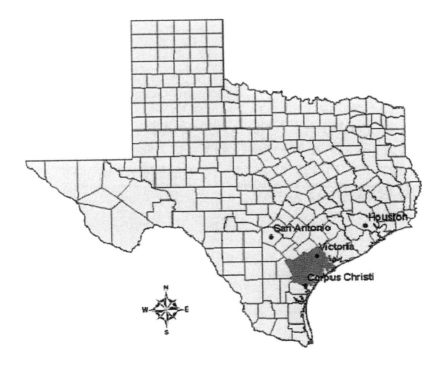

Figure 1. Coastal Bend Region of Texas

The coastal bend region ranks high in the production of sorghum and cotton. The industrial base is primarily comprised of petrochemical industries. The ubiquitous usage of organic chemicals in agricultural and industrial activities has raised concerns regarding exposure to organic chemicals. South Texas has a fairly high incidence of neural tube defects (NTD) and other endocrine-related ailments often suspected to be caused from high exposures to organic chemicals (Newsweek, 1992). South Texas is also a common stopping ground for many migratory birds during their winter sojourn. Therefore, in addition to human health issues, exposure to ecological receptors including several endangered species is also of concern.

Recent field campaigns have detected a variety of organic chemicals in many streams and lakes. Organic compounds detected include – aliphatics and aromatics in the air, industrial estrogens such as nonylphenol, triazine pesticides, plasticizers such as benzene dicarboxylic acid and other legacy organochlorine compounds such as DDT and metabolites in sediments and surface waters (USGS, 1999; Garcia et al., 2001).

Understanding the fate and transport of organic chemicals in South Texas environments is necessary for several standpoints including: 1) assessing exposure to human beings via different pathways; 2) quantifying exposure to migratory birds and wildlife; 3) identifying sound agricultural management practices and 4) fostering environmentally friendly growth and industrial development.

Therefore, the broad goal of this research is to develop a suite of models and methodologies that will enable better elucidation of the fate and transport of organic chemicals in South Texas environments. In particular, the focus is to couple state-of-the-art information technologies like Geographic Information Systems (GIS) and fuzzy set theory with multimedia modeling schemes to develop a knowledge base for potential exposure routes. As a first step in that direction, the proposed research focuses on characterizing the fate and transport of benzene in Choke Canyon Reservoir.

2.   STUDY AREA

2.1   Site Overview

The Choke Canyon reservoir (Figure 2) is an approximately 26000 acres impoundment in the Frio river watershed constructed in the early 80's to meet the growing water demands of the city of Corpus Christi and its surrounding areas. The maximum depth of the lake is approximately 95 ft with 10 -15 feet fluctuations. The inflow to the reservoir is from San Miguel Creek and Frio Rivers and the outflow is into Nueces River. It is monitored by the United States Geological Survey (USGS). The outflow from the reservoir is mandated to be at least 33 ac-ft/day (Agreed Order, 1995). There are roughly 18000 visitors per month and recreational activities at the lake include camping, boating, fishing and wildlife viewing.

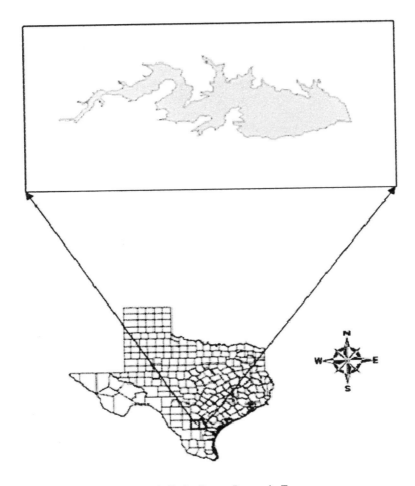

Figure 2. Choke Canyon Reservoir, Texas

2.2   Characterization of Benzene Sources

A refinery operated by Valero, Inc. is the only major benzene emitting facility in the vicinity of the lake. Emissions reported by this facility in the USEPA toxic release inventory (TRI) between the periods of 1996 – 2002 were used to obtain estimates for benzene releases from the refinery. Over 99.5% of these emissions were in vapor form into the air and a very small amount was discharged into the water. Given that the discharge into the water was very small (~ 30 lbs/yr) all of it was assumed to enter the lake to simulate conservative water emissions. It is however important to recognize that not all of the benzene emitted from the refinery in the air will actually traverse over Choke Canyon reservoir and assuming all the emissions will pass over the lake might be too conservative. Hence, it is important to apportion the amount of emissions that will actually pass over the lake (e.g., Mackay and Hicke, 2000).

A GIS analysis was carried out to identify the angle between the two extreme points of the reservoir and the stack at Valero refinery. Using distances obtained from GIS analysis, this angle was computed to be approximately 53° in the west - north quadrant (Figure 3). Therefore winds blowing from east/southeast directions will carry hydrocarbon emissions from the stack over the lake. Wind speed/wind direction data collected by the Crop Weather Program at Texas A&M University at Beeville, TX was used to identify the annual frequency of winds blowing in the east/south east direction. The wind rose plot (Figure 4) indicates that winds are in this direction for about 65% of the time with wind speeds ranging from 1 – 10 knots. Thus, 65% of the annual emissions from the refinery were assumed to pass over the lake. No losses were assumed to occur between the stack and the lake and this assumption is reasonable given the proximity of the release (< 5 miles), typical wind speeds (~7 km/hr) and half-life of benzene in air (~ 17 hours).

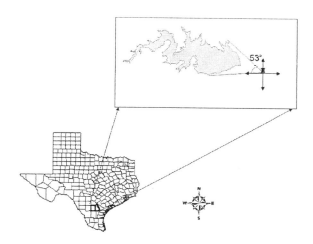

Figure 3. Identification of Wind Directions that can carry benzene from the Refinery over the lake

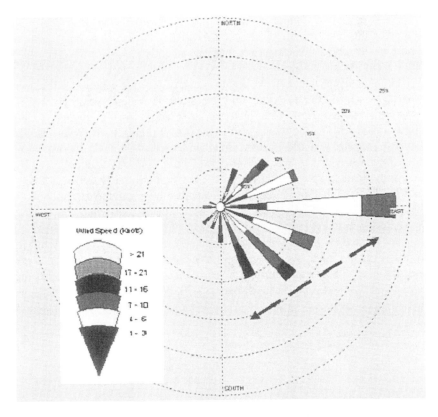

Figure 4. Wind Rose Plot from the nearest weather station at Beeville, TX. (The arrowed line indicates the winds that can transport emissions over the lake)

In addition to point releases, non-point releases are also important. Major non-point releases of benzene include: 1) vehicular emissions in the nearby highways; 2) emissions within the lake from recreational vehicles (boats); and 3) emissions from camping and other activities. Emission inventories for these categories were not readily available or carried out by park service or other regulatory personnel. Therefore indirect approaches to estimate emissions from these sources were used. The national parks service (NPS) has recently completed a comprehensive emission inventory at Lake Meredith, TX (NPS, 2003). Emission estimates for these non-point categories were assumed to be a linear function of visitors to the park. The ratio of visitors to Choke Canyon reservoir and Lake Meredith was used to estimate non-point source emissions at Choke Canyon reservoir from Lake Meredith emission inventory numbers. All non-point source emissions were assumed to occur in air and roughly 5-10% of the hydrocarbon emissions were assumed to be benzene. Non-point emissions into water were assumed to be negligible as benzene is highly volatile and not very soluble. Also as benzene has a relative small sediment-water partitioning coefficient, sources of benzene in sediments and mass-transfer between water and sediment compartments were assumed to be negligible.

3.   MATHEMATICAL MODEL DEVELOPMENT

Based on potential sources of benzene, and the scope of the present investigation, the Choke Canyon reservoir was conceptualized to comprise of two well-mixed compartments (namely air and water) and is depicted in Figure 5. The inflows and outflows of air and water were assumed to be at steady-state and represented using their average values. This assumption is consistent with typical multimedia modeling formulations used in similar studies (e.g., Mackay 2001).

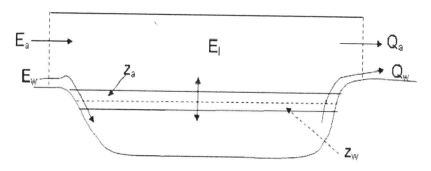

Figure 5. Two Compartment Conceptual Model for Choke Canyon Reserv oir

3.1   Mass Balance Expressions

Two mass-balance expressions are necessary as the transport of benzene in air and water is simultaneously considered here. The mass balance expression can be qualitatively described as follows:

$$
\begin{bmatrix} Accumulation \\ In\ the\ Compartment \end{bmatrix} = \begin{bmatrix} Inflow\ to\ the \\ compartment \end{bmatrix} - \begin{bmatrix} Outflow\ from\ the \\ compartment \end{bmatrix} - \begin{bmatrix} Net\ Losses \\ Within \end{bmatrix}
$$

(1)

The loss term in Equation (1) will include emissions from non-point sources generated within the lake; degradation of benzene in the lake due to abiotic and biotic reactions; and the mass-transfer of benzene between air and water compartments. As the concentration of benzene in air and water is hypothesized to be low, the degradation reactions were assumed to follow first-order kinetics and were represented using a lumped first-order coefficient based on half-life values suggested in Mackay (2001).

A stagnant two-film model was used to model the air-water mass-transfer of benzene (Mackay, 2001). According to the two-film theory, there is no net accumulation of benzene at the air-water interface and the total resistance to mass-transfer is equal to the sum of individual resistances a molecule of benzene experiences as it moves from the bulk water compartment to the interface and from the interface into the bulk air compartment or vice versa, assuming the mass-transfer at the air-water interface is a diffusion-type process. The flux of benzene from water to air can be written as:

$$F_{w-a} = v_w C_w = \frac{D_w}{z_w} C_w \tag{2}$$

$$F_{a-w} = v_a C_a = \frac{D_a}{z_a H} C_w \tag{3}$$

where F is the flux (mg/cm^2-sec.); C is the concentration (mg/cm^3); D is the coefficient of molecular diffusion (cm^2/s); H is the Henry's law constant (dim); v is the mass-transfer (piston) velocity (cm/s); and z is the characteristic length (cm). The subscripts a and w are used to denote air and water compartments respectively. Note the units presented above are for illustrative purposes only and any set of consistent units can be used with Equations (2) and (3).

Based on the above assumptions, the mass balance expressions for air and water can be written as:

$$V_a \frac{dC_a}{dt} = Q_a C_{a,in} - Q_a C_a + E_a + v_w C_w - v_a C_a - \lambda_a C_a \tag{4}$$

where V_a is the volume of the air compartment; Q_a is the air flowrate; C_a as stated before is the concentration in air compartment and $C_{a,in}$ is the inflow concentration; E_a is the emissions within the air compartment; v_w and v_a are piston velocities given by Equations (2) and (3); and λ_a is the lumped first-order rate constant in air. Similarly, the mass-balance expression for water can be written as:

$$V_w \frac{dC_w}{dt} = Q_w C_{w,in} - Q_w C_w + E_w - v_w C_w + v_a C_a - \lambda_w C_w \quad (5)$$

The variables in Equation (5) have similar meanings to those in Equation (4). V_w is the volume of the water compartment; Q_w is the water flowrate; C_w is the concentration in water compartment and $C_{w,in}$ is the inflow concentration into water; E_w is the emissions within the water compartment; v_w and v_a are piston velocities given by Equations (2) and (3); and λ_w is the lumped first-order rate constant in water.

The initial concentrations of benzene in air and water are also necessary to complete the mathematical description and these initial conditions are stated as follows:

$$C_a(@t = 0) = C_{a,i} \quad and \quad C_w(@t = 0) = C_{w,i} \quad (6)$$

3.2   Mathematical Solution

Equations 4 – 6 represent two coupled ordinary differential equations (ODEs). It is possible to solve this set of equations analytically, if the coefficients of the concentration terms are all constant. An approximate numerical solution can be obtained even when the coefficients are not constant (but functions of time). The ODEs (equation 4 -6) can be reduced to algebraic equations using Euler's method as follows:

$$C_a^t \approx C_a^{t-1} + \frac{\Delta t}{V_a} \left[\left(Q_a C_{a,in} + E_a + v_w C_w \right) - \left(Q_a + v_a + \lambda_a \right) C_a \right]_{-1} \quad (7)$$

$$C_w^t \approx C_w^{t-1} + \frac{\Delta t}{V_w} \left[\left(Q_w C_{w,in} + E_w + v_a C_a \right) - \left(Q_w + v_w + \lambda_w \right) C_w \right]_{-1} \quad (8)$$

Equation (7) and Equation (8) can be solved starting $t = 1$ and marched till $t = T$ (the end of simulation period). Note all terms on the right-hand side are computed at the previous time step and as such the concentration at the present time step is a function of input values at previous time steps. The initial conditions are used to solve the concentration at the first-time step. The accuracy of Equation (7) and Equation (8) improves as the time-step Δt

becomes smaller and smaller. However, this would imply increased number of computations for a given simulation time period T.

3.3   Simplifications using Equilibrium and Steady-state Transport Assumptions

It is often advisable to employ mathematical models in an iterative fashion starting with simple conceptualizations and progressively increasing the complexity. The values for piston velocities (especially the characteristic lengths z_w and z_a over which mass-transfer occurs) are not readily available for Choke canyon reservoir. If the air-water mass-transfer is assumed to be instantaneous then equilibrium assumption can be invoked. When equilibrium is assumed between the phases, the ratio of concentrations of benzene in air and water stays constant and is equal to the Henry's law constant. Mathematically,

$$H = \frac{C_a}{C_w} \quad \Rightarrow \quad C_a = H C_w \quad \Rightarrow \quad \frac{dC_a}{dt} = H \frac{dC_w}{dt} \qquad (9)$$

Adding Equation (4) and Equation (5) cancels out the mass-transfer terms on the right hand side and results in the following expression:

$$V_w \frac{dC_w}{dt} + V_a \frac{dC_a}{dt} = \left(Q_a C_{a,in} + Q_w C_{w,in} + E_a + E_w\right) - \left(Q_a + \lambda_a\right)C_a - \left(Q_w + \lambda_w\right)C_w \qquad (10)$$

Using relationships specified in Equation (9), Equation (10) can be simplified as:

$$\left(V_w + V_a H\right)\frac{dC_w}{dt} = \left(Q_a C_{a,in} + Q_w C_{w,in} + E_a + E_w\right) - \left(Q_w + \lambda_w + Q_a H + \lambda_a H\right)C_w \qquad (11)$$

Only the initial concentration of benzene needs to be specified to complete the mathematical description.

$$C_w(@\,t = 0) = C_{w,i} \tag{12}$$

Thus, using equilibrium assumption, a system of two ordinary differential equations can be reduced to a system of one ordinary differential equation and one algebraic equation which is considerably easier to solve.

Measurements of benzene either in the aqueous phase or in the vapor phase were not available at Choke Canyon reservoir to solve Equations 11 and 12. Therefore the concentration of benzene in the water was assumed to be at steady-state to further simplify the mathematics. The steady-state assumption implies benzene concentrations don't change with time. This assumption is reasonable if the emissions don't fluctuate significantly in time. While some temporal fluctuations are to be expected, the uncertainties in benzene concentrations obtained by solving a transient problem (Equation 7 - 8 or 11) using imprecise (unknown) initial conditions may far exceed the variability around a steady-state value. Hence, the steady-state value was used here to obtain a first-cut estimate. The steady-state benzene concentration in water can be obtained by setting the left-hand side of Equation (11) to zero and solving for C_w as:

$$C_{w,ss} = \frac{\left(Q_a C_{a,in} + Q_w C_{w,in} + E_a + E_w\right)}{\left(Q_w + \lambda_w + Q_a H + \lambda_a H\right)} \tag{13}$$

The subscript ss indicates that the obtained concentration represents a steady-state value. The steady-state concentration of benzene in air can be obtained from steady-state concentration in water ($C_{w,ss}$) and Henry's Law constant (H) using Equation (9).

3.4   Uncertainty Assessment Using Fuzzy Logic

Statistical approaches like Monte Carlo simulation and First-Order Second Moment (FOSM) analysis are commonly used to assess how uncertainty and variability in input parameters affect the accuracy of the estimated model outputs. Essentially, these approaches treat input parameters as random variables and characterize the variability using probability distribution functions (pdf). FOSM assumes the variables are normally distributed with known mean and variance, while any arbitrary distribution can be specified with Monte Carlo simulation.

Deriving appropriate site-specific distributions for various input parameters was not possible in this study due to lack of sufficient data.

Commonly a uniform distribution can be used if any value in a given range is equally likely. This assumption highlights our ignorance with regards to statistical distribution for various inputs which in turn manifests in our output values. In other words, assuming uniform distributions for all parameters would imply any value for output is equally likely and as such of little practical use.

Fuzzy set theory or possibilistic approach provides an alternative means for assessing parametric imprecision. In this approach, input variables are considered to be fuzzy numbers rather than random variables (in a statistical sense). The fuzzy inputs are characterized using a membership function (Figure 6) which represents a modeler's subjective point of view regarding an input. The α-cut signifies the level of confidence on part of the modeler as to whether the input value lies in a particular range. For example, in Figure 6, the modeler is 100% sure that the parameter a lies in the interval $[a,b]$ and 50% sure that the parameter lies between $[a_1,b_1]$. Fuzzy set theoretic approaches are increasingly being used in environmental applications and their applicability has been reviewed by Uddameri, (2004). The uncertainty propagation using fuzzy set theory can be carried out using fuzzy calculus rules suggested by Dubois and Prade (1988) which entails the following steps:

1. Specify a α-cut
2. Obtain ranges for all inputs corresponding to the specified α-cut
3. Use the values of inputs from step 2 to compute minimum and maximum values for the output
   a. An optimization algorithm may come in handy here
4. Repeat step 1 – 4 at different α-cuts
   a. Typically 4 -5 different α-cuts are evaluated
5. Use output values obtained at different α-cuts to construct the membership function for the output.
   a. This membership function characterizes the imprecision in the output due to imprecision in the inputs.

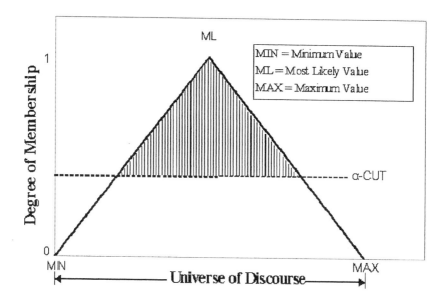

Figure 6. Schematic of a Fuzzy Membership Function

The α-cut procedure described above is fairly generic and can be applied to any arbitrary membership function. However, triangular membership functions are commonly used in practice as minimum; most likely and maximum values for an input can be identified by a modeler (Cox, 1995).

4.   MODEL APPLICATION

Typical values for various physical properties such as volume of air and water compartments, point and non-point source loadings and air and water flow rates considered representative of steady-state (average) conditions at Choke Canyon reservoir were tabulated from a variety of sources and are presented in Table 1 along with expected range of variability. Chemical properties of Benzene and ranges of variability were based on data presented in Mackay, (2001).

Table 1. Model Inputs for Choke Canyon Reservoir

Model Inputs	Symbol	Range of Values		
Physical Properties		Low	Likely	High
Flowrate of Air (cu.ft/month)	Qa	3.85E+11	3.85E+12	3.85E+13
Flowrate of water (cu.ft/month)	Qw	1.14E+09	1.14E+10	1.14E+11
Loading in air (lb/month)	QaCain + Ea	67.434	337.17	1685.85
Loading water (lb/month)	QwCwin + Ew	0.055	0.275	1.375
Volume of water (cu-ft)	Vw	1.84E+11	2.30E+11	2.76E+11
Volume of Air (cu-ft)	Va	1.17E+11	2.33E+11	3.50E+11
Chemical Properties[1]				
Chemical Name			Benzene	
Henry's Law Constant	Dim	0.18	0.2	0.22
Half Life (air)	hours	3.4	17	85
Half Life (air)	months	0.004722	0.023611	0.118056
Half Life (water)	Hours	34	170	850
Half-life (water)	Months	0.047222	0.236111	1.180556

[1]Likely values from Mackay (2001); Low and High values were obtained by assigning specified variability

4.1   Steady-State Estimates

The steady-state benzene concentration in water was obtained to be 1.94 ng/L by substituting data presented in Table 1 into Equation (13). Then using Equation (9) the steady-state vapor concentration was equal to 0.42 ng/L. Although the magnitude of concentration in the air phase is lower than that in water, over 99.9% of the mass was contained in the air phase (the lower concentration is due to a larger atmospheric compartment). The estimated concentration values are low and the results indicate that under currently assumed conditions, exposures to benzene at Choke Canyon reservoir are not significant and do not pose a significant risk.

4.2   Sensitivity Analysis

Regression-based sensitivity analysis was carried out by computing output values for different values of inputs and carrying out a linear regression between inputs and the output. Over 75% of the variance was explained by the regression equation suggesting that this approach to calculating sensitivity is reasonable. The results presented in Figure 7 indicate that the concentration of benzene in water is more sensitive to air-related inputs than water-related inputs. This result is to be expected since most of the benzene enters the system through air emissions; also benzene is highly volatile and has a greater affinity to atmospheric compartment. The concentration of benzene in air (not shown here) is also affected more by air-related inputs than water inputs and this again is to be expected as concentration is the product of concentration in the water and Henry's law constant.

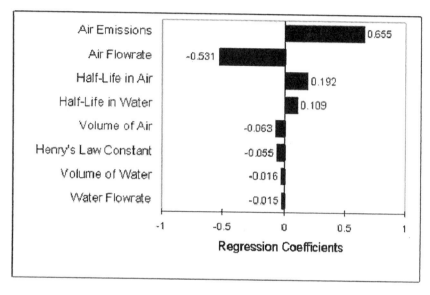

Figure 7. Identification of Sensitive Model Parameters via Linear Regression

4.3   Fuzzy Uncertainty Analysis

Fuzzy uncertainty analysis was carried out by assuming triangular membership functions for all inputs, the maximum and minimum values for the outputs at different α-cuts were obtained by using SOLVER add-in available in MS-EXCEL to carry out the nonlinear minimization and maximization problems.   The membership functions for benzene concentrations in water and air are presented in Figure 8 and Figure 9.   As can be seen the membership functions for the concentrations are not triangular in shape even though triangular membership functions were assumed for all inputs.   This behavior is to be expected because fuzzy multiplication and division are nonlinear operations (Kaufmann and Gupta, 1985).   The results indicate that for assumed ranges of input variability, the concentration of benzene in water ranges between 0.6 ng/L – 3.4 ng/L and the concentration of benzene in air ranges from 0.11 ng/L – 0.74 ng/L.   The variability around the most likely value is more symmetric for air concentrations and slightly skewed towards the higher end for water concentrations.   The variability is high for water concentrations and reflects the limited information we have about this parameter. In general, the range of variability obtained using the fuzzy approach will be higher than that obtained using Monte Carlo simulations. This result occurs because multiplication operations between two small numbers further reduce the probability of occurrence of an output. From a conceptual standpoint, by specifying probability distributions in a Monte Carlo simulation we profess more knowledge (i.e., we not only know most likely values and variability but also know the frequency with which we can expect to see occurrences of an input). On the other hand no such statistical claims are made with fuzzy membership functions. Therefore, fuzzy approaches are better suited when our ignorance regarding input parameters is high.

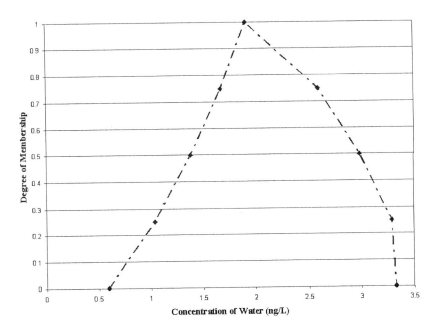

Figure 8. Membership Function for Benzene Concentrations in Water Depicting Parametric Variability

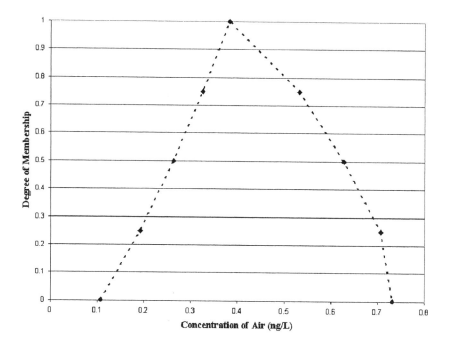

Figure 9. Variability in Predicted Benzene Concentration in the Air due to Input Variability

5. SUMMARY AND CONCLUSIONS

The objective of this study was to develop a two-compartment mathematical model to assess the fate and distribution of benzene (or similar aromatics) in the vicinity of Choke Canyon reservoir, TX. Data for model inputs were complied from several sources. A Geographic Information System (GIS) was to identify the fraction of total emissions that actually pass-over the reservoir. The amount of available data did not allow development of site-specific probability distribution functions to statistically characterize uncertainty and variability in the inputs. Therefore, fuzzy set theoretic approaches were used to assess imprecision in estimated steady-state water and air phase benzene concentrations in the lake. The analysis suggests that for the assumed conditions, exposure to benzene either through vapor or aqueous pathways is most likely insignificant. GIS and fuzzy set theory were seen to be useful tools to manage data and characterize uncertainty.

ACKNOWLEDGMENTS

This work was partially supported by Texas A&M University-Kingsville, University Research Council (URC) award to the first author.

REFERENCES

Alderite, M. (1996). "Editorial." Environmental Health Perspectives.

Census (2000). www.census.gov (accessed 10/2004).

Dubois, H. and R. Prade (1988). Possibility Theory. NY, Plenum Press.

Garcia, S. S., C. Ake, et al. (2001). "Initial Results of Environmental Monitoring in the Texas Rio Grande Valley." *Environmental International* 26: 465-474.

Kaufmann, A. and M. Gupta (1985). An Introduction to Fuzzy Arithmetic. San Diego, CA, Academic Press.

Mackay, D. (2001). Multimedia Environmental Modeling - A Fugacity Approach. Chelsea, MI, Lewis Publishers.

Mackay, D. and B. Hickie (2000). "Mass Balance Model for Source Apportionment, Transport and Fate of PAH in Lac Saint Louis, Qubec." *Chemosphere* 41: 681-692.

Newsweek (1992). article - June 8th: p52.

Norwine, J. R. (1995). "The changing climate of Texas: predictability and implications for the future."

NPS (2003). Emissions Inventory for Lake Meredith, Texas, National Parks Service.

Order (2001). An Agreed Order Between Texas Commission on Environmental Quality and City of Corpus Christi regarding Lake Corpus Christi and Choke Canyon Reservoir. Corpus Christi, TX. 2003.

Uddameri (2004). A Review of Fuzzy Set Theoretic Approaches for Hazardous Waste Management. *Contaminated Soils and Sediments* vol. 19. P.T.K. a. E.J. Calabrase. Amherst, MA, Kluwer Academic Publishers.

USDA (2002). "Agriculture Census of the United States."

USGS (1999). Chemical Quality of Sediment Cores from Laguna Madre, Laguna Atascosa and Arroyo Colorado, Texas. Austin, TX, United States Geological S urvey.

PART IV: MTBE AND OXYGENATES

CHAPTER 9

THE MtBE REMOVAL EFFECTIVENESS OF AIR SPARGING, TESTED ON AN INTERMEDIATE SCALE LABORATORY APPARATUS

Claudio Alimonti[1] and Daniele Lausdei[2]

[1]*Università di Roma "La Sapienza", v. Eudossiana 18, 00184 Roma, Italy;* [2]*Golder Associates S.r.l., v. Sante Bargellini 4, 00157 Roma, Italy*

Abstract: Among all the *in situ* groundwater remediation treatments designed to remove MtBE and other gasoline components, *Air Sparging (AS)* has been widely used since it is one of the best established, econom ical and reliable technologies for the remediation of volatile compounds dissolved in groundwater. However although *AS* has been successfully applied at several contaminated sites, the airflow distribution in saturated media and the interactions of various physical, chemical and microbial processes during *AS* operations are still not well understood. This experimental study was designed to investigate the effectiveness of *AS* in removing dissolved MtBE from a saturated media, performing a removal test under confined and controlled conditions in an intermediate scale tank (m 1 x 1 x 1,2). The experimental conditions focused on the study of the stripping process driven by air injection. Stripping is considered to be the most effective removal process driven by *AS* in the initial period of its *in situ* application. The study confirmed that *in situ AS* has significant potential for remediating groundwater contaminated by MtBE, by showing that the stripping action, driven by air injection contributes significantly to the overall removal action of the *AS* technology.

Key words: site remediation; *in situ* technology; groundwater treatment; saturated media; stripping.

1.   INTRODUCTION

In the last decades, groundwater contamination by methyl *ter* butyl ether (MtBE) due to fuel spills or leakage from underground storage facilities has become a highly sensitive environmental issue. Among all the *in situ* groundwater remediation treatments, aimed at removing MtBE and other gasoline components, *Air Sparging* (*AS*) has been widely used since it is one of the most well-established, economical and reliable technologies for the remediation of volatile compounds dissolved in the groundwater of shallow aquifers. The broad use of *AS* systems in contaminated groundwater treatment is due to the simplicity of the plant installation, operation and maintenance and, consequently, to its low associated costs (D.O.D., 1994).

Even though *AS* has been successfully applied at several contaminated sites (Ducco et al., 2001), the airflow distribution in saturated media and the interactions of various physical, chemical and microbial processes during *AS* operations are still not well understood (Suthersan S. S., 1999). This is mainly due to the difficulty of obtaining, at reasonable costs, enough information on the phenomena occurring within the contaminated saturated zone to be treated by air injection (Braida and Ong, 2001).

Golder Associates S.r.l., in partnership with the University of Rome "*La Sapienza*", is currently developing a research program to investigate the effectiveness and the efficiency of *AS* in dissolved MtBE removal from a saturated media, performing removal tests under confined and controlled conditions within an intermediate scale tank. The experiment was designed to study the stripping process driven by air injection, since stripping is considered to be the most effective of the removal processes driven by *AS* in the initial period of its *in situ* application.

2.   LABORATORY APPARATUS

The laboratory apparatus used to perform removal tests consists of the following components as illustrated in Figure 1:
- the steel TANK (1), dimensions 1 x 1 x 1,2 m, containing 1 m³ of soil;
- the SATURATION UNIT, including:
    - a *saturation vessel* (2), placed onto the external scaffolding, where the water is stocked prior to flow within the tank and then saturate the soil;
    - a nylon *saturation line* (3), to have the stocked water flow to the top of the soil inside the tank;
- the INJECTION UNIT, including:
    - an *air compressor* (4) to provide air to the system;

- a *flowmeter* (5), for the on-line measurement of the injected airflow;
- a *manometer* (6), for the on-line measurement of the injected air pressure;
- a "Rilsan" type *injection line* (7), joining the air compressor to the sparge point;
- a *sparge point* (8) to release injected air within the saturated media;
- the VAPOR RECOVERY UNIT, including:
- an "Oregon" type flexible *tubing* (9), conveying the vapors from the saturated soil directly to the filter vessel;
- a filter *vessel* (10), placed on an external scaffolding, containing activated carbon filter to separate and adsorb volatilized MtBE from the airflow outlet of the tank.

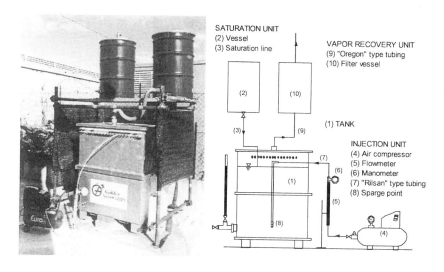

Figure 1. Laboratory experimental apparatus and its related flowsheet.

Several sampling points (15) were evenly distributed within the saturated media in order to withdraw samples of solution at fixed times to monitor the MtBE removal process from water. All the samples of solution were then analyzed by gascromatography.

The sampling points' position was chosen to maximize the information on the spatial distribution of MtBE concentrations within the saturated media. Fourteen sampling points were located on three levels, whilst the fifteenth (point F in Figure 2) was placed at one of the corners on the bottom of the tank, away from the zone reached by the injected air. All sampling points were connected to the sampling ports located on the front side of the tank using nylon pipes.

The sampling points distribution within the tank is represented in Figure 2, where *north*, *east*, *south* and *west* are conventional names for the four sides of the tank and 1, 2 and 3 are, respectively, the horizontal levels where the sampling points were placed.

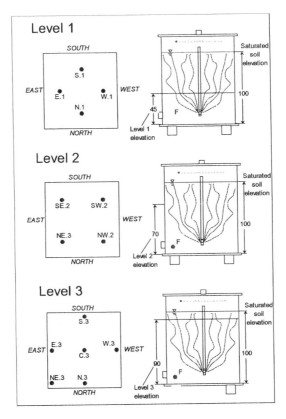

Figure 2. Sampling points distribution within the saturated porous media (the d imensions are expressed in centimeters).

The characterization of the saturated media was necessary prior to start the removal test in order to obtain adequate input data for the system modeling. Through experimental determinations, the intrinsic soil properties (grain size distribution, specific weight, water content and organic carbon fraction) were measured, as well as the general properties of the settled media, once the tank was filled and saturated (porosity and hydraulic conductivity).

In Figure 3, the grain size distribution curve and the other characteristics of the soil used in removal test are summarized.

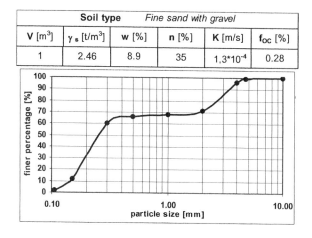

Soil type		Fine sand with gravel			
V [m³]	γ_s [t/m³]	w [%]	n [%]	K [m/s]	f_{oc} [%]
1	2.46	8.9	35	$1,3*10^{-4}$	0.28

Figure 3. Grain size distribution curve and characteristics of the soil used in removal test.

3. EXPERIMENTAL PHASES DESCRIPTION

The experimental activities were split in five main operating phases, characterized by a sequence of elementary actions, in order to simplify their operation and also to make them repeatable (Table 1).

Table 1. Sequence of the operating phases, and related elementary actions, undertaken to perform the removal test

	OPERATING PHASES	ELEMENTARY ACTIONS
1	LABORATORY APPARATUS STRUCTURAL UNITS SET UP	• External scaffolding assemblage • Saturation, injection and vapor recovery units set up • Activated carbon filter set up
2	SATURATED MEDIA ARRANGEMENT	• Soil characterization and tank filling • Sampling points and withdrawal lines installation • Injection point installation • Soil saturation • Saturated media characterization
3	SATURATED MEDIA CONDITIONING	• Preparation of the contaminant solution • Media conditioning through several cycles of saturation and drainage
4	INJECTED AIR REGULATION	• Air flow and pressure of injection regulation
5	REMOVAL TEST EXECUTION	• Initial sampling

OPERATING PHASES	ELEMENTARY ACTIONS
	• Air Sparging system start up • Sampling of solution and temperature measurements • Samples delivery to the chemical laboratory

The most "critical" operating phases, among the ones listed in Table 1, are concisely described in the following paragraphs.

3.1   Saturated media conditioning

At the end of the saturation operations, a 3 cm thick layer of water was left on top of the soil and a peristaltic pump was set to circulate water within the media. The pump withdrew water from a port located at the bottom of the tank conveying it on top of the soil, through the port at the end of the *saturation line* (Figure 1).

A certain amount of pure MtBE (1.75 g) was injected using a syringe directly in the pump pipe, first to obtain a uniform distribution of MtBE in the 3 cm layer of water, then to let it uniformly percolate through the porous media. The peristaltic pump completed 3 continuous[1] cycles of saturation and drainage to uniformly spread the solution within the saturated media. At the end of the saturation and drainage cycles, the MtBE concentration within the saturated media reached an average concentration (excluding point F) of about 3.4 mg/l, as shown in Figure 4. This concentration was slightly lower than the theoretical concentration of 5 mg/l, probably because some MtBE evaporated from the 3 cm layer of solution left on top of the soil before starting the saturated media conditioning process.

---

[1] With the exception occurred after few minutes of operating, when the bottom port was unconnected for a while, to allow the 3 cm of excess water to flow out of the tank.

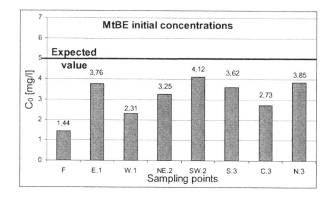

Figure 4. Dissolved MtBE concentration distribution within the saturated media at the end of its conditioning.

3.2   Injected air regulation

Both the flow of the injected air and its pressure were controlled through a valve located along the *injection line*, right after the air compressor. It was observed that the soil was lifted for injected airflow greater than 1.2 Nm^3/h whilst injected air could not exceed the saturated soil resistance and then flow into the media for injected airflow lower than 0.3 Nm^3/h.

The regulation valve was then set to have an injected airflow of 0.6 Nm^3/h, with a pressure of 1.5 bar.

3.3   Removal test execution

Right after the initial sampling of the solution, the removal test began with the air injection start-up, followed by several scheduled sampling rounds over twenty days. During the test, the number and the frequency of solution samplings were voluntarily limited to avoid media desaturation before the completion of the test. Only 8 sampling points, among the 15 shown in Figure 2, were chosen to monitor the MtBE removal process within the saturated media. At the end of the removal test 74 samples of solution had been collected through 11 monitoring rounds with a logarithmic-like distribution over time. At each withdrawal of solution an infrared thermometer measured the temperature of each sample.

All samples of solution were delivered to the chemical laboratory of Raw Material Department of University of Rome "*La Sapienza*" to be analyzed by gascromatography.

4.   ANALYTICAL RESULTS

In Table 2 are reported all the analytical results of MtBE concentrations measured on solution samples taken from each sampling point during the removal test. The chemical analyses carried out on the solution samples did not reveal the presence of any other analyte but MtBE, indicating the absence of any process of MtBE degradation.

Table 2. Analytical results of MtBE concentrations measured at each sampling point during the removal test.

Time		Temp	F	E.1	W.1	NE.2	SW.2	S.3	C.3	N.3
[h]	[d]	[°C]	[mg/l]	[mg/l]	[mg/l]	[mg/l]	[mg/l]	[mg/l]	[mg/l]	[mg/l]
0	0	22.8	1.44	3.76	2.31	3.25	4.12	3.62	2.73	3.85
6	0.25	23.4	2.02	2.83	2.95	3.11	4.60	4.22	3.56	3.88
12	0.5	19.6	3.38	3.79	3.89	4.12	5.58	1.51	4.19	1.93
24	1	22.0	n.a.*	2.59	1.81	3.71	2.32	1.29	1.79	1.08
49	2	18.4	2.69	3.10	0.71	4.98	4.76	0.72	1.10	0.97
75	3	16.8	n.a.*	1.41	0.18	2.98	2.29	n.a.*	0.46	0.24
100	4	22.0	2.90	1.42	0.07	2.55	2.30	n.a.*	0.66	0.23
145	6	23.1	n.a.*	0.99	0.08	2.56	2.07	n.a.*	0.56	n.a.*
193	8	23.5	6.78	0.43	0.06	1.87	1.28	n.a.*	0.51	n.a.*
265	11	23.0	3.74	0.45	0.06	1.79	1.30	n.a.*	0.49	n.a.*
385	16	23.7	4.82	0.07	0.02	0.80	0.78	n.a.*	0.31	n.a.*
481	20	24.7	3.82	0.04	0.01	0.34	0.42	n.a.*	n.a.*	n.a.*

*n.a.: not available

In Figure 5, the trends of MtBE concentration are plotted to illustrate the MtBE removal process from the saturated media. A general effectiveness of the *AS* to remove MtBE from the contaminated saturated media can be observed.

Where the media was directly reached by the injected airflow (Figure 2: all sampling points except point F), the MtBE concentration trend after an initial fluctuation shows a strong tendency to attenuate. The concentration fluctuations generally occurred at the early stage of air injection, within the first two days of operation. Then, the concentrations asymptotically decreased towards zero.

The MtBE concentrations at point F, located at a corner of the tank and away from the sparging point (Figure 2), do not reveal a reduction similar to the other sampling points, remaining almost equal to the average initial concentration.

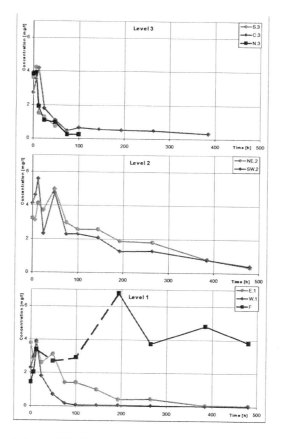

Figure 5. Trend of MtBE concentration measured during the removal test.

5. DISCUSSION OF RESULTS

It is well-known that the *AS* technology removes contaminants from groundwater through three main mechanisms (Suthersan S. S., 1999):

- *stripping*, based on the mass transfer of dissolved contaminants to the vapor phase through the air-water interface created by the injection of air within the water-saturated media;
- *direct volatilization*, based on the mass transfer from the trapped and/or adsorbed phase below the water table to the vapor phase, through the air-water interface created by the injection of air within the water-saturated media;

- *aerobic biodegradation*, based on the disruption of the contaminants molecule, through the action of microorganisms that use (and consume) oxygen as electron donor.

Among the three contaminant removal mechanisms mentioned above, the experiment solely highlighted the action of *stripping*.

In fact, the absence of any kind of substance but the MtBE, revealed at the chemical analysis, meant that *biodegradation* was totally inhibited, probably caused by the use of tap water to saturate the soil (generally treated with chlorine for drinkable usage) and/or by the short duration of the test. The *direct volatilization* could not be evaluated during the test, since MtBE was present within the media only in a dissolved phase.

Assuming *stripping* is the dominant mass removal mechanism, Henry's law provided a basis for analyzing the trend and the distribution of MtBE concentration within the saturated media.

Since the partitioning of the dissolved contaminant occurred at the discrete air-water interface created by the air injection (bubbles and/or air channels), lower MtBE concentrations were expected in the immediate vicinity of the air paths within the saturated media creating a concentration gradient directed to the air-water interfaces. Therefore, a mass transfer through the mechanism of diffusion occurred within pore water providing new dissolved contaminant mass to be removed by stripping.

Hence, the density and distribution of air paths play a significant role in mass transfer effectiveness, as well as the residence time of the injected air traveling through these discrete paths. The dissolved MtBE concentrations measured at different times and sampling points, during the removal test, were affected by the local availability of interfacial surface area and the MtBE concentration in vapor phase within the air paths.

Since the Henry constant is strongly affected by temperature any possible relation between the remarkable fluctuation of MtBE concentration, noticed in some sampling points, and the analogue and simultaneous fluctuation of temperature was evaluated (Figure 6).

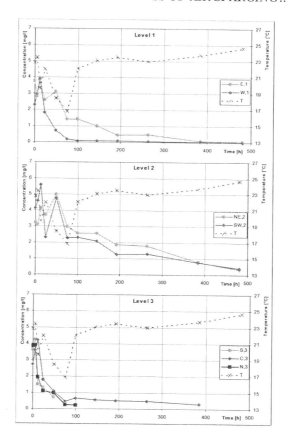

Figure 6. Trend of solution temperatures measured at the sampling points.

As shown in Figure 6, an inverse correlation between concentration and temperature peaks was observed only for some sampling points (E1, NE2 and SW2) and for short intervals during the test. The temperature does therefore seem to have a negligible influence on the general MtBE removal process.

The regularity of fluctuations registered at some sampling points (E1 and SW2), as well as some sudden increases of MtBE concentration (reaching values sometimes higher than the initial concentrations, as in W1, NE2 and C3), has been interpreted as the evidence of a pore water mixing, due to the air injection that locally displaced pore water within the saturated media.

This phenomenon, enhanced by the limited dimensions of the tank, likely played an important role in dissolved MtBE removal. In fact the convective movement of displaced water through the media enhanced the availability of air-water interfaces, reducing the limitation of the diffusion mechanism for mass transfer.

The high external temperature (28÷30 °C) registered during the removal test enhanced the evaporation of the solution from the top of the soil, even though the tank was protected from sunlight. This prohibited the taking of solution samples from the shallowest sampling points after the first 48 hours of operation. However, during this initial period a significant decrease of MtBE concentration was observed in these points.

6.   CONCLUSIONS

The applicability of the *AS* technology for MtBE removal from water is usually considered to be limited by the physico-chemical properties of MtBE. The purpose of the described experiment was to evaluate the effectiveness of *AS* in MtBE removal from water.

A removal test was carried out in an laboratory apparatus that replicates, at a smaller scale, an *AS* system applied to saturated porous media. The "intermediate" scale of the laboratory apparatus allowed knowing and controlling all boundary conditions during the execution of the test. Several sampling points were located at fixed positions within the media and used to withdraw samples of solution at predefined time intervals. In this way it was possible to follow the evolution of the removal process both in time and space.

Analysis of the chemical results on the water samples reveals that the *AS* is clearly effective at removing MtBE from water. The experimental results clearly showed a progressive decrease of MtBE concentrations in collected samples taken from points directly impacted by the injected airflow, whilst MtBE concentrations stayed constant in the sampling point located outside the zone of influence of the airflow.

The influence of geometry, water temperature and bulk water movement was considered for the evaluation of the effectiveness of *AS* removal from MtBE contaminated water. In particular, convective movement and consequent mixing of the bulk water produced by the air injection seem to affect mass removal in a significant way.

This study thus confirmed that *in situ AS* has a significant potential for remediating groundwater contaminated by MtBE, showing that stripping processes driven by air injection constitute a strong contribution to the overall removal action of the *AS* technology.

REFERENCES

Braida, W.J. and Ong, S.K. 2001. Air sparging effectiveness: labo ratory characterization of air-channel mass transfer zone for VOC volatilization. *Journal of Hazardous Materials*, B87, pp. 241-258. May 2001.

D.O.D. (U.S. Department of Defense). 1994. Remediation Technologies Screening Matrix and Reference Guide. *Federal Remediation Technologies Roundtable, second edition*, vol. 1. October 1994.

Ducco, F., Gigli, P., Lausdei, D. and Mangherini, E. 2001. Comportamento dell'MtBE nella bonifica di siti contaminati da idrocarburi. *Siti Contaminati*, n. 4, pp. 46-49. September 2001 (in italian).

Nichols, E. and Drogos, D. 2000. Strategies for Characterizing Subsurface Releases of Gasoline Containing MtBE. *American Petroleum Institute, publication* n. 4699. February 2000.

Suthersan, S.S. 1999. *In situ* Air Sparging. *Remediation engineering: design concepts*. ed. Suthan S. Suthersan, Boca Raton, CRC Press LLC.

CHAPTER 10

MAINE'S EXPERIMENT WITH GASOLINE POLICY TO MANAGE MTBE IN GROUNDWATER

John M. Peckenham[1], Jonathan Rubin[2], and Cecilia Clavet[3]

[1]*Senator George J. Mitchell Center for Environmental and Watershed Research, University of Maine, 102 Norman Smith Hall, Orono, ME 04469;* [2]*Senator Margaret Chase Smith Center for Public Policy and Department of Resource Economics and Policy, University of Maine, Orono, Maine;* [3]*Department of Resource Economics and Policy, University of Maine, Orono, Maine*

Abstract: The gasoline additive MtBE has become one of the most commonly detected contaminants in groundwater nationwide and has caused much concern in the state of Maine. In 1998 the Maine Department of Human Services conducted a statewide survey of groundwater wells and MtBE was detected in 16% of private and public wells tested. These findings resulted in the state regulatory agencies deciding to opt out of the reformulated gasoline (RFG) program in 1999. Subsequently, the average concentration of MtBE in gasoline dropped from ~15% to 2% by volume to protect water resources. This major policy change provided a microcosm to study the economic and environmental effects of this gasoline additive. In order to test the effect of this policy on water quality, groundwater samples were analyzed over a period of six years (1998-2003) from 19 wells distributed across a sand and gravel aquifer in Windham, Maine. MtBE continues to occur in detectable concentrations in 30 to 40% of the study wells despite Maine's decision to opt out of the RFG program in 1999. Although recent detected concentrations are lower than in previous years, this study confirms MtBE's temporal and spatial persistence in the environment. Reducing MtBE concentrations in gasoline may not be sufficient to eliminate its occurrence in groundwater. The economic perspective is that MtBE increases the cost of groundwater remediation, as compared to MtBE-free gasoline. Economic data for spills are being analyzed to assess if reducing MtBE concentration in gasoline has affected remediation cost. Preliminary results suggest that MtBE increases costs, even when present in low concentrations in gasoline.

1. INTRODUCTION

The 1992 implementation of the 1990 Clean Air Act mandated oxygenated gasoline containing methyl *tert*-butyl ether (MtBE) in certain areas in the country to meet part of the Federal requirement of reducing VOC concentrations in Maine's air by 15 percent. The State of Maine elected to use reformulated gasoline (RFG) that contained at least 11 percent MtBE by volume (Maine Department of Human Services, 1995). In December 1994, RFG was introduced in southern Maine, in Kennebec, Sagadahoc, Androscoggin, Cumberland, York, Lincoln, and Waldo Counties and by 1995, oxygenated gasoline use was common throughout the northeast. Soon after the widespread use of gasoline containing MtBE began, water-quality surveys began detecting MtBE in groundwater, often without the other sparingly soluble gasoline compounds—benzene, toluene, ethyl benzene, and xylenes (BTEX)—usually found in groundwater near gasoline spills (Maine Department of Human Services, 1998).

MtBE has been detected in groundwater in Maine as early as 1985 (Garrett et al., 1986). However, it was not until 1998 that evidence of widespread, low-concentration MtBE contamination in the groundwater was documented, when several widely-publicized occurrences of contamination in drinking-water wells prompted the State of Maine to conduct a random sampling of 951 private wells and nearly all the 830 non-transient public water supplies in the state of Maine. The sampling found MtBE in groundwater in all parts of Maine, including areas not required to use RFG (Maine Department of Human Services, 1998). This study and additional studies by the U.S. Geological Survey (USGS) across the country documented low concentrations of MtBE in groundwater, especially in urban areas (Zogorski et al., 1998) and in over 8 per cent of public drinking water supplies (Gullick and LeChevallier, 2000). The widespread occurrence of MtBE has prompted discussion about possible nonpoint sources of MtBE in groundwater, especially precipitation and atmospheric deposition (Squillace, Pankow et al., 1996; Squillace, Zogorski, et al. 1996; Pankow at al., 1997; Lopes and Bender, 1998; Baehr at al., 1999; Moran et al., 1999). Few studies to date, however, have examined the persistence of MtBE, or how MtBE concentrations in groundwater respond to changes in gasoline formulations.

In this paper we report on additional occurrences of MtBE in groundwater from 1998 during RFG use to 2003, four years after RFG use ceased. In 1998, the USGS-Maine District in cooperation with Senator George J. Mitchell Center at The University of Maine, the Maine Department of Environmental Protection and the Town of Windham, Maine (Figure 1) began a study of MtBE in the Windham aquifer (Nielsen and Peckenham, 2000). That study was designed to collect information on the occurrence and distribution of low concentrations of MtBE (defined as less than 2 μg/L) in the aquifer, a shallow aquifer used for drinking water in an area in which RFG was used. Additional sampling and analysis occurred in August 2001 and September 2003. Key objectives were to determine: (1) if MtBE was still detectable in groundwater post-RFG; (2) if the rates of detection and spatial patterns have changed between 1998 and 2003; (3) if the spatial patterns relate to potential MtBE sources; and (4) if the concentrations of MtBE changed uniformly over time. The remediation costs of using MtBE were analyzed to determine if the type of fuel blend, RFG or non-RFG gasoline, affected the costs of remediation.

1.1   Previous Investigations

In 1998, the State of Maine conducted a comprehensive survey of MtBE in groundwater, which consisted of sampling 951 randomly selected domestic water wells and 793 of the 830 public supply wells (Maine Department of Human Services, 1998). MtBE was detected at concentrations greater than 0.1 μg/L in 15.8 percent of the residential wells and in 16.0 percent of the public supply wells. More than 90 percent of the detections of MtBE were below 1.0 μg/L. The data were interpreted with respect to well type, association with a recent gasoline spill nearby, population density, and inclusion in the RFG area. No significant differences in MtBE occurrence were observed by well type (drilled bedrock wells compared to unconsolidated surficial wells) or among wells where there were known recent gasoline spills compared to those where there were no known spills (State of Maine, 1998). The State tested the risk of an MtBE occurrence associated with population density and whether or not the well was in a mandatory-RFG area. Both factors were statistically significant, when controlled for the other, in determining the risk of MtBE detections in domestic and public-supply wells. The reason for the smaller percentage of MtBE detections in higher density areas for public-supply wells where RFG was not required was not evident. MtBE detection rates for areas with more than 1,000 people/mi^2 were similar to those for the National Ambient Water Quality Assessment (NAWQA) studies (State of Maine, 1998).

Known and suspected sources of MtBE in groundwater include point and non-point sources. Point sources include pipelines, storage tanks (above- and below-ground), accidental spillage, homeowner disposal, spillage during fueling, and waste motor oil. Heating-oil spills also have been identified as another source of some MtBE in groundwater (Robbins et al., 1999). Suspected non-point sources include atmospheric deposition (recharge from precipitation or direct transport to groundwater through gaseous diffusion), vehicle evaporative losses, and urban runoff (Moran et al., 1999; Baehr et al., 1999). Point sources will cause locally elevated concentrations of MtBE and other related gasoline compounds. Point sources have the potential to generate very high concentrations of MtBE in water, up to 50,000 mg/L (Barker et al., 1991). The geochemical characteristics of point sources are: local effects with a limited area of contamination, steep concentration gradients with marked changes in concentrations away from source area, and substantial time-series changes in concentrations over time.

Non-point sources, most significantly direct precipitation and subsequent recharge of contaminated rainfall, and stormwater runoff primarily containing gasoline flushed from land surfaces, have been suggested as the source of some trace concentrations of MtBE in groundwater (Zogorski et al., 1998; Pankow et al., 1997). Concentrations of MtBE in groundwater from non-point sources are low, well below 20 µg/L (Moran et al., 1999).

Investigators have developed models to predict the behavior of aqueous phase MtBE once it starts to infiltrate the unsaturated zone (Johnson et al., 2000). Pankow et al. (1997) modeled various scenarios of recharge in a hypothetical aquifer to test the possibility that MtBE in precipitation could have traveled into shallow groundwater during the time frame of its use as a gasoline additive. Assuming no attenuation, they found that it took just 14.3 in/yr of recharge to saturate groundwater 6.5 feet below the water table with atmospheric levels of MtBE in 5 years. Baehr et al. (1999) developed a model to predict the concentration of MtBE at the water table, taking into account diffusion, recharge flux, thickness of the unsaturated zone, and several decay functions. This model concentration of MtBE at the water table is most sensitive to the thickness of the unsaturated zone and that the reason why MtBE is not universally detected in shallow groundwater could be explained by variations in saturated thickness and recharge. Factors that may affect the degradation rate of MtBE, such as aquifer material composition, also would contribute to the heterogeneity of MtBE detection.

MtBE was detected in more than 43 percent of stormwater samples collected for permitting requirements for urban stormwater runoff in areas with RFG use (Delzer et al., 1996). The source of MtBE in the stormwater samples may be either gasoline washoff from paved surfaces or MtBE in the precipitation (Lopes and Bender, 1998). Modeling the behavior of MtBE and other BTEX compounds found in stormwater runoff, at concentrations of MtBE above 1.0 µg/L showed that the source of MtBE was most likely dissolved gasoline entrained during the storm event and not precipitation. This was determined primarily by the co-occurrence of MtBE and other BTEX compounds in the stormwater runoff.

The costs associated with MtBE in groundwater are not well characterized. The University of California-Davis study (1998) concluded that MtBE adds 140 to 180 percent of the cost of a gasoline spill. This doesn't include costs associated with private wells that may have low concentrations of MtBE with or without an association with a gasoline spill.

1.2   Windham Aquifer Hydrology

The Windham aquifer depicted in Figure 1 is a thick, glacial sand and gravel aquifer in Cumberland County, southern Maine. It is mapped as a significant sand and gravel aquifer (Neil, 1998), with expected yields ranging from 10 to more than 50 gal/minute (38 to 189 L/m). The aquifer consists primarily of marine deltaic sand and gravel, with localized interbeds of marine silt/clay deposit. A buried esker lies in the northern part of the aquifer (Gerber, Inc., 1997). Thickness of the aquifer ranges from approximately 10 feet to more than 120 feet (3 to 37 m), and depth to the water table in the aquifer ranges from 5 feet to more than 50 feet (1.5 to 15 m). Hydraulic conductivities calculated from slug tests ranged from less than 1 foot/day to more than 400 feet/day (0.7 to 122 m/d) (Caldwell, 2002). Average hydraulic conductivities used in a calibrated model of the aquifer (Gerber, Inc., 1997) were 250 feet/day (76 m/d) in the buried esker; 44 feet/day (13 m/d) in the sand and gravel; 1.5 feet/day (0.46) in the till surrounding the aquifer; and less than 0.1 feet/day (0.03 m/d) in the marine silt/clay. Flow paths based on observed and modeled heads indicated flow to the east-southeast and discharge along the southerly margins (Gerber, Inc., 1997; Caldwell, 2002). Some recharge also may come from Little Sebago Lake, a heavily developed lake that borders the Windham aquifer on the northeast.

Temperature data Maine provide an average annual temperature of 40.7°F (4.83°C) (Northeast Climate Center, 2004). Annual temperatures were consistently greater than the mean for 1998 (43.3°F/6.3°C), 1999 (45.0°/7.2°C), 2001 (42.5°F/5.8°C), 2002 (41.6°/5.3°C); but not 2000 (40.8°/4.9°C) or 2003 (40.4°/4.7°C). Average annual precipitation near Windham is 42.28 inches (1.07 m). Precipitation varied around the mean for the study years 1998 (44.73 inches/ 1.13 m), 1999 (45.05 inches/1.14 m), 2000 (42.64 inches/ 1.08 m), 2001 (29.59 inches/0.75 m), 2002 (40.36 inches/ 1.02 m), and 2003 (44.80 inches/ 1.14 m).

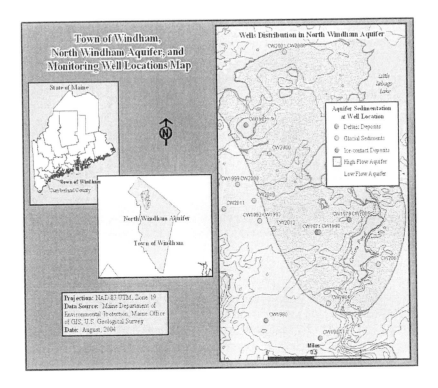

Figure 1. Windham, Maine Locus Map

The potential for precipitation to recharge the Windham aquifer depends on the amount of precipitation, amount of groundwater recharge, and average yearly and monthly temperatures. An estimate of the maximum amount of yearly recharge to sand and gravel aquifers can be calculated using average annual temperature to estimate evapotranspiration and precipitation (Lyford and Cohen, 1988). The estimated average annual temperature of 43°F (6.1°C) for the Windham, Maine, region translates into an annual evapotranspiration of 19.5 inches (0.49 m). This evapotranspiration combined with an annual average precipitation of 44 inches (1.12 m) for the Windham area, yields a maximum potential recharge to the Windham aquifer of 22 in/yr (0.56 m). In the Windham area, some runoff is collected in detention basins that route runoff from paved areas into groundwater recharge zones and some is surface runoff. The potential for MtBE to be transported into the aquifer is high.

2. METHODS

2.1 Groundwater samples

An existing network of shallow wells (Figure 1) installed by the USGS and distributed throughout the groundwater flow system in the Windham aquifer was used for this study (Nichols and Silverman, 1998). These wells are completed in different types of glacial sediments. Samples were collected in July and August 1998 in 31 wells; subsequent sampling rounds in November and December 1998, April and May 1999, and August 2001, included a smaller number of wells (Table 1). Well depths ranged from 19 to 132 feet (5.8 to 40 m), with screened interval depths ranging from 8 to 130 feet (2.4 to 39.6 m). Screened intervals were mostly 5 or 10 feet (1.5 to 3.0 m) long, with a few wells having longer screens, up to 30 feet (9.1 m). Depth to water in the wells at the time of sampling ranged from 4.5 to 61.1 feet (1.37 to 18.6 m). All the wells were completed in unconfined parts of the aquifer; the sampling points ranged from water-table position to more than 110 feet (33.5 m) below the water table. Seven pairs of nested wells were sampled to evaluate vertical distributions of MtBE. The dominant land cover near each well was determined from the NLCD land-cover classification: urban, undeveloped, or low-density residential.

USGS National Water Quality Assessment protocols for sampling groundwater wells (Koterba et al., 1995) were followed in sample handling, quality assurance/quality control (QA/QC), sampling equipment, and cleaning. The USGS protocols were modified, however, to follow the U.S. Environmental Protection Agency (USEPA) low-flow (minimal drawdown) sampling procedures (Puls and Barcelona, 1995). A small number of wells did not yield enough water to meet the minimum drawdown requirements. These wells were pumped dry and sampled the next day. A detailed description of the sampling methods is contained in Nielsen and Peckenham (2000). The collection method was modified in 2001 to minimize cross-contamination by using disposable polyethylene bailers following the purging of the wells.

Samples were analyzed for fuel oxygenates and volatile organic compounds using modified USEPA Methods (SW-846) and specialized laboratory methods including Standard Methods (5000 and 6000) and Church *et al.* (1997). Detections limits as low as 0.1 μg/L were obtained. Due to contamination of a blank sample, detections of MtBE in the 1998 round of samples (11 detections) were recoded as a "less than" value larger than the observed analytical result. For example, a detection in the first round of 0.42 μg/L would have been recoded to <0.5 μg/L, to make sure that the database did not contain detections of MtBE from potentially compromised samples.

MtBE was also detected in subsequent sampling rounds in 9 of the 11 wells with in which it was detected in the first round.

The costs of remediating petroleum spills (any fuel type) are tracked by the Maine Department of Environmental Protection (MDEP). Historical data are available back to circa 1970 and the completeness of the data improved markedly in the late 1980's. Over 30,000 spill reports were reviewed to locate gasoline spills with a corresponding record of activities and costs. Spill information and costs are located in separate databases, so each spill had to be cross-referenced by hand. Approximately 1 percent of the spills had sufficient information for cost analysis and analytical data on MtBE in soil or water. Additional cultural data were derived from the 2002 U.S. Census.

The cost variables were determined to be either continuous or discrete. The continuous variables included MtBE concentrations and length of monitoring, as well as, demographics (age, population, and income). Discrete variables included media affected, spill causes, spill locations, and time period relative to RFG use. An additional set of discrete variables were used to divide the state into three geographical regions, from north to south: Presque Isle, Bangor, and Augusta.

Statistical analyses of MtBE concentrations to test for significant differences between years were performed using SYSTAT ver.10. Stepwise multiple-regression analyses of spill costs were performed using SAS.

2.2   Persistence of MtBE in the North Windham Area

MtBE was detected in the majority of the water samples from the Windham aquifer. This result is based on analyses of 90 individual samples collected from 31 wells (Table 1). A cumulative total of one to four water samples were collected from each of the wells on different sampling dates in 1998, 1999, 2001, and 2003. Sixteen of the 31 wells sampled (52 percent) had at least one detection of MtBE (Table 1). MtBE was detected in 38 individual samples (42 percent of all samples), and the median concentration of MtBE was 1.85 µg/L (excluding the samples that were recoded as "less thans" from the first sampling round). MtBE was below the detection limit in 52 samples (58 percent); however, 13 of these samples had elevated detection limits at greater than 0.2 µg/L. Two of the three highest concentrations detected, 14.0 µg/L and 6.64 µg/L, were collected from the same well, CW 2012. This well is located in an urbanized area, close to a gasoline station. The single highest concentration, 38.7 µg/L, was detected in well CW 2004 located in a light-industrial and commercial area.

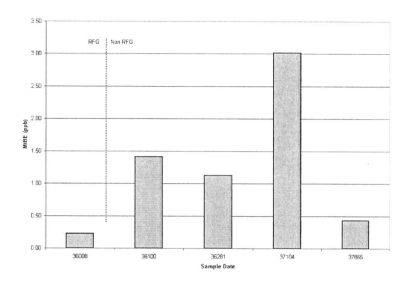

Figure 2. Temporal Changes in Mean MtBE Concentrations

Table 1. MtBE Concentrations in Groundwater Monitoring Wells.

Sample Date		RFG		post-RFG		
		Aug-98	Nov-98	May-99	Aug-01	Sep-03
Well ID	USGS ID	MtBE (ug/L = ppb)				
ARL-1	CW 1971	<0.2		<0.2	<0.1	<1
CPW-1	CW 1979	<0.2		0.20	0.50	0.18
MTW-1	CW 1980	<0.2			0.49	
CHA-2	CW 1985	<0.5	<0.2	<0.2	<0.1	0.42
KEL-1	CW 1987	<0.2		<0.2	<0.1	
ARL-2	CW 1990	<0.2		0.23	<0.1	<1
Bent-1	CW 1992	<0.2		<0.2	<0.1	<1
Bent-2	CW 1993	<0.2		<0.2	<0.1	<1
Key 1	CW 1999	<0.2		<0.2	<0.1	<1
SSW-1	CW 2000	<0.4	0.46	0.35	0.16	<1
MPW-1	CW 2001	<0.8	<0.2	<0.2	<0.1	<1
UWW-1	CW 2003	<1.0	0.72	0.47	<0.1	0.22
TCBY-1	CW 2004	<0.7	3.14	0.60	38.7	0.34
CPW-2	CW 2005	<1.8	2.73	1.54	4.22	0.73
MPW-2	CW 2008	<0.2		<0.2	5.65	0.38
Key 2	CW 2009	<0.2		<0.2	<0.1	<1
MDW-1	CW 2010	<0.2		1.89	0.33	<1
BRW-1	CW-2011	<0.4	<0.2	<0.2	<0.1	<1
CBW-1	CW 2012	<0.75	3.96	14.0	6.64	4.60

The detection of MtBE was accompanied by the detection of the two sparingly soluble BTEX compounds- benzene, or toluene in less than 10 percent of the samples. This observation is consistent with other findings of MtBE along with some BTEX in trace concentrations in Maine (State of Maine, 1998). Since 1998 the median concentration for detections has risen from 0.35 µg/L in 1998 to 4.20 µg/L in Summer 2001 and fallen to 0.40 µg/L in Summer 2003 (Figure 2). Over the same period the mean MtBE concentration increased from 1998 to 2001 then decreased in 2003. This trend has been driven by a few high concentrations from the same group of three wells (CW 2004, CW 2005, and CW 2012).

Eight of the wells sampled (CW's 1979, 1990, 2000, 2004, 2005, 2008, 2010, and 2012) have exhibited a trend of increasing MtBE concentrations through 2001, then a decline in 2003. These wells are located in undeveloped areas (CW 2000 and CW 2008), rural residential neighborhoods (CW 1979, CW 1990, and CW 2005) and in commercial districts (CW 2004 and CW 2010). Two wells (CW 1985 and CW 2003) exhibited low concentrations of MtBE that increased in 2003. Both of these wells are located in lightly developed areas. Well CW 1985 is adjacent to a former public water supply well and CW 2003 is located on an undeveloped parcel within a rural residential neighborhood.

2.3   Cost Analysis

The cost-analysis regressions were performed in two groups. The first group used 133 observations from the Bangor and Presque Isle regions. In this group, nine variables had a significant relationship with total remediation costs:

   highest concentration detected;
   number of properties monitored;
   monitoring time period;
   medium affected = groundwater;
   cause = other;
   cause = corrosion;
   cause = mechanical failure;
   location = gasoline terminal; and
   median income.

The effects on costs are summarized in Table 2. The variables-highest detected concentration, number of wells monitored, cause is corrosion or other, were significant in each step. The effect on cost table shows the percent change in cost for each additional microgram per liter (µg/L) of

MtBE. For example, as MtBE concentrations increased by 1 µg/L, total costs increased by 0.001 per cent. Not surprisingly, mechanical failures caused the most expensive spills to remediate. In one of the regression steps, the variable- medium affected was groundwater, was significant and negative. This effect may be due to the inclusion of large spills that affected soil and not groundwater. In another regression step, median income had a significant negative effect. This may reflect an association of activities using motor fuels that varies by economic class. Overall, these significant variables explained about 42 per cent of the variations in cost. Monitoring the timeframe indicates that costs increase for longer periods of site activity.

Table 2. Model Expected Values-Dataset 1

Variable	Expected Percent Change on Total Cost				
	1	2	3	4	5
Highest Detected Concentration	0.001**	0.001**	0.001**	0.001*	0.001*
Number of Properties Monitored	14.14**	11.38**	10.78**	10.70**	10.80**
Monitoring Timeframe		1.44**	1.16**	1.12**	1.20**
Medium Affected: Groundwater		-59.15*			
Cause: Other		251.27*	372.07**	388.37**	540.39**
Cause: Corrosion	258.73*	454.40**	639.42**	610.21**	814.37**
Cause: Mechanical Failure	587.57**	790.56**	907.62**	886.24**	1081.00**
Location: Gasoline Terminal			189.97**	186.21**	207.85**
Median Income					-0.01*

** Significance level = 0.05, * Significance level = 0.1

The second group used 174 observations from the Bangor, Presque Isle, and Augusta regions. In this group, seven variables had a significant relationship with total remediation costs:

geographical region = Presque Isle;
number of wells affected;
medium affected = surface water;
cause = mechanical failure;
location = gasoline terminal;
per cent of population > 59 years; and
median income.

The effects on costs for the second group are summarized in Table 3. The variables- highest detected concentration, number of wells affected, were significant in each step. Mechanical failures and location at gasoline terminals also caused the most expensive spills to remediate. In one step, the variable- medium affected is surface water, was significant and positive. This effect may be due to the inclusion of a large surface spill to a lake or river. In another step, median income had a significant negative effect, as did an increase in older population. This may reflect an association of activities using motor fuels that varies by economic class and age. Finally, the costs to remediate spills in the Presque Isle region were significantly more expensive compared to Bangor or Augusta. This region of the state is more remote and higher travel costs are likely incurred. The model fit for the second group explained only 15 per cent of the variations in costs.

Table 3. Model Expected Values-Dataset 2

Variable	Expected Percent Change on Total Cost				
	1	2	3	4	5
Region: Presque Isle	177.91*	189.17**	204.18**	211.15**	228.50**
Impacted Wells	34.45*	32.28*	44.92**	44.22**	43.35**
Medium Affected: Surface water	161.63*				
Cause: Mechanical Failure			193.07*		280.38**
Location: Gasoline Terminals			216.95**	224.42**	202.27**
Percent Over 59 years					-0.28*
Median Income					-0.01*

** Significance level = 0.05, * Significance level = 0.10

The gasoline formulations RFG or post-RFG had a small positive, but inconsistent, effect on costs. The regression coefficients were not statistically significant. Remediation costs do not appear to be affected any differently with respect to RFG and post-RFG formulations.

3.   SUMMARY AND CONCLUSIONS

A study was conducted on the temporal occurrence and spatial distributions of the fuel oxygenate MtBE (methyl *tert*-butyl ether) in a glacial sand and gravel aquifer in southern Maine. Ninety samples were collected from 31 different wells in the Windham aquifer, in North Windham, Maine, for analysis of MtBE between July 1998 and September 2003. MtBE was detected in 42 percent of the samples and in 52 percent of the individual wells sampled. In addition, 92 percent of wells having detectable concentrations of MtBE were in an area of the aquifer designated as a "high-yielding" aquifer (Neil, 1998).

The median concentration in wells with detectable MtBE was increased from 1998 to 2001 and then decreased in 2003 to concentrations similar to 1998. MtBE was detected in association with any land use from undeveloped to residential to commercial. The strong association of MtBE with aquifer transmissivity suggests that geological vulnerability is an important control on the distribution of MtBE in groundwater. The mechanisms by which MtBE enters the aquifer were not identified in this study. The Maine State study of MtBE tentatively identified very small spills of gasoline associated with filling of lawn-care equipment fuel tanks, recreational vehicle tanks, and automobile tanks as likely being responsible for the majority of the low concentrations of MtBE found. If true, MtBE should be detectable in similar urban to suburban settings that are situated over transmissive aquifers.

Changing fuel blends (RFG to post-RFG) had no significant effect on the costs of remediating gasoline spills. Initially, it appeared that changing from RFG was associated with an increase in the ambient concentrations of MtBE in groundwater in Windham, Maine. Data collected in 2003 suggest that the mean concentration of MtBE has declined while the detection frequency remains stable. Thus the policy decision to opt out of RFG has had no effect on remediation costs by site and it has had a negligible effect on the rate of MtBE detections in groundwater in the study area. Since post-RFG fuels may contain up to four percent MtBE and it is very soluble, total eradication of MtBE in groundwater may only be possible if fuels contained no MtBE.

ACKNOWLEDGEMENTS

This work was supported the U.S. Geological Survey-Water Resources Research Institute grants to the University of Maine. Additional support for this project was provided by funding from the U.S. Geological Survey-Water Resources Division-Maine District, Maine Department of Environmental Protection, the town of Windham. Therese Anderson provided laboratory assistance. Peter Garrett, Steve Kahl and Catherine Schmitt provided critical assistance in reviewing early drafts of this paper.

REFERENCES

Baehr, A.L., Charles, E.G., and Baker, R.J., 2000, Methyl tert -butyl ether degradation in the unsaturated zone and the relation between MtBE in the atmosphere and shallow groundwater: *Water Resources Research*, v. 37, p. 223-233.

Baehr, A.L., Stackelberg, P.E., and Baker, R.J., 1999, E valuation of the atmosphere as a source of volatile organic compounds in shallow groundwater : *Water Resources Research* , v. 35, no. 1, p. 127-136.

Barker, J.F., Gillham, R.W., Lemon, L., Mayfield, C.I., Poulsen, M., and Sudickey, E.A., 1991, Chemical fate and impact of oxygenates in groundwater —Solubility of BTEX from gasoline—Oxygenate compounds: Wash ington, D.C., American Petroleum Institute Publication 4531, 90 p.

Caldwell, J. M., 2002, Groundwater levels and water-quality data from monitoring wells in Windham,, Maine, water years 1997-2001 U.S. Geological Survey Open-File Report 02-145, 19 p.

Church, C., L. Isabelle, J. Pankow, D. Rose, and P. Tratnyek, 1997, Method fo r determination of methyl tert -butyl ether and its degradation products in water, *Env. Sci. & Tech.*, Vol. 31, no. 12, pp. 3723 -3726.

Delzer, G.C., Zogorski, J.S., Lopes, T.J., and Bosshart, R.L., 1996, Occurrence of the gasoline oxygenate MtBE and BTEX compounds in urban stormwater in the United States, 1991-1995: U.S. Geological Survey Water -Resources Investigations Report 96 -4145, 6 p.

Garrett, P., M. Moreau, and J. Lowry, 1986, MtBE as a groundwa ter contaminant, Proc. NWWA/API Conf. on Petrol. Hydrocarbons, November 1986, Houston, TX.

Gerber, Robert G. Inc., 1997, Windham groundwater resource evaluation, Phase 2: Freeport, Maine, Robert G. Gerber, Inc., Freeport, Maine, report dated May 15, 1997, 26 p. plus tables, figures, appendixes, sheets.

Grady, S.J., 1997, Distribution of MtBE in groundwater in New England by aquifer type and land use, *in* American Chemical Society Division of Environmental Chemis try preprints of extended abstracts, 213th, San Fran cisco, Calif., April 1997, v. 37, no. 1, p. 392 -394.

Gullick, R.W. and M.W. LeChevallier, 2000, Occurrence of MtBE in drinking water sources: *Jour. Amer. Water Works Assoc.,* v. 92, p. 100-113.

Helsel, D.R., and Hirsch, R.M., 1992, Statistical methods in water resources: New York, New York, Elsevier Sci ence Publishing Company, Inc., 522 p.

Johnson, R., J. Pankow, D. Bender, C. Price, and J. Zogorski, 2000, MtBE: To what extent will past releases contaminate community water supply wells?: *Env. Sci. Tech., v. 34, p.* 2A-7A.

Koterba, M.T., Wilde, F.D., and Lapham, W.W., 1995, Groundwater data -collection protocols and procedures for the Nationa l Water-Quality Assessment Pro gram—Collection and documentation of water quality samples and related data: U.S. Geological Survey Open -File Report 95-399, 113 p.

Lopes, T.J., and Bender, D.A., 1998, Nonpoint sources of volatile organic compounds in urban areas—relative importance of land surfaces and air: *Environmental Pollution* , v. 101, p. 221 -230.

Lyford, F. P., and Cohen, A.J., 1988, Estimation of water available for recharge to sand and gravel aquifers in the glaciated Northeastern United States, *in* Randall, A.D. and Johnson, A.I., eds., 1988, Regional aquifer systems of the United States –the northeast glacial aquifers: American Water Resources Association Monograph Series 11, 156 p.

Maine Department of Human Services, 1995, An assessment of the health affects of reformulated gasoline in Maine, a report presented by the task force on health effects of reformulated gasoline: Augusta, Maine, April 1995, 35 p.

Maine Department of Human Services, 1998, The presence of MtBE and other gas oline compounds in Maine's drinking water —Maine MtBE drinking water study, preliminary report, Octo ber 13, 1998: Augusta, Maine, Maine Bureau of Health, Department of

Human Services; Bureau of Waste Man agement and Remediation, Departm ent of Environmental Protection; Maine Geological Survey, Department of Conservation, 15 p., 8 figures.

Moran, M.J., Zogorski, J.S., and Squillace, P.J., 1999, MtBE in groundwater of the United States–Occurrence, potentia l sources, and long-range transport, *in* Water Resources Conference, American Water Works Association, Norfolk, Va., Sept. 26 -29, 1998 [Proceedings]: American Water Works Association, Denver, Colo.

Neil, C.D., 1998, Significant sand and gravel aquifers in the North Windham quadrangle, Maine: Maine Geological Survey, Open -File Map 98-158, scale 1:24,000.

Nielsen, M. G. and J. M. Peckenham, 2000, Methyl tert-Butyl Ether (MTBE) in groundwater, air, and precipitati on in an urbanized area in Maine , USGS Water Res. Invest. Report 00 - 4048, 28 p.

Nichols, W.J., and Silverman, P.N., 1998, Hydrologic data for the Presumpscot River Basin, Cumberland and Oxford Counties, Maine 1995 to 1996: U.S. Geological Survey Open - File Report 98-265. 53 p.

Pankow, J.F., Thompson, N.R., Johnson, R.J., Baehr, A.L., and Zogorski, J.S., 1997, The urban atmosphere as a nonpoint source for the transport of MtBE and other volatile organic compounds (VOCs) to shallow ground water: *Environmental Science and Technology* v. 31, no. 10, p. 2821 -2828.

Puls, R.W., and Barcelona, M.J., 1995, Low -flow (minimal drawdown) groundwater sampling procedures: U.S. Environmental Prot ection Agency Research Brief EPA/540/S-95/504.

Robbins, G. A., Henebry, B.J., Schmitt, B.M., Bartolomeo, F.B., Green, A., and Zack, P., 1999, Evidence for MtBE in heating oil: *Groundwater Monitoring and Remediation,* v. 19, no. 2, p. 65-69.

Squillace, P.J., Pankow, J.F., Korte, N.E., and Zogorski, J.S., 1996, Environmental behavior and fate of methyl *tert*-butyl ether (MtBE): U.S. Geological Survey Fact Shee t FS-203-96, 6 p.

Squillace, P.J., Pankow, J.F., Korte, N.E., and Zogorski, J.S., 1997, Review of the environmental behavior and fate of methyl *tert*-butyl ether: *Environmental Toxicology and Chemistry*, v. 16, no. 9, Sept. 1997, p. 1836 -1844.

Squillace, P.J., Pope, D.A., and Price, C.V., 1995, Occur rence of the gasoline additive MtBE in shallow groundwater in urban and agricultural areas: U.S. Geological Survey Fact Sheet FS-114-95, 4 p.

Squillace, P.J., Zogorski, J.S., Wilber, W.G., and Price, C.V., 1996, Preliminary assessment of the occurrence and possible sources of MtBE in groundwater in the United States, 1993-1994: *Environmental Science and Tech nology*, v. 30, no. 5, p. 1721 -1730.

University of California, Davis, 1998, Health and Environmental Assessment of MtBE , Report to the Governor and Legislature of the State of California.

Zogorski, J.S., Delzer, G.C., Bender, D.A., Squillace, P.J., Lopes, T.J., Baehr, A.L., Stackelberg, P.E., Landem eyer, J.E., Boughton, C.J., Lico, M.S., Pankow, J.F., Johnson, R.L., and Thompson, N.R., 1998, MtBE —Summary of findings and research by the U.S. Geological Survey, *in* Annual Conference of the Amer ican Water Works Association — Water Quality, June 21 -25, 1998, Dallas, Tex., [Proceedings]: AWWA, Den ver Colo., p. 287-309.

PART V: RADIONUCLIDES

CHAPTER 11

DEVELOPMENT OF RADON ENRICHMENT IN SOIL GAS OVER QUARTZ-MICA SCHIST IN VIRGINIA

Douglas Mose, George Mushrush, Charles Chrosniak and Paul DiBenedetto
Center for Basic and Applied Science, 20099 Camp Road, Culpepper, VA 22701 and Department of Chemistry, George Maso n University, Fairfax, VA 22030 -4444

Abstract: A major portion of northern Virginia is underlain by a quartz -muscovite soil, on average about 10 meters thick, that has developed on a bedrock of polymetamoirphic schist. The schist formed f rom an ancient clay -rich sediment eventually recrystallized several times, as the modern Appalachian rocks were heated deep in the Earth and subsequently exposed by erosion. The total gamma radioactivity and the permeability of the schist are higher than average, and combine to generate a radon -rich soil-gas that can be brought into homes by the pressure differential normally present in local homes that commonly are well insulated and have basements. More than half of the homes, based on three-month measurements, exceed the U.S. Environmental Protection Agency recommended maximum for indoor radon of 4 pCi/L. Fortunately, while the area is experiencing a rapid inc rease in new home construction, it is possible to avoid types of home construction susceptible to, and areas of, high soil -gas radon and high permeability.

Key words: Radon, Uranium, Aeroradioactivity

1.   INTRODUCTION

Because of health implications, the early work on indoor radon has given rise to a broad range of research characterizing ^{222}Rn and progeny occurrence and control in inhabited structures. With the tendency for the gas to concentrate in buildings where air exchange is limited, radon is becoming identified as a major form of indoor air pollution. There is growing belief that exposure to radon gas poses one of the nation's most significant radiological health problems in the form of an increased risk of developing lung cancer. State-wide compilations reveal that about 15% of the homes in the United States have indoor radon concentrations in excess of 4 pCi/L, the U.S. Environmental Agency recommended Maximum Contamination Level for the buyers of new homes (White et al., 1992; Alexander et al., 1994; Marcinowski et al., 1994).

The correlation between breathing indoor radon and lung cancer arises because the very short half lives of radon and its daughters allow the disintegrations to occur on the epithelial lining of the lungs before the mucociliary clearance system has a chance to move the particles out of the lungs. The body has repair mechanisms which can handle damage from radiation. However, if the rate at which the cells are damaged or destroyed is greater than the body's ability to repair them, the damage will be cumulative.

Uranium in soil and rock is the source of most radon to which people are exposed. The importance of soil as a source of indoor radon combined with the increasing evidence of unacceptably high radon concentrations in a significant fraction of houses has raised the question of whether one might predict on a geological basis where high indoor radon levels might be found. The potential for high indoor radon concentrations depends on several factors: radium content of the soil, moisture content of the soil, permeability of the soil, the season, and the weather. Radon gas enters the atmosphere by crossing the soil air interface. It has been estimated that the emanation rate is 0.42 pCi/m^2 per second from soil in the U.S. (NCRP, 1984).

The following report is based on indoor radon measurements obtained using eight consecutive three-month intervals in the basements of homes constructed in native soil over the Peters Creek Schist, which by earlier studies is known to be a set of homes with the highest indoor measurements in northern Virginia. It will be shown that the observed indoor radon concentrations are related to the season and to the type of home heating system, and that soil radiation, measured by airborne surveys, is a useful method to separate low-radon homes from high-radon homes.

2. METHOD

Indoor radon values were obtained from 242 homes using seasonal measurements during winter periods (November-January), spring periods (February-April), summer periods (May-July) and fall periods (August-October). Each home in this study set is in terrain underlain by the Peters Creek Schist (Figure 1), where the soil is about 10 meters thick. In some areas the soil is almost 20 meters thick, and in some other areas of stream erosion and road construction, the schist is exposed on the surface.

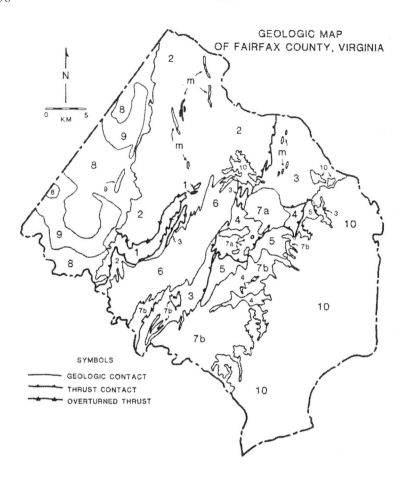

Figure 1. Generalized geological map of Fairfax County in Northern Virginia showing the
areas underlain by the major geological units

1 = metamorphosed lava flows, 2 = Peters Creek Schist, 3 -6 = sequence of metamorphosed
sedimentary strata, 7a and 7b = metamorphosed granitic rocks, 8 = red sandstones, 9 = post -
metamorphic lava flows, 10 = recent deposit s of sand and clay strata.

The conclusions are based on three-month indoor radon measurements
using alpha-track radon monitors analyzed by Tech/Ops Landauer
Corporation of Illinois. Tech/Ops develops the film in the monitors,
measures the "tracks" produced by the decay of radon progeny near the film

surface and calculates the average amount of indoor radon recorded by the film. Estimates of uncertainty for the alpha-track monitors are related to the measurement interval, so intervals of three months were used in this study. At the 90% confidence level, the alpha-track monitors carry an uncertainty of about +/- 25% for the three-month measurement interval (Mose et al., 1990).

Earlier studies (Mose et al., 1996) showed that the homes built in the soils that have fomed over the polymetamorphic rocks in northern Virginia generally have a large portion that exceeds 4 pCi/L. The Peters Creek Schist is a quartz-rich polymetamorphic rock, which originated as submarine landslides deposited in a large fan-shaped accumulation on oceanic crust (Drake and Morgan, 1981, 1983). Because of its mineralogy, the soil formed on the schist has relatively high permeability (Sextro et al., 1987). This is important because high permeability soils facilitate rapid radon migration by convective flow, allowing radon from a large volume of soil to reach the basement of a home before the radon decays. Others have concluded that high permeability, and high uranium (and radon) content, causes the schist to have high radon potential (Otton et al., 1988).

3. RESULTS AND DISCUSSION

The indoor radon measurements obtained over the four seasons of study in northern Virginia homes constructed over the Peters Creek Schist (Table 1) were obtained from a basement location or on the lowest level of their home (approximately 90% of the study homes had a basement). The comparison between each season shows, as have other studies, that winter is most often the time of greatest radon concentrations, probably because these homes are normally closed to outside air. Conversely, the summer is the time of lowest radon concentration due to the generally increased air exchange rate caused by open windows and doors. Spring or fall radon measurements tend to most closely approximate the average for the entire year.

Table 1. Summary of basement indoor radon measurements from 242 homes constructed on soil over the Peters Creek Schist in Fairfax County.

Measurement Season	Average Rn in pCi/L	Median Rn in pCi/L	% Over L4 pCi/L
Winter	5.6 pCi/l	4.1 pCi/l	57 %
Spring	5.4 pCi/l	3.6 pCi/l	46 %
Summer	4.0 pCi/l	3.4 pCi/l	37 %
Fall	5.3 pCi/l	4.1 pCi/l	52 %

Homes equipped with an oil or gas heating system tend to have lower levels of indoor radon than homes equipped with an electrical heating system (Table 2). This may occur because during the winter and during heating days in the spring and fall homes heated with a combustion system experience a brief low-pressure interval when the furnace ignites, driving air out of the basement in the form of chimney gasses. The momentary low-pressure interval is immediately eliminated as outside radon-free air is drawn into the home, around windows and doors. The indoor radon levels of homes with oil or gas heating systems are also lower in the summer than in homes with electrical heating systems. This unexpected result is probably because of the whole-home air cooling ability of the heat pump, which is the most commonly used electrical heating system in the study area. During the summer, homes with the whole-home service of heat pumps are often closed to keep the warm but also radon-free outside air from entering the home.

Table 2. Summary of basement indoor radon from homes that use oil and gas heating and from homes that use electrical heating, using 242 homes constructed on soil over the Peters Creek Schist in Fairfax County.

Measurement Season	Heating System	Average Rn in pCi/L	Median Rn in pCi/L	% Over 4 pCi/L
Winter	Oil/Gas	4.7	3.7	47%
	Electrical	6.4	3.6	47%
Spring	Oil/Gas	4.2	4.0	51%
	Electrical	6.3	5.0	61%
Summer	Oil/Gas	3.7	3.6	38%
	Electrical	6.6	4.4	54%
Fall	Oil/Gas	4.4	3.7	36%
	Electrical	10.9	4.1	50%

It has been noted that measurements of surface radiation, taken from an airplane, seem to be very good indicators of high radon potential areas (Gundersen and Schumann, 1996; Revzen et al., 1988). The radiation signal of the Earth's surface in the northern Virginia study area is rather better known than perhaps anywhere else in Virginia (Daniels, 1980; Figure 2). To make this map, total-count gamma-ray radioactivity measurements were flown with a ground clearance about 500 feet along east-west flight lines. The flight lines were spaced at only ½ mile apart over Fairfax County. At this altitude the effective area of response of the scintillation equipment is approximately 1000 feet in diameter, and the signal is generated by the uppermost 1 foot into the ground. The combination of close flight lines and area of response is such that essentially all of the ground surface was examined for its radioactivity signal.

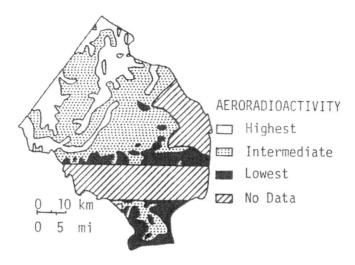

Figure 2. Generalized aeroradioactivity map of Fairfax County showing the realtive total - gamma radioactivity of the soil. Areas identified as Highest Radioactivity exhibited over 400 counts/second, Int ermediate exhibited 200 -300 cps, and lowest exhibited less than 200 cps.

The data used to make a total-gamma map are total gamma-ray radioactivity, derived from [214]Bi (a radioactive daughter of uranium, produced by the decay of [222]Rn), thorium and potassium. To a rough approximation uranium, thorium and potassium tend to have a similar geochemical behavior. Potassium-rich rocks tend to be enriched in uranium and thorium. In any event, as will be shown below, there is a good correlation between indoor radon and the total gamma-ray signal, suggesting that the component of the total signal due to [222]Rn is relatively constant percentage of the total signal.

The aeroradioactivity map for Fairfax County (Figure 1) was examined to evaluate its use as a predictor of indoor radon of homes built over the Peters Creek Schist. The comparison between indoor radon levels is rather good (Table 3). For every season, homes constructed where the aeroradioactivity signal is in the 300-500 c.p.s. range average twice as much indoor radon as the homes constructed in soils in the 100-300 c.p.s. range. In short, indoor radon increases as the aeroradioactivity increases.

Contaminated Soils- Radionuclides

Table 3. Comparison of total gamma aeroradioactivity to indoor radon , using 242 homes constructed on soil over the Peters Creek Schist in Fairfax County.

Aeroradioactivity Ranges		Average Radon	Median Radon	% over 4 pCi/L
Winter	200 and less c.p.s.	4.8 pCi/L	4.0 pCi/L	55%
	>200-300 c.p.s.	4.8	3.9	49
	>300-400 c.p.s.	7.5	4.9	60
	>400-500 c.p.s.	9.9	9.3	100
Spring	200 and less c.p.s.	4.8	4.4	50
	>200-300 c.p.s.	4.7	3.3	39
	>300-400 c.p.s.	6.9	4.7	53
	>400-500 c.p.s.	8.2	8.8	89
Summer	200 and less c.p.s.	4.4	3.9	47
	>200-300 c.p.s.	3.6	2.8	29
	>300-400 c.p.s.	4.5	3.8	44
	>400-500 c.p.s.	5.2	5.0	60
Fall	200 and less c.p.s.	4.7	4.0	50
	>200-300 c.p.s.	4.4	3.7	45
	>300-400 c.p.s.	6.3	4.5	60
	>400-500 c.p.s.	6.0	5.8	78

Perhaps the most significant advantage of the Fairfax County aeroradioactivity survey is that the survey covers essentially all of the county surface. This makes such surveys a cost-effective improvement over an analogous land-based study in which a geologist would examine the radon emanation characteristics of all the area underlain by each geological unit. A comparable land-based study based mainly on geological units and soil types would probably be always less useful because, as can be demonstrated by comparing geological maps with aeroradioactivity maps for the study area, most rock units vary in surface radioactivity.

Since it was apparent that the type of home heating system and the local aeroradioactivity could both be used to separate high radon from low radon potential homes, the homes over the Peters Creek Schist were grouped into one subset that should have lower radon potential (oil/gas heat and low aeroradioactivity) and another that should have higher radon potential (electrical heat and high aeroradioactivity). Using these subsets (Table 4), the homes of lower potential were found to be have indoor radon measurements that were about half as great as the high radon potential subset. Approximately 35% of the low radon potential homes exceeded 4 pCi/L over the entire year, compared to approximately 70% of the high radon potential homes.

Table 4. Comparison between low radon potential homes (oil/gas heat and aeroradioactivity less than 300 c.p.s) and high radon potential homes (electrical heat and aeroradioactivity of 300 c.p.s. or higher).

Measurement Season	Heating System and Aeroradioactivity	Average Radon	Median Radon	% Over 4 pCi/L
Winter	Oil/Gas & <300 c.p.s.	3.8 pCi/L	3.5 pCi/L	43%
	Electrical & 300+ c.p.s.	8.2	5.7	71
Spring	Oil/Gas & <300 c.p.s.	4.4	3.6	45
	Electrical & 300+ c.p.s.	7.7	5.9	67
Summer	Oil/Gas & <300 c.p.s.	3.4	2.9	30
	Electrical & 300+ c.p.s.	6.5	5.3	67
Fall	Oil/Gas & <300 c.p.s.	4.5	3.8	38
	Electrical & 300+ c.p.s.	7.3	5.9	76

4.   CONCLUSIONS

In the present study, the indoor radon concentrations can be related to the season, the type of home heating system, and the aerial radiation survey of a region. Winter measurements are the highest, spring and fall indoor radon values were comparable, and the summer season tends to be the time of lowest values. Homes with electrical heating systems tend to have higher indoor radon than homes with oil or gas furnaces as do homes constructed in areas with higher total-gamma surface radiation.

It is obvious that with some care a potential home buyer can select a home with construction and location factors that can be associated with a lower-than-average indoor radon concentration. It is also clear that owners of presently occupied homes can be alerted to a potential indoor radon problem if they are informed about factors that contribute to a radon hazard.

REFERENCES

Alexander, B., Rodman, N., White, S.B., and Phillips, I. 1994. Areas of the United States with elevated screening levels of Rn222. *Health Physics.* 66, 50-54.

Daniels, D.L.. 1980. Geophysical-geological analysis of Fairfax County, Virginia . *U.S. Geol. Survey Report* 80-1165, 64 pp.

Drake, A.A., Jr., and Morgan, B.A. 1981. The Piney Branch Complex, a metamorphosed fragment of the central Appalachian ophiolite in northern Virginia. *American Journal of Science.* 281, 484-508.

Drake, A.A., Jr., and Morgan, A.A. 1983. Reply: Melanges and the Piney Branch Complex, a metamorphosed fragment of the central Appalachian ophiolite in northern Virginia , *American Journal of Science* . 283, 376-381.

Gundersen, L.C.S., and Schumann, R.R. 1996. Mapping the radon potential of the United States, examples from the Appalachians, *Environmental International.* 22, S829-S837.

Marcinowski, F., Lucas, R.M., and Yeager, W.M. 1994. National and regional distributions of airborne radon concentrations in U.S. homes. *Health Physics.* 66, 699-706.

Mose, D.G., Mushrush, G.W., and Chrosniak, C.E. 1990. Reliability of inexpensive charco al and alpha-track radon monitors, *Natural Hazards.* 3, 341-355.

Mose, D.G., Mushrush, G.W., and Chrosniak, C.E. 1996. Environmental factors governing indoor radon, *Journal of Environmental Science and Health* . A31, 553-577.

NCRP (National Council on Radiation Protection and Measurements). 1984. Exposures from the uranium series with emphasis on radon and its daughters. Report No. 77, 131 pp.

Otton, J.K., Schumann, R.R., Owen, D.E., Thurman, N., and Duval, J.S. 1988 . Map showing radon potential of rocks and soils in Fairfax County, Virginia , U.S. Geological Survey Miscellaneous Field Studies Map MF -2047.

Revzan, K., Nero, A., and Sextro, R. 1988. Mapping surficial radium content as a pa rtial indicator of radon concentrations in US houses. *Radiation Protection Dosimetry* . 24,179.

Sextro, R.G., Moed, B.A., Nazaroff, W.W., Revzan, K.L., and Nero, A.V. 1987. Investigations of soil as a source of indoor radon , in Hopke, P.K., ed., Radon and Its Decay Products- Occurrence, Properties, and Health Effects: Washington, American Chemical Society Symposium Series. Report 331, 10 -29.

White, S.B., Bergsten, J.W., Alexander, B.V., Roadman, N.F., and Phillips, J .L. 1992. Indoor radon concentrations in a probability sample of 43,000 homes across 50 states, *Health Physics.* 62, 41-50.

CHAPTER 12

INFLUENCE OF HOME SIZE ON THE RISK FROM SOIL-GAS AND WATERBORNE INDOOR RADON

Douglas Mose, George Mushrush, George Saiway and Fiorella Simoni
Center for Basic and Applied Science, Culpeper, VA 22701 and Chemistry Department, George Mason University, Fairfax, VA 22020

Abstract: In a recent study of about 700 homes in Virgin ia and Maryland, three-month measurements of airborne radon derived from soil-gas combined with indoor radon derived from potable water in the home ranged from about 10-40 pCi/L. The radon in the potable water ranged from less than 100 to 8000 pCi/L/L. The home sizes ranged from about 20,000 to 100,000 cubic feet. In a study set composed of all the homes using water with low concentrations of waterborne radon, no correlation was observed between indoor radon and home size. In study sets of increasingly high waterborne radon, a correlation can be seen between waterborne radon and indoor radon. As the waterborne radon increases, smaller homes ten d to have more indoor radon than larger homes. In terms of the risk of developing lung cancer , the greatest risk is experienced by people using well water while living in small homes.

Key words: indoor radon, waterborne ra don, potable water, lung cancer

1.   INTRODUCTION

Indoor radon concentrations are quite variable, even in the basement or above the base-level slab, where radon enters through soil-facing surfaces. Indoor radon appears to be dependent on the geological material under the home (Mushrush et al., 1989a), and on the home construction (Mushrush et al., 1989b). Due to the complex interaction of several factors, it has proven difficult to the predict indoor radon concentration in any particular home, but estimates of averages for groups of homes based on geological material and home construction are reasonably accurate.

Indoor radon is a mixture of soil-gas radon that comes through soil-facing walls, base-level slabs, and waterborne radon that escapes from the potable water in homes. The only difference is that the radon in well water originates from somewhat deeper geological material beneath the home. Radon is are always found in natural soil and in natural water, but because radon easily dissolves into water, its concentration in water (particularly in undergroundwater) is usually many times greater than the concentration in soil through which the water moves.

Several studies, starting with reports of Kahlos and Asikainen (1980), have suggested that radon dissolved in the home water supply can substantially contribute to the indoor radon of the home. Several subsequent studies have estimated that a water supply with a radon concentration of 10,000 pCi/L would add about 1 pCi/L to the indoor radon due to the outgassing of the domestic water supply (Cothern, 1987; Prichard, 1987). Three booklets published by the EPA (USEPA 1987a, 1987b, 1987c) state that radon-enriched domestic water must have about 10,000 pCi/L of radon to contribute 1 pCi/L to indoor radon (a 10,000:1 ratio). The purpose of this research is to examine the relationship between indoor radon concentration and waterborne concentration, and to determine if home size influences the relationship.

2.   PROCEDURE

As part of an ongoing study, seasonal measurements have been made of soil radon, and drinking water radon. Most of the study homes are in Fairfax County in Northern Virginia, and the immediately adjacent Montgomery County in southern Maryland. At the present time, over 1200 Virginia and Maryland homes have been examined in our study, each over an entire year, using a series of four alpha-track indoor radon monitors (Tech/Ops Landauer Corporation, Type SF). The alpha-track monitors were provided to the homeowners during each season to determine seasonal variations. The winter data in the following tables are a compilation of three-month exposures during November, December and January. The spring exposure included February, March and April, the summer exposure included May, June, July, and the fall exposure included August, September and October.

Participants in the indoor radon project were offered a test kit with which to measure drinking water radon. About 700 homeowners (half of the participants in the indoor radon project) joined the radon-in-water study. The homeowners who participated filled out a questionnaire about the size of their homes and their home water supply (e.g., from where is the water obtained, if from a well how deep and how far away is the well, is the water treated, and if so how, etc.). The participant homeowners were each provided with a syringe, a capped vial with 5 mL of toluene-based liquid scintillation fluid, and directions about how to collect 10 mL of drinking water from a commonly used water tap. Although only a single water sample was analyzed for its radon content from each home during each season, radon concentration in groundwater is thought to show variation below a factor of two (Prichard and Gesell, 1981). Studies now in progress will serve to determine if this level of variation is found in the Virginia and Maryland study area, but preliminary data suggest that the variation is indeed below a factor of two.

3. RESULTS AND DISCUSSION

During the winter when homeowners close their homes to conserve heated air, the basement indoor radon tends to be at its greatest concentration (Table 1). Basements with concrete block walls tend to have higher indoor radon concentrations than basements with poured concrete walls (Table 2), probably because the commonly observed cracks that develop in the mortar between the blocks allow soil derived radon to move into the home. Homes with electrical heating systems tend to have greater concentrations of indoor radon than homes with combustion heating systems (Table 3). This is probably because homes with electrical heating systems have no momentary intervals of decreased air pressure that can pull radon-poor outside air into the home. The momentary "inhalation" of outside air that occurs in homes with combustion heating systems is caused by the rapid exhaust of combustion gasses.

Table 1. Summary of basement indoor radon concentrations

Season and County	Number of Homes	Average Radon pCi/L	Median Radon pCi/L	% Over 4 pCi/L	% Over 10 pCi/L	% Over 20 pCi/L
Winter						
Fairfax County	844	4.4	3.2	38	6	1
Montgomery City	293	4.9	3.1	40	11	2
Spring						
Fairfax County	829	4.1	3.0	33	5	1
Montgomery City	242	4.7	3.1	36	9	2

Season and County	Number of Homes	Average Radon pCi/L	Median Radon pCi/L	% Over 4 pCi/L	% Over 10 pCi/L	% Over 20 pCi/L
Summer						
Fairfax County	927	3.3	2.5	24	2	0
Montgomery City	323	3.6	2.7	27	4	2
Fall						
Fairfax County	898	4.2	3.2	35	4	1
Montgomery City	307	4.4	3.2	36	9	1

Table 2. Comparison between basement indoor radon and basement wall construction.

	Concrete Block Walls				Poured Concrete Walls			
	Winte	Spring	Summe	Fall	Winter	Sprin;	Summe	Fall
Fairfax County VA								
Average Rn(pCi/l)	4.4	4.1	3.1	3.9	4.2	4.0	3.1	4.5
Median Rn(pCi/l)	3.3	3.0	2.5	3.1	3.0	2.9	2.4	3.2
% Over 4 pCi/	138%	33%	23%	35%	35%	30%	20%	32%
% Over 10 pCi/l	6%	5%	1%	3%	6%	6%	3%	6%
Number of homes	515	502	548	532	249	262	277	261
Montgomery County MD								
Average Rn(pCi/l)	5.3	4.8	3.7	4.4	4.3	4.3	3.4	4.5
Median Rn(pCi/l)	3.3	2.9	2.6	3.2	3.0	3.6	3.0	3.7
% Over 4 pCi/	142%	35%	27%	36%	39%	41%	27%	40%
% Over 10 pCi/	111%	8%	5%	6%	7%	6%	1%	9%
Number of homes	214	176	237	226	56	49	66	63

Table 3. Comparison between basement indoor radon and type of home heating system.

	Fall Season		Winter Season		Spring Season		Summer Seasor	
	Fuel	Electric	Fuel	Electric	Fuel	Electric	Fuel	Electri
Fairfax County VA								
Average Rn(pCi/l)	4.0	6.3	3.7	4.7	2.8	3.9	3.7	5.0
Median Rn(pCi/l)	2.9	3.9	2.8	3.1	2.3	3.1	3.0	3.7
% Over 4 pCi/	33%	47%	29%	40%	17%	33%	29%	46%
% Over 10 pCi/l	5%	13%	3%	9%	0%	5%	1%	10%
Number of homes	419	241	407	257	442	272	444	253
Montgomery County MD								
Average Rn(pCi/l)	5.1	5.5	4.7	6.2	3.8	4.2	4.4	4.9
Median Rn(pCi/l)	3.1	5.6	2.9	5.8	2.8	3.6	3.1	4.5

	Fall Season		Winter Season		Spring Season		Summer Season	
% Over 4 pCi/	37%	61%	32%	69%	28%	44%	36%	48%
% Over 10 pCi/	11%	11%	8%	8%	4%	8%	6%	6%
Number of homes	196	28	166	25	215	36	209	31

During the indoor radon study, one of the homes with a private water well was being monitored to examine variations in drinking water radon. Indoor radon varies seasonally, and this particular home was one of our first that was being studied to see if its drinking water radon varied in a fashion similar to that of its indoor radon. The indoor radon varied from about 3 to 12 pCi/L (basement average was about 9 pCi/L; first floor average was about 4 pCi/L) and the drinking water radon varied from about 3200 to 3700 pCi/L (average was about 3500 pCi/L). To examine the outgassing of radon from the home water supply, a continuous indoor radon monitor (Femto-Tech Indoor Radon Monitor) was placed in this home, in a first floor 50 square foot bathroom with a shower. The shower was left on for several hours, and the bathroom door was left closed. This is similar to experiments reported by McGregor and Gourgon (1980), Hess et al., (1982), and Kearfott (1989).

As shown in Table 4, there was a significant increase in the radon content of the bathroom. The degree to which this shower would increase the overall radon concentration of the home can be estimated. For example, if the air handling system circulated the radon enriched air throughout a 2500 square foot home, (with a volume of about 25,000 cubic feet), the increase would have been about 1/50 of the bathroom concentration, which would be about 7 pCi/L. If the shower had been used for 1/20 of 5 hours (a shower duration of about 15 minutes), one shower per day would add about 0.35 pCi/L to the indoor radon of the home, which is 1/10,000 of the average radon-in-water concentration.

Table 4. Compilation of airborne radon increase due to outgassing of ashower in a 50-square-foot bathroom.

Indoor Radon Test	Hour-long test intervals for indoor radon before water shower was turned on			Hour-long averages for indoor radon during the time that water shower was turned on (indoor radon in pCi/l)				
	3rd Hour	2nd Hour	1st Hour	Hour 1	Hour 2	Hour 3	Hour 4	Hour 5
Test 1	15 pCi/	16 pCi/l	8 pCi/l	71	190	dnt*	dnt*	dnt*
Test 2	27 pCi/l	5 pCi/l	4 pCi/l	24	134	226	293	dnt*
Test 3	35 pCi/l	4 pCi/l	5 pCi/l	29	134	238	317	350

*data not taken

Obviously, the 10,000:1 ratio of waterborne radon to indoor radon calculated above is not always true. If each day there were two showers, one cycle of a dishwasher and one cycle of a clothes washer, the indoor radon concentration would be increased by more than 1/10,000 of the radon-in-water level. While interesting, it seemed intuitively true that a more important point to consider is the relationship between the size of the home, waterborne radon concentrations and in indoor radon concentrations.

Our collection of measurements consisted of homes with indoor radon concentrations that ranged from about 1 to 40 pCi/L, water in this collection that ranged from less than 100 to 8,000 pCi/L, and home sizes that ranged from about 20,000 to over 100,000 cubic feet. At these relatively low water radon concentrations, the 10,000:1 ratio mentioned above implies that we would not see a positive correlation between waterborne radon and indoor radon in this collection of homes. If we did not take home size into consideration, we did not see a positive correlation during the winter when the homes are generally more tightly sealed to conserve heat (Figure 1), spring, summer or fall. We did see a positive correlation when home size was considered.

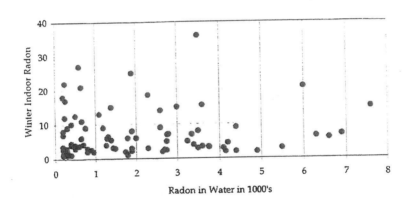

Figure 1. Comparison between basement indoor radon measurements during the winter and radon in the drinking water of homes with private wells.

To determine the importance of home size, the homes were divided into groups determined by the radon concentration of the home water supply. In homes where the domestic water supply is provided by a municipal water system which contains essentially no dissolved radon, the indoor radon concentration was not related to the home size (Figure 2). The average indoor radon in a small home was similar to the average found in a large home. However, Figures 3-5 show that as the waterborne radon concentration increased, the effect of the size of the home increased. As we look at groups of homes with a greater amount of waterborne radon, is became clear that smaller homes generally have higher indoor radon.

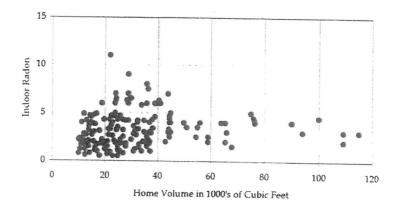

Figure 2. Comparison between the annual indoor radon in Fairfax County and Montgomery County homes with the volume of the home, using homes that have no measurable radon in their home water supply.

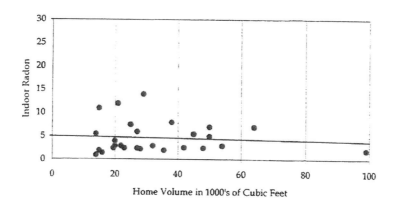

Figure 3. Comparison between the annual indoor radon in Fairfax County and Montgomery County homes with the volume in the home, using only homes that have 100 -500 pCi/L of radon in well water.

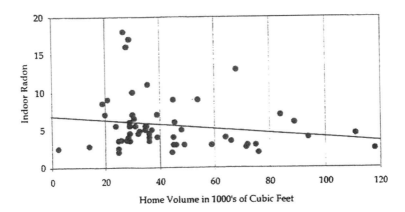

Figure 4. Comparison between the annual indoor radon in Fairfax County and Montgomery County homes with the volume of the home, using only homes that have 500 -5000 pCi/L of radon in well water.

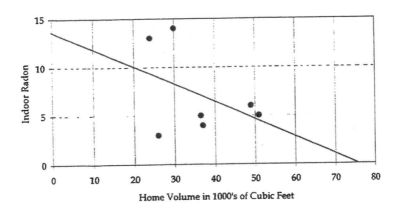

Figure 5. Comparison between the annual indoor radon in Fairfax County and Montgomery County homes with the volume of the home, using only homes that have more than 5000 pCi/L of radon in well water.

What is important to note is that waterborne radon concentrations of less than 10,000 pCi/L exert a measurable effect on indoor radon, but the effect is observed in the smaller homes. Conversely, when home volumes reach 100,000 cubic feet (a large home), the indoor radon concentration is relatively low, even if the waterborne radon concentration is relatively high.

In short, waterborne radon is most likely to present a problem for people with water wells and small homes. This may represent a population group that tends to be less able to make changes in either their homes or their water supply.

4.   CONCLUSIONS

The study of indoor radon and drinking water radon in Virginia and Maryland showed that the indoor radon concentrations were almost all between 1 and about 40 pCi/L, and the waterborne radon in homes supplied by a private well ranged from about 100 pCi/L to about 8,000 pCi/L. Differences in weather, basement wall construction and home heating system influenced indoor radon derived from the soil. We also found that the extent to which the waterborne radon concentration affected the concentration of indoor radon was determined by the volume of the home, in combination with the radon content of water used in the home.

As in most investigations, thoughtful precautions and additional data are often useful. In our study, we provide indoor radon monitors and we encourage homeowners with private water supplies to obtain a measurement of the radon concentration in their water. In homes with an indoor radon concentration that the homeowner viewed as representing an unreasonable risk, we suggested that they reduce the airborne radon with the use of sub-slab ventilation. Similarly, in homes with unreasonably high radon concentrations in the drinking water, we suggested the installation of a charcoal water purification system.

Unfortunately for the general population, the perception of what represents an unreasonable risk depends on many factors, only one of which is generally agreed standards. At this stage of public awareness, airborne radon clearly garners more attention than waterborne radon. We suggest, pending the gathering of additional data of the type compiled in this report, that the risks of radon in both air and water be equally emphasized by public health officials.

REFERENCES

Cothern, C.R. 1987. Estimating the health risks of radon in drinking water. *J. Amer. Water. Works Assoc.,* 79, 153 158.

Hess, C.T., Weiffenbach, C.V., and Norton, S.A. 1982. Variations of airborne and waterborne Rn 222 in houses in Maine. *Environ. Int.,* 8, 59 66.

Contaminated Soils- Radionuclides

Kahlos, H., and Asikainen, M. 1980. Internal radiation doses from radioactivity of drinking water in Finland, *Health Phys.* 39, 108 111.

Kearfott, K. J. 1989. Preliminary experiences with 222Rn gas in Arizona homes, *Health Phys.* 56, 169 179.

McGregor, R.G., and Gourgon, L.A. 1980. Radon and radon daughters in homes utilizing deep well water supplies, Halifax County, Nova Scotia, *J. Environ. Sci. and Health.* 15, 25 35.

Mushrush, G.W., Mose, D.G., and Chrosniak, C.E. 1989a. Indoor radon in northern Virginia: Seasonal changes and correlations with geology. *Episodes.* 12, 6 9.

Mushrush, G.W., and Mose, D.G. 1989b. The effect of home constructi on on indoor radon in Virginia and Maryland, *Environ. Int.* 14, 395 402.

Prichard, H.M. 1987. The transfer of radon from domestic water to indoor air: *J. Amer. Water Works Assoc.* 79, 159 161.

Prichard, H.M., and Gesell, T.F. 1981. An estimate of population exposure due to radon in public water supplies in the area of Houston, Texas, *Health Phys.* 41, 599 606.

USEPA (U.S. Environmental Protection Agency). 1987a. Radon reduction methods, a homeowner's guide (second edition). Publication OPA 87 011, 10 pp.

USEPA (U.S. Environmental Protection Agency Publication) 1987b. Removal of radon from household water: Publication OPA 87 010, 21 pp.

USEPA (U.S. Environmental Protection Agency 1987c). 1987c. Radon reference manual. Publication 520/1 87 20. 140 pp.

PART VI: REGULATORY

CHAPTER 13

HOW INTERSTATE COLLABORATION CAN IMPROVE SITE CLEANUPS: TRIAD AND THE ITRC

Ruth R. Chang[1] and Stuart J. Nagourney[2]

[1]*California Environmental Protection Agency, Department of Toxic Substances Control, Hazardous Materials Laboratory, 700 Heinz Avenue, Suite 100, Berkeley, CA 94710;* [2]*New Jersey Department of Environmental Protection, Office of Quality Assurance, PO Box 424, Trenton, NJ 08625*

Abstract: The standard practice currently used by state and federal regulatory agencies for remedial action has been proven to be very costly, time consuming and labor intensive. To improve project quality and to save resources, the U.S. Environmental Protection Agency has initiated a Triad approach that integrates systematic project planning, dynamic work strategies and real-time measurement technologies for the management of environmental projects. The central principle of the Triad approach is managing decision uncertainty. Experience from several previous investigations has shown significant savings in time and costs by using Triad approach, while providing more reliable scientific data for decision-making. The Sampling, Characterization and Monitoring Team of the Interstate Technology Regulatory Council has completed a Technical and Regulatory Guidance document summarizing the principles of the Triad approach and the scientific and technical requirements to employing this paradigm shift. The advantages of adapting this innovative approach in hazardous waste site investigation and remediation will be highlighted.

Key words: Triad approach, project management, data quality, managing uncertainty

1.   INTRODUCTION

Currently, there are more than 100,000 sites in the United States which require remediation. In most cases, site cleanups are under the jurisdiction of State or Federal regulatory agencies. Each year the U.S. Environmental Protection Agency (U.S. EPA) and the State regulatory agencies spent several billion dollars collecting environmental data for site cleanup, risk assessment and regulatory compliance. However, in several historical cases that regulatory agencies found data collected at the conclusion of the investigations were insufficient for making decision. Thus, more environmental sampling and analysis were required to provide additional information. This multistage investigative process has proven to be very costly, time-consuming and labor intensive. Inflexible project plans often require analytical data generated from regulatory approved methods used in fixed laboratories. Under these operating conditions, site restoration usually takes years to complete. The U.S. EPA has introduced a new concept that integrates systematic project planning, dynamic work strategies and real-time measurement technologies to improve remedial efficiency. This combined process is called the "Triad" approach. To implement a Triad approach, regulators need to have an innovative-thinking, flexible work plan and accept field data for decision making. This new paradigm shift may not be consistent with some state policies and presents a challenge to several regulatory agencies. The Interstate Technology Regulatory Council (ITRC), a state-led coalition that brings together regulators, industry, academia and the public, seeks innovative solutions to overcome regulatory barriers in accepting novel concepts and emerging technologies. The ITRC guidance document *"Technical and Regulatory Guidance for the Triad Approach: A New Paradigm for Environmental Project Management"* (ITRC, 2003) developed by the Sampling, Characterization and Monitoring Team (SCM) is the basis for this presentation. Comprehensive information on the implementation of the Triad approach can be obtained from the U.S. EPA, Technology Innovation Office the Triad Resource Center web site: http://www.triadcentral.org

2.   THE TRIAD APPROACH

As the environmental media are fundamentally heterogeneous at macro and micro scales, collecting site-representative samples for decision making presents a challenge to the analytical communities and regulatory agencies. Various environmental factors, such as heterogeneity of chemical contaminations, environmental fate and transport, spatial and temporal effects, and metrological conditions, can contribute to the heterogeneity of environmental matrices. Unless a large number of samples are collected for the analysis, it is impossible to depict the contaminated profile of a real world. Consequently, due to time and resource constraints, the analytical data obtained from limited numbers of tiny samples, analyzed in a fixed laboratory were extrapolated to predict a site-contaminated profile. When the sample analyzed does not represent the site conditions, it is due to a sampling error. Sampling errors contribute to the inaccurate Conceptual Site Model (CSM) which in turn lead to erroneous decisions about risk and cleanup strategies. In addition to sampling errors, there are other uncertainties associated with the sampling and analysis process. The "Triad" approach is a process to reduce uncertainty levels in order to meet the project goal. The Triad approach encompasses three major components: systematic project planning, dynamic work strategies and real-time measurement technologies (Figure 1). The central principle behind the Triad approach is to manage data uncertainty for decision making.

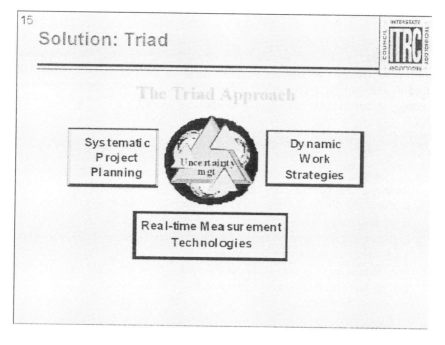

Figure 1. The Triad Approach

2.1   Systematic Project Planning

Systematic project planning is essential to establishing clear objectives for all environmental projects and is the key for the successful implementation of Triad. The project team should include the regulators, the responsible parties, the technical staffs, and the stakeholders. The project team should work together starting at the beginning phase of the project planning, gathering site information and regional data to develop a conceptual site model (CSM). The CSM provides information on site history, physical characteristics, and environmental data indicating how contaminants are distributed throughout the site, along with contaminant's fate and transport and exposure pathways. The CSM is the basis for the design of sampling plans for data collection. As new data become available, the CSM is updated for the next phase of data generation. The primary product of the Triad approach is an accurate CSM that can support decisions about exposure to contaminants, site cleanup and restoration.

Relative to the conventional operation, Triad requires more comprehensive, systematic project planning before data collection takes place. A core technical team with multidisciplinary principles and expertise must be organized to direct site activities during the project implementation phase. Systematic project planning to select the correct analytical methods and to develop an appropriate quality control (QC) program is critical in managing uncertainties to accomplish the project goal.

2.2   Dynamic Work Strategies

The word "dynamic" describes the flexibility or adaptability of the intended flow of work activities. Dynamic work strategies give experts working in the field the flexibility to make decisions and change directions based upon information acquired on site. Since the project cost is proportional to time invested, the focus is on completing the field work with as little mobilization as possible, and on modifying the field activities as quickly as possible as the investigation proceeds, to complete the project. Table 1 summarizes the difference between the dynamic work plan and the static work plan used in the traditional site mitigation.

Table 1. Dynamic Work Plan vs. Static Work Plan

	Static Work plan	Dynamic Work plan
Sampling location	Sampling locations are specified in the work plan.	Sampling locations are based on results of field measurement
Analytical Method	Regulatory approved laboratory-based methods	Real-time measurement technologies
Conceptual Site Model	Modifying CSM after completion of field work	Updating CSM as field data available
Decisions	Work plan for decisions are made before field work.	Work plan for decisions are updated according to real time measurement to manage uncertainties.

Implementation of dynamic work plan requires close involvement of project team and process to expedite data evaluation and decision making. To manage the large data sets, the environmental data management system is a useful tool to support the investigation process. Documents supporting a dynamic work strategy should contain:

- decision logic that adapts the investigation approach to change conditions
- mechanisms for rapid communication and decision making
- real-time data management

2.3   Real-time Measurement Technologies

Real-time measurement is the cornerstone of triad data acquisition. Triad uses real-time measurement to delineate the site through high density and spatial sampling. Real-time technology and dynamic work plan are carried hand-in- hand to support real-time decision making. Relative to the laboratory-based measurements, field analytical methods are less expensive and give short sample turnaround time. This is the way to acquire a large number of samples to overcome matrix heterogeneity. Figure 2 illustrates that by collecting a larger number of less-expensive samples, a more complete site characterization can be accomplished. High analytical data quality is seldom needed to refine the CSM. However, without a reliable CSM to support the representativeness of high-quality analytical data points, those data may mislead the result and cause decision error.

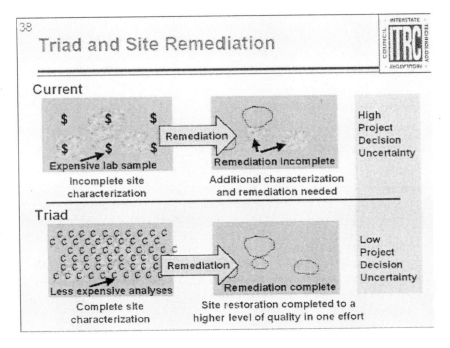

Figure 2. Triad and Site Remediation

On the other hand, field analysis is not a replacement for conventional laboratory analysis. It would never be expected to achieve a one-to-one correspondence between the laboratory data and the field data. These deviations can be due to the different analytical principles, measuring different analytical variables, or different sample supports. The role of laboratory analysis is to identify those contaminates not currently amenable to field analysis, such as insufficient sensitivity and selectivity, and to confirm the effectiveness and reliability of the field data. Field analysis complementary with the laboratory analysis is the solution to optimize the efficacy of site investigations. High numbers of less-expensive field analysis are used to develop the CSM to manage sampling uncertainties, while expensive laboratory analyses are used to manage analytical uncertainties. The solution is to complement the field analysis with the laboratory analysis to produce collaborative data sets in managing all sources of data uncertainties important to decision making (Figure 3).

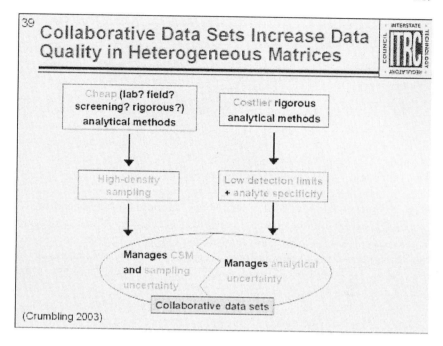

Figure 3. Collaborative Data Sets Increase Data Quality

Real-time measurement techniques commonly used in site characterization and remediation can be classified into geophysical, analytical and geological techniques. Information to select appropriate field analytical technologies can be found in the U.S. EPA, Technology Innovation Office, Field Analytic Technologies Encyclopedia web site: http://fate.clu-in. org.

- Geophysical techniques: electrical and electromagnetic, borehole techniques, seismic, etc.
- Analytical techniques: immunoassay, portable GC, portable GC/MS, XRF, *in situ* Probes, Open –path Spectroscopy, etc.
- Geological techniques: direct-push down-hole video, cone penetrometer, etc.

3.   QUALITY CONTROL, DATA QUALITY AND UNCERTAINTY

3.1   Quality Control Program

The goal of the QC program is to produce data of known quality to achieve project decision goals. The Triad approach usually uses multiple field analytical technologies in conjunction with fixed laboratory analytical methods. Therefore, the quality control (QC) program shall consider both laboratory and field analytical techniques including geophysical, analytical and geological techniques. Selection of appropriate field methods through mixing and matching analytical techniques is commonly practiced in the Triad. As specified in the National Environmental Laboratory Accreditation Conference (NELAC) for the implementation of a performance-based system, a pilot study to demonstrate method applicability for the analyte of concern in the matrix of concern should be part of the QC program in the project planning (NELAC, 2005).

3.2   Data Quality and Uncertainty

It has long been recognized that the largest source of data uncertainty is sampling variability associated with the heterogeneity of environmental matrices (Homsher 1991, Jenkins et al. 1997). The uncertainties of environmental data are introduced during the sampling and analysis steps. (Variables that associate with sampling and analysis are presented in Figure 4.) Sample support, sampling design, sample preservatives and subsampling are the variables contributed to sampling errors; while sample preparations, cleanup procedures, determinative methods, and result reporting are the variables contributed to analytical errors. Inappropriate handling of any of these variables can contribute to data uncertainty.

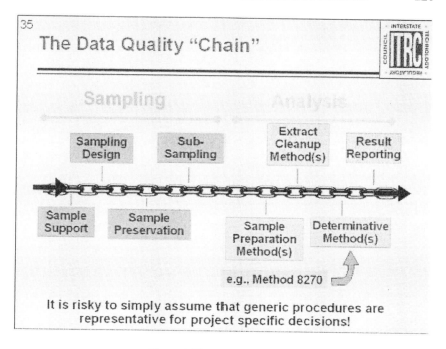

Figure 4. The Data Quality Chain

There are several common misconceptions in the analytical communities as well as in the regulatory agencies: (1) Environmental samples are analyzed to generate data and data are used for decision (Figure 5); (2) Screening methods produce data for screening purposes and uncertain for decision making, and (3) Definitive methods produce definitive data for decision making (Figure 6).

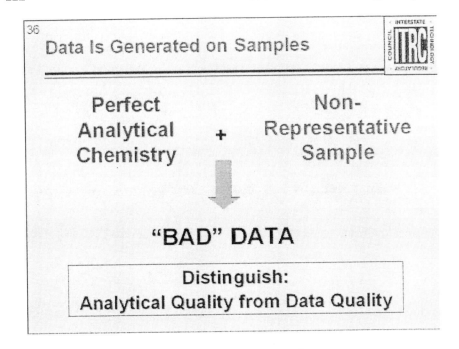

Figure 5. Data is Generated by Samples

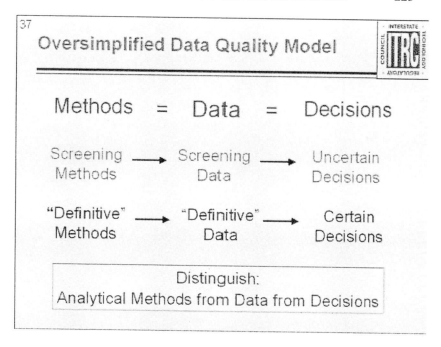

Figure 6. Oversimplified Data Quality Models

As shown in Figure 2, high-quality and expensive analytical data, if not generated from representative samples, can produce a bad data set that leads to incorrect decisions and fails to support project decisions. Therefore, the regulators and the responsible parties should have a clear concept to distinguish quality analytical data from data quality used for decisions.

4.   REGULATORY ISSUES AND INVOLVEMENT

For implementation of the Triad approach, regulators need to employ innovative thinking, a flexible work plan and to accept the data from field measurements. Accepting the field data for regulatory purposes is of great concern because of the potential issues in legal defensibility. For some states, data acceptability is linked to laboratory certification, which does not currently apply to field analytical methods. As in any large organization, regulatory procedures become institutionalized overtime. To deviate from these routine business practices, is a challenge to the management of many organizations. The barriers that prohibit the implementation of the Triad are identified in six areas:

- Organizational or institutional barriers
- Concerns with real-time measurement technologies
- Conflicts with states law
- Lack of regulatory guidance in implementing Triad
- Lack of established action levels or cleanup criteria
- Uncertainty associated to specific decisions

In consideration of these issues, a survey was carried out among SCM team members who participated in the development of the Triad guidance plan. These states include California, Delaware, Kentucky, Missouri, New Jersey, Oklahoma, Vermont, South Carolina, and Wisconsin. The outcomes of this survey regarding individual state policies in implementing Triad are documented for reference (ITRC, 2003). It revealed that institutional barriers, in general, are the greatest challenge. No specific regulatory barriers have been identified to prohibit the implementation of Triad in these states.

The New Jersey Department of Environmental Protection (NJDEP) has recently taken formal steps to endorse Triad for environmental project management. Several activities have been initiated to support theTriad Implementation Plan. These activities include creating an interdisciplinary project team, training managers and staff working on Triad projects and writing a Triad implementation guide for the NJDEP. A certification program is set up to certify four categories of field methods (immunoassay, GC, GC/MS, and XRF). Certification processes will involve: review of applicant qualifications, standard operating procedures and on-site audit. The goal is to blur distinctions between field data and laboratory data.

To evaluate the performance of novel technologies, several environmental technology verification and certification reviews are to be sponsored by the State and Federal agencies. These include U.S. EPA Environmental Technology Verification (ETV) and Superfund Innovative Technology Evaluation (SITE) programs and several State certification programs to evaluate the performance of novel technologies in California, Massachusetts, Washington, New Jersey, and Canada.

5.   CASE STUDIES

So far, the Triad approach is mostly used by the Department of Energy (DOE) and the military in cleanup of hazardous waste sites. The applications include characterization of several large DOE sites contaminated with a wide variety of pollutants (metals, organic and radioactive wastes), military installations requiring expedited decisions, small brownfields and research centers contaminated with known or unknown contaminants. Successful case studies have been reported in California, New Jersey, Vermont, Florida Washington, and Ohio (ITRC, 2003). The following are two case studies that quantified the advantages of Triad in terms of time and costs.

5.1   McGuire Air Force Base C-17 Hanger in New Jersey:

This case study involved a chlorinated solvent contamination in McGuire Air Force Base C-17 Hanger Site in New Jersey. The project team included the regulators and an experienced technical team who planned and executed Triad approach. A series of real-time measurement technologies, including CPT, MIP, DSITMS, XRF and the aid of global positioning system (GPS), geographical information system (GIS), and contaminant analysis software were employed. The site was completely characterized within three weeks, with an estimated time savings of 18 to 24 months, and cost savings of $1.34 million.

5.2   Camp Pendleton, CA Site 1114

During a base-wide groundwater survey, the U.S. Navy Public Works Command (PWC) in San Diego suspected a Perchloroethene (PCE) source zone at Camp Pendleton, CA. Under the operation of Triad, a large data set was collected to localize a very small PCE source in groundwater. Regulators agreed to close the site without further long-term monitoring. The US Navy PWC estimated time savings of 3 years and cost savings of $2.5 million.

6.   SUMMARY

The Triad approach must encompass three essential elements: systematic project planning, dynamic work strategies and real-time measurement technologies. Dynamic work strategies and/or real-time measurement technologies alone are not the "Triad" approach. In summary:

- The goal of the Triad is to improve investigation effectiveness
- The central concept of the Triad is managing uncertainties for project decisions with social, economic and political considerations
- A technical team with multidisciplinary expertise must be involved throughout project planning and implementation
- Systematic project planning focuses on identifying project goals
- Data collections emphasis on the conceptual site model and sample representativeness
- Applications of field technologies increase sample density to overcome matrix heterogeneity and sampling error
- Reduced mobilizations avoid repeated planning, field execution and analytical cost.
- QC programs should include field and analytical methods to ensure data of known quality to achieve project goals
- The bottom line is significant savings of time and costs.

REFERENCES

Crumbling, D.M. January 2003. "The Triad Approach to Address Data Quality Issues" Presentation given at the Intestate Technology & Regulatory Council Sampling, Characterization and Sampling Team Meeting, Tampa, Fla.

Homsher, M.T., Haeberer, P.J., Marsden, R.K., Mitchum, D ., Neptune and Warren, J. 1991. "Performance-Based Criteria, a Panel Discussion" *Environmental Lab,* October/ November.

ITRC, 2003. Technical and Regulatory Guidance for the Triad : A New Paradigm for Environmental Project Management. prepared by The Interstate Technology Regulatory Council, Sampling, Characterization and Mentoring Team, December 2003.

Jenkins, T.F., Walsh, P.G., Thorne, S., Thiboutot, G., Ampleman, T.A., Ranney, and Grant, C.L. 1997. Assessment of Sampling Error Associated wi th Collection and Analysis of Soil Samples at a Firing Range Contaminated with HMX, Special Report 97-22.U.S> Army Corps of Engineers/Cold Regions Research and Engineering Laboratory, National Technical Information Service. Available online at http://www.crrel.usace.army.mil/ techpub/CRREL Reports/reports/SR97 22.pdf

NELAC, 2004, National Environmental Laboratory Accreditation Conference, Chapter 5 Quality Systems, Appendix C Demonstration of Capabilities. NELAC Standards effective July 1, 2004.

PART VII: REMEDIATION

CHAPTER 14

ORGANOCLAY/CARBON SYSTEMS AT MILITARY INSTALLATIONS

George Alther

Biomin, Inc., P.O. Box 20028, Ferndale, MI 48220

Abstract: Organoclays have been used as a pre-polisher for activated carbon, or post-polisher for oil/water separators and DAF units, for the removal of small amounts of oil, grease, PCB, PNA, BTX and other organic hydrocarbons of low solubility for the cleanup of groundwater and wastewater. The end user can save 50% or more of his operations costs by removing large hydrocarbons which plug the pores of activated carbon beforehand, allowing carbon to remove the last 5 ppm or less of volatile compounds. Organoclays can remove 7 times as much oil and other organic hydrocarbons of low solubility, as does carbon.

This article describes what organoclay is, how it is used, and presents several case histories of large systems at military bases and other places.

1.   WHAT ARE ORGANOCLAYS?

Organoclays, or organically modified clays, are a blend of a cationic surfactant and bentonite. This blend creates a new product, a nonionic surfactant with a solid base. The quaternary amine chains are, on the positive end, ion exchanged onto the bentonite clay, while the neutral end extends into the water column. By means of the partition process this chain will fixate non-polar organic compounds. In contrast to activated carbon, where the organic compounds are adsorbed into pores, which soon become fouled, the partitioning phenomena takes place outside of the clay particles, eliminating the problem of fouling. (Alther, 2002a)

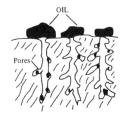 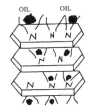 

Activated carbon Granule – Pore spaces of activated carbon, blinded by emulsified oil.

Clay Platelets, modified with quaternary amine, removal of emulsified oil on the clay surface.

Activated Carbon downstream of organoclay, ready to remove the more soluble compounds.

Figure 1. Mechanism of oil droplet removal from water by activated carbon and organoclay .

The organoclay is blended with anthracite to extend its performance life. The vessels used for the organoclay are the same as those used for activated carbon. The treatment train is thus the same, i.e. oil/water separator, possibly preceded by an equalization tank, a 5-micron bag filter, organoclay vessel, and activated carbon vessel.

2.   ORGANOCLAY TECHNOLOGY

Tables 1 and 2 and Figure 2 summarize the results from the column experiment studying the sorptive capacity of organoclay to an aqueous-oil solution. The oil used was a vegetable oil. A 30-inch long (76.2 cm) by 3-inch diameter (7.62 cm) column was constructed from poly-vinyl-chloride (PVC) and filled with the sorbent material to be studied. A peristaltic pump forced an aqueous-metal solution containing 680 mg/L vegetable oil up through the column to displace void-space air and ensure maximum contact with the sorbent material. Samples were collected periodically at the outflow of the column where the organic composition was analyzed using a chemical oxygen demand (COD) analysis.

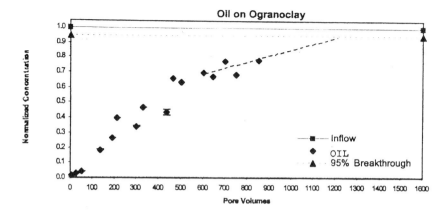

Figure 2. Shows a column test, which was conducted at the laboratory of the University of Virginia, Civil Engineering Department.

Table 1. Sorbent mass, porosity, flowrate and residence time information for organoclay column experiment.

Sorbent	Mass Sorbent		Porosity	Flow Rate		Residence
	(kg)	(lb)		(mL/min)	(gal/hr)	(min)
Organoclay /Anthracite	0.141	0.31	0.3	15.45	0.23	8

Table 2. 95% breakthrough for organoclay sorbent materials given in pore volumes and minutes along with estimated mass of oil sorbed per mass of sorbent in mg/kg, lb/lb and percent basis.

Sorbent	Breakthrough		Mass Sorbed		Mass Sorbed/Mass Sorbent		
	PV	min	(g)	(lb)	(g/kg)	(lb/lb)	(% by sorbent)
Organoclay	1150	9200	65.8	0.14	475	0.475	47.5

The results show that the organoclay/anthracite blend removed nearly 50% of its weight in oil.

Figure 3 shows some practical results which were derived from a set of jar tests, describing the capacity of non-ionic organoclay for the removal of 3 separate oils with varying densities. A sample of bituminous activated carbon was included for comparison. The organoclay is far superior to activated carbon for the removal of oils from water. A combination of organoclay, followed by activated carbon, is far superior for the removal of BTEX, PNAs and PC-B from water. (Alther, 2002b; Alther 1996).

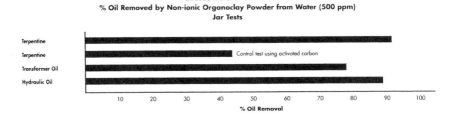

Figure 3. Jar Test Results showing the removal capacity of non -ionic organoclay from terpentine, transformer oil and hydraulic oil.

3.   CASE HISTORY, HILL AIR FORCE BASE, UTAH

This is a typical case for the kinds of cleanup operations for storm water and wastewater, which are needed at such facilities. Water to be cleaned is derived from a number of operations, including storm water, water from airplane washing, paint stripping, de-greasing, and electroplating. These operations add such solvents as Methylene Chloride; 1,1-Dichloroethane; Chloroform; 1,1,1-Trichloroethane; Trichloroethane; 1,1,2-Trichloroethene; Tetrachloroethane; and oil and grease. BTEX's from gasoline, jet fuel and diesel fuel are present as well. The electroplating operation adds heavy metals such as Chromium 3, Nickel, Zinc, Copper and Cadmium. The NPDES discharge limitations in 1990 where: O&G: 100 ppm; Total VOCs: 1.92 ppm. Today a number of these chlorinated compounds are considered hazardous. Of the heavy metals, the discharge standards have become much stricter, around 1 ppm or less. In order to remove the chlorinated compounds, BTEX's and metals, the oil and grease must be removed first. For this reason, a unique water-treatment system was installed, with a flow rate of 350 gpm: A dissolved air flotation system to remove oil and grease and sediments was followed by an alum flocculation system to remove the metals. The DAF brought the O&G content down to 10 ppm, with spikes as high as 50 ppm. Next followed several tanks filled with organoclay/anthracite, filled with a total of 90,000 lbs of organoclay/anthracite. This was followed by a series of bituminous activated carbon tanks, concluding the treatment train with an air stripper. The organoclay not only removes oil and solvents but also heavy metals such as Cd, Cu, Ni, Pb and Zn.

Organoclay/anthracite removes oil and grease from airplane washwater.

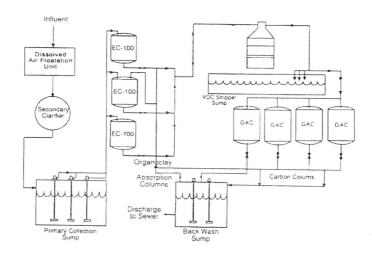

Figure 4. Process Flow Diagram, Hills Airforce Base, Utah

Operation: Electroplating
                     Degreasing
                     Paint Stripping
Contaminants: Grease
                     Chlorinated Hydrocarbons Oil (10-50 ppm)
                     Total: 10-50 ppm
                     Heavy Metals
Flow through rate: 350 gpm
Approximate cost of oil and grease removal: $0.55 – 0.65/1000 gal waste water
Cost reduction by using EC-100 plus carbon: 50%

4.   CASE HISTORY: BIO-SLURPING

This method combines vacuum-enhanced recovery and bio-venting in one singular step. Bio-slurping includes remediation of residual soil contamination in the Vadose Zone, and the recovery of free product. One pump extracts soil gas, free product and groundwater simultaneously. On the surface, the water is then treated conventionally with oil/water separators to recover free product, activated carbon and air strippers, followed by coconut carbon to remove the VOCs. At Andrews Air Force Base the objective was to remove LNAPLs from an aquifer. Once the water was pumped out and stored in a settling tank, the presence of 1000 ppm oil was detected. After a settling time of 4 hours, 200 ppm of oil was still left. The water was then passed through a small filter vessel filled with 250 lb organoclay/anthracite, followed by activated carbon. The outflow after the organoclay/anthracite contained 5.1 ppm oil. The activated carbon removed the remainder of this oil and LNAPLs, the water was clean enough to meet discharge standards.

5.   CASE HISTORY, DEWATERING

A city in California installed a storm-drain system in an area of shallow groundwater. The groundwater was contaminated with crude oil and diesel fuel, which had leaked from pipelines which where installed some 60 years ago. At first, activated carbon alone was used to remove these hydrocarbons. However, once construction began, it turned out that the hydrocarbon levels where much higher than anticipated, and the activated carbon was spent within 48 hours. It became necessary to install a 5000 gpm system within a very short time.

A thorough analysis suggested that the following treatment train was necessary: An oil/water separator, 5 high flow sediment filters, thirty 72-cubic-foot media vessels which included 120,000 lbs of organoclay/anthracite. Two systems where set up a 1/2 mile apart of 1800 gpm and 600 gpm flow capacity each.

The set-up of these systems was complicated. A linear trench extended over a mile length as construction progressed. The area also had to accommodate local traffic. It was decided to place the equipment along a railroad easement, which had height restrictions due to electrical wires. The project lasted for two months, and required no change from organoclay.

Figure 5. Complete Filtration System

Figure 6. 1600 gpm System

Figure 7. Sediment Filters and String Wound Filters

6.   CASE HISTORY STORM WATER CLEANUP

A Major refinery in California was in violation of local discharge standards and was forced to upgrade its wastewater treatment system. The refinery collected parking lot and plant runoff, which was contaminated with oil, grease and sediments. The runoff was mixed with process water and the mixture was pumped through a dissolved air flotation unit. To break any emulsions and float as much oil as possible, polymers where added to the mix. The oil was skimmed off before the water entered a settling tank. After sediments in the water settled, it was discharged into an open waterway.

The Regional Water Quality Control Board's standard for oil and grease was 5 ppm. The refinery met the standards occasionally, but spikes often overtaxed the system. This resulted in frequent, heavy fines. To bring the refinery into compliance, the following system was designed and constructed: Two particulate filters with 50 micron cartridges, followed by two particulate filters removed O&G, but also solvents, BTEX, and heavy metals. The organoclay/anthracite removes around 0.1% of metals by its weight, and 50-60% of its weight in oil. Removing compounds such as toluene, xylene and ethyl benzene allows the carbon to remove benzene without being desorbed due to the presence of these less soluble compounds.

The cost of organoclay treatment was around $0.60/1000 gallons water. The organoclay and activated carbon where changed out annually. Using organoclay rather than carbon alone saved the taxpayer 50% of operations costs. The cost of hauling this water to a waste water treatment facility would have been prohibitive, in the millions of dollars, since the hauling costs for hazardous waste containing water are around $0.60-1/1000 gallons.

7.   CONCLUSIONS

These four case histories show the utility of organoclay/anthracite for the removal of mechanically emulsified oil from water. Using activated carbon alone would raise the treatment costs to prohibitive levels.

Figure 8. Stormwater Cleanup at a Refinery

REFERENCES

Alther, G.R., 1996. Organically modified Clay removes Oil from Water. *Waste Management*, Vol 15, No.8, pp. 623 -628, Elsevier Science, NY.

Alther, G.R., 2002 (a) Removing Oil from Water with Organocl ays. *Journal AWWA*, American Water Works Association. Pp. 115 -121 AWWA. Denver, CO.

Alther, G R , 2002 (b) Using Organoclays to enhance Carbon Filtration. *Waste Management*, 22, pp. 507-513. Elsevier Science Ltd. London, U.K.

CHAPTER 15

REMEDIATION OF PETROLEUM-CONTAINING SOIL AND GROUNDWATER AT A FORMER RAIL YARD LOCOMOTIVE FUELING AREA

Scott R. Compston[1], Bruce R. Nelson[1], Scott A. Underhill[1], Andrew R. Vitolins[1], and Leann M. H. Thomas[2]

[1]Malcolm Pirnie, Inc., 15 Cornell Road, Latham, NY 12110; [2]Canadian Pacific Railway, 501 Marquette Avenue, Suite 804, Minneapolis, MN 55402

Abstract: A multi-faceted remedial program was implemented at a former locomotive fueling area (FLFA) at a rail yard in upstate New York to address diesel - affected soil and groundwater . Main line tracks running through the FLFA prohibited removal of affected soil and, consequently, an *in situ* remedy was developed. The remedy combines air sparging to provide oxygen to intrinsic diesel-degrading microorganisms and to volatilize petroleum compounds, and soil vapor extraction to actively remove volatilized diesel compounds fr om the subsurface. System components include vapor extraction and air sparging wells within the FLFA and low-flow biosparging wells between the FLFA and down gradient properties. The biosparging wells create an oxygen barrier to migrating diesel compound s. Based on vapor extraction flow rates and the concentration of volatile organic compounds (VOCs) in extracted air, an estimated 1,000 pounds of petroleum mass have been removed by the vapor extraction system to date. Mass removal and biological activit y is strongly correlated with seasonal fluctuations in subsurface temperature, which varies by more than eight degrees Celsius in the treatment zone over the course of a year. Analyses of microbial biomass in the treatment area indicate that diesel - degrading organisms increased by four orders of magnitude in unsaturated soil and by three orders of magnitude in saturated soil within five months of system start up. Regulated VOC and semi -volatile organic compound (SVOC) concentrations in soil decreased in s ubsurface soil to below detection limits in most locations in approximately 24 months. Concentrations of petroleum compounds in groundwater have been reduced to less than regulatory standards over the majority of the site and have declined 84 percent on average. This integrated *in situ* approach to the treatment of diesel -impacted soil and groundwater has greatly reduced cleanup costs and cleanup time for the site.

Key words: soil remediation, diesel fuel, air-sparging, soil vapor extraction, railroad

1.   INTRODUCTION

A long-used rail yard with a history of operation spanning more than 130 years is located in upstate New York on the flood plain of the Susquehanna River. Former activities at the yard included rail car and engine repair and locomotive refueling. The majority of operations at the yard have ceased. Though most of the railroad track has been removed, east-to-west main-line tracks and some spurs still extend through the yard. The former locomotive fueling area (FLFA) previously contained four-to-six fueling stations for fueling diesel locomotives. Based on discussions with rail-yard employees, fueling operations at the FLFA began around 1950 and continued through the 1980s. The fueling stations were removed when fueling activities ceased at the site.

Petroleum compounds as well as light non-aqueous phase liquid (LNAPL) were initially discovered in the subsurface of the site during the 1980s. A subsurface site investigation (SSI) in the late 1990s further delineated the extent of petroleum in the subsurface as well as subsurface conditions at the site. Figure 1 shows the general site layout and sampling and remedial system locations.

1.1   Subsurface Properties

The surficial geology of the site is mapped as glacial lacustrine sands and outwash sand and gravel deposits which are typical of the Susquehanna River Valley in upstate New York (Caldwell, et al., 1987). Consistent with the mapped deposits, geologic materials encountered during the subsurface investigation in the area of the remedial effort consisted primarily of fine-to-coarse sand and gravel. A layer of silt was also encountered in several soil borings at a general depth of 16 to 18 feet below ground surface (bgs). Generally, groundwater was present in the subsurface at between 8.5 and 13 feet bgs. However, continued monitoring of groundwater has indicated that the water table fluctuates by as much as four feet seasonally. Based on *in situ* hydraulic conductivity tests conducted during the SSI, the average horizontal hydraulic conductivity of the subsurface materials ranges from 1×10^{-3} centimeters per second (cm/sec) to 1×10^{-1} cm/sec. Given a hydraulic gradient of 0.003, an average horizontal hydraulic gradient of 1×10^{-2} cm/sec, and an effective porosity of 0.25 (Fetter, 1988), the average horizontal groundwater velocity at the site is approximately 0.3 feet per day. The direction of groundwater flow at the site is toward the Susquehanna River, located approximately 2,500 feet south of the site.

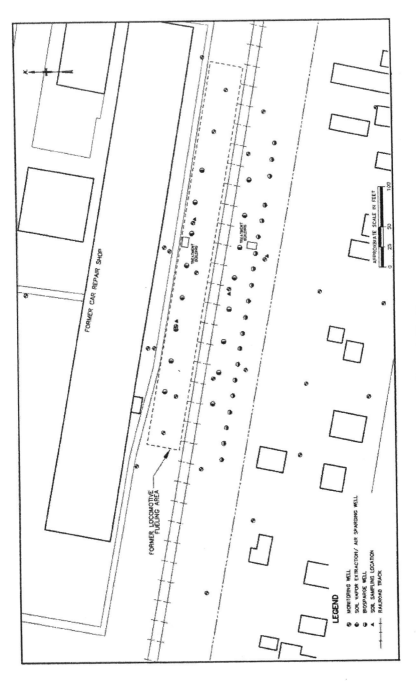

Figure 1. Site Plan

1.2   Extent of Petroleum in the Subsurface

An area of petroleum-containing soil was present within the FLFA extending east to west for approximately 480 feet along the main line tracks, and to approximately 250 feet south (down gradient) of the FLFA. The petroleum compounds extended beneath the main line tracks and onto adjacent properties to the south. Within the FLFA, petroleum-containing soil was present vertically from approximately 2.5 feet bgs to a maximum depth of approximately 14 feet bgs. In areas down gradient of the FLFA, petroleum–containing soils were generally limited to the zone of vertical fluctuation of the water table (i.e., smear zone), which ranges from approximately 8.5 feet bgs to approximately 13 feet bgs seasonally across the site.

Regulated volatile organic compounds (VOCs) and semi-volatile organic compounds (SVOCs) were detected at concentrations in excess of New York State groundwater standards and/or guidance values in groundwater samples collected from monitoring wells installed within, and immediately down gradient (south) of the FLFA, including, to a limited extent, down gradient properties. Furthermore, LNAPL was present in monitoring wells in the FLFA at measured thicknesses ranging from 0.01 feet (in a majority of the wells where LNAPL was detected) to a maximum thickness of 0.87 feet.

1.3   Remedial Strategy Evaluation and Selection

The presence of main-line railroad tracks running through the petroleum-affected area eliminated excavation of petroleum-containing soil as a viable remedial alternative. As a result, an *in situ* remedial strategy was required that would both effectively treat the full extent of petroleum-impacted soil and groundwater while not interfering with railroad operations. During the SSI, efforts were made to determine if biodegradation of diesel compounds was on-going at the site. Most Probable Number (MPN) estimates of bacteria populations with the ability to degrade diesel compounds indicated that diesel-degrading bacteria were present in the subsurface at the site and were present in greater numbers in the petroleum-affected area than in unaffected areas of the site (between 4.8×10^2 and 4.8×10^4 MPN/g versus 4.8×10^1 MPN/g). Soil samples collected in the affected area contained carbon, nitrogen, and phosphorus in sufficient concentrations to support bacterial growth.

Groundwater samples were collected from all monitoring wells installed both within and outside the petroleum-affected area and were analyzed for nitrate, sulfate, ferrous iron, dissolved oxygen, and reduction-oxidation (redox) potential to evaluate potential natural attenuation processes degrading the petroleum compounds. The two primary microbiological processes involved in the natural attenuation of petroleum compounds are aerobic and anaerobic metabolism. Aerobic metabolism is most efficient in attenuating petroleum compounds. However, in an oxygen-limited environment, such as groundwater, aerobic metabolism results in the depletion of oxygen, after which anaerobic processes predominate. Anaerobic metabolism is evidenced by the use of alternative electron acceptors such as nitrate, ferric iron (iron^{3+}), and sulfate, if reducing conditions (Redox < (+) 100) are occurring. The preferred order of electron acceptor use is: oxygen, nitrate, ferric iron, and sulfate. Therefore, if petroleum compounds are actively being degraded in the FLFA, it is expected that concentrations of nitrate and sulfate would be lower and ferrous-iron concentrations, the byproduct of the reduction of ferric iron, would be higher in the petroleum-affected area than in the unaffected area.

Analyses of the groundwater samples confirmed this trend within and down gradient of the FLFA. Sulfate and nitrate concentrations provided the strongest evidence for natural attenuation. Sulfate concentrations ranged from 10.3 mg/l and 50.6 mg/l in the unaffected area west of the FLFA and were not detected in monitoring wells within and immediately down gradient of the FLFA. Though nitrate concentrations were low across the site, nitrate was not detected in samples collected from within or down gradient of the FLFA. Ferrous-iron concentrations were not consistently greater in samples collected within the FLFA. However, the two highest concentrations measured on-site were collected from wells in the FLFA. Combined with a reducing environment, the lack of dissolved oxygen measured in the FLFA, and results of the MPN analyses, it was concluded that microbial degradation of the diesel compounds was likely occurring in the subsurface of the site.

Based on the findings of the SSI, the remedial strategy selected for the site included installing an air sparging system and soil vapor extraction system. As previously mentioned, aerobic metabolism is typically the more efficient means by which petroleum compounds may be utilized by microorganisms. The air sparging system would provide a constant supply of oxygen to microorganisms to allow for aerobic metabolism of petroleum compounds and reduce or eliminate metabolic rate restrictions caused by insufficient oxygen. Further, the air-sparging system would also volatilize petroleum compounds while the soil vapor extraction system would extract petroleum vapor from the soil for removal in an activated carbon treatment system. Details of remedial system design are provided below.

2. MATERIALS AND METHODS

2.1 Pilot Test

Pilot tests were conducted in December 2001 to determine the effectiveness of both the soil vapor extraction (SVE) and air sparging (AS) components of the remedial plan. One SVE-AS combination well was installed in an area where soil and groundwater sampling during the SSI indicated petroleum was present. The combination SVE-AS well was constructed by drilling a borehole to 22 feet bgs, which fully penetrated the petroleum affected area. The 1.5-inch diameter AS well was installed in the bottom of the boring and screened from 22 to 20 feet bgs. Filter pack sand was placed in the annular space and extended one foot above the AS well screen. Two feet of bentonite plug and five feet of grout were used to seal the borehole above the AS well. The two-inch diameter SVE well was installed in the borehole above the grout and included six feet of screen from 12 feet bgs to six feet bgs. Filter pack sand was emplaced in the annular space between and extending one foot above the SVE well screen and the borehole was sealed at the surface with two-feet of bentonite plug and three feet of grout. Six monitoring points were installed around the well in a radial pattern at distances ranging between five and 25 feet.

During the SVE pilot test, air was extracted from the SVE-AS extraction well at a rate of approximately 10 cubic feet per minute (cfm). Vacuum pressures were measured at the monitoring points during the test. The concentration of VOCs in the off-gas from the extraction point was monitored using a photo-ionization detector (PID). The air sparging test was conducted following completion of the SVE pilot test. During the test, air was injected into the SVE-AS well air sparging point at a rate of approximately 10 cfm and a pressure of approximately three pounds per square inch (psi). Volatile organic vapor and dissolved oxygen (DO) concentrations were measured in each monitoring point before, during, and after the air sparging pilot test using a PID and an in-well DO meter, respectively.

The results of the SVE test indicated that vacuum influence was observed in monitoring points a distance of 25 feet or greater from the extraction well at an extraction rate of 10 cfm. Using a vacuum measurement of 0.1 inches of water as an indicator of the actual radius of influence (ROI) (USEPA, 1995), the ROI of the extraction well was approximately 34 feet. Volatile organic vapor concentrations measured in the extraction well off-gas prior to treatment ranged from 84 parts per million (ppm) to 104 ppm. These data indicated that the SVE pilot well was effective at removing petroleum mass from the unsaturated subsurface during the test.

The air sparging test results showed that volatile organic vapor and dissolved oxygen concentrations in the monitoring points increased significantly during the air sparging test. Volatile organic vapor concentrations increased from less than one ppm prior to testing to greater than 100 ppm in all monitoring points during the sparging test. This indicated that the sparging system was effectively removing petroleum mass from the saturated zone at least 25 feet from the sparge point. Additionally, dissolved oxygen concentrations increased by a minimum of three milligrams per liter (mg/l) in all monitoring points within two hours of the start of the test. With the exception of one monitoring point, dissolved oxygen concentrations decreased to their pre-test levels within one day after the end of the test. The remaining monitoring point decreased to the pre-test level after approximately two days.

Based on the data collected during the pilot testing, it was determined that the SVE-AS remedial approach would actively treat the petroleum compounds in the subsurface through volatilization and biological degradation. The actual ROI of the pilot vapor extraction well was approximately 34 feet. Effective extraction well spacing was evaluated using an equation developed by the Naval Facilities Environmental Service Center for similar projects:

$$L=2rcos(30)$$

where "L" equals the distance between extraction wells and "r" equals the radius of influence. Using this equation, the maximum distance between extraction wells was calculated to be no greater than 58 feet.

2.2   System Layout and Design

To treat the full extent of petroleum-affected soil and groundwater in the area of the FLFA, two treatment systems, each north and south of the main line tracks, were installed. Each system includes a treatment building, containing SVE and AS blowers, as well as system controls and activated carbon treatment canisters for SVE off-gas, and eight SVE-AS wells that were constructed as previously described. The SVE-AS wells were installed to 22 feet bgs or to the top of a layer of silt underlying the western portion of the site. The SVE-AS wells were installed parallel to the main line tracks in the petroleum-affected area. To account for variability in subsurface conditions and system air flow rates, the SVE-AS system well spacing for the site was set at 40 feet.

In addition to the eight SVE-AS wells, the south system operates 20 biosparge (BS) wells. The line of biosparge wells were placed along the down gradient site property boundary creating a positive dissolved oxygen barrier to migrating petroleum compounds. Similar to the AS wells, the BS wells were installed with a 2-foot screen to the top of the silt layer or a depth of 22 feet bgs, whichever was shallower. Biosparge wells were constructed of 1.5-inch diameter PVC screen and riser and were aligned linearly along the site's southern property boundary between the FLFA and down gradient adjacent properties. The spacing between the biosparge wells is twenty feet.

Main manifold lines extend east and west from the treatment buildings and carry injected and extracted air to and from the SVE/AS well points. Soil vapor extraction, AS, and BS wells were connected to the main manifolds by 1.5-inch schedule 40 PVC laterals. Each lateral is equipped with a sampling port and an in-line 1.5-inch ball valve, allowing for the adjustment of air flow to or from each individual well. The sampling port consists of a 0.5-inch PVC ball valve attached to the 1.5-inch PVC lateral by a reducing tee. By opening the ball valve sampling port, air flow or vapor concentrations can be measured for the individual well. Air flows are measured by inserting the probe of a TSI Model 8346 Velocicalc® air velocity meter into the 1.5-inch lateral via the sample port. Vapor concentrations are measured by inserting Teflon® tubing through the sample port into the lateral, extracting vapor with a Gilian® GilAir-3 Personal Sampling pump and measuring volatile concentrations in the extracted vapor with a Mini-Rae® PID.

The designed extraction rate for each SVE well is 30 cubic feet per minute (cfm). The designed flow rate for each air sparging well is 10 cfm while each biosparging well receives between one and two cfm.

System operation monitoring has been performed on an approximate monthly basis since the system was started in May 2002. System monitoring includes vapor and flow measurements at individual extraction points, flow measurements at sparging points, and LNAPL and water level measurements at SVE and selected monitoring wells. Additionally, operating parameters are recorded from instrumentation inside the treatment building. Groundwater and soil samples are collected at the site on a semi-annual basis.

3.   TREATMENT RESULTS

3.1   Soil

Soil samples are collected for VOC and SVOC analyses semi-annually from four locations at the site, two from the area of influence of the north system, and two from the area of influence of the south system. The samples are collected from depths identified as containing the greatest evidence of petroleum-impacts as noted during the SSI. Concentrations of VOCs in samples collected from the north system's area of influence have declined from approximately 22,000 micrograms per kilogram (μg/kg) in October 2002 to non-detectable concentrations in both sampling locations in June 2004. Concentrations of VOCs in samples collected from the south system's area of influence have declined from 41,000 μg/kg and 23,000 μg/kg in October 2002, to non-detectable concentrations and 90 μg/kg, respectively in June 2004. No VOCs were detected at concentrations exceeding regulatory standards. Total SVOC concentrations in samples collected from the north system's area of influence have declined 99 percent and 14 percent while samples collected from the south system's area of influence have declined 98 percent and 95 percent. The smaller decrease of 14 percent is associated with a sampling location that initially contained an order of magnitude lower concentrations of SVOCs.

3.2   Groundwater

Groundwater samples collected from wells within and down gradient of the area of influence of the treatment systems have contained decreasing concentrations of VOCs and SVOCs since the initiation of the remedial program. Total detectable VOC concentrations have decreased between 100 percent and 19 percent and the average decline for wells within and down gradient of the treatment systems is 84 percent. Four wells, which prior to initiation of the treatment program contained groundwater with concentrations of one or more VOC analytes in excess of regulatory groundwater standards, contained no VOCs in excess of those standards. Total SVOC concentrations have decreased by between 100 percent and 52 percent in samples collected from wells within and down gradient of the treatment area. The average SVOC reduction in these wells is 84 percent. Two wells, which prior to initiation of the treatment program contained groundwater with concentrations of one or more SVOC analytes in excess of regulatory groundwater standards, contained no SVOCs in excess of those standards.

3.3   Soil Vapor Extraction System Performance

Estimates of mass removal are based on VOC concentrations and vapor extraction rates measured at the sampling ports on each SVE well lateral. As of July 2004, it is conservatively estimated that approximately 1,000 pounds (approximately 150 gallons) of petroleum mass have been removed by the SVE systems, approximately 64 percent was removed by the north system.

Figure 2 shows cumulative mass removal from system start-up through July 2004. As shown, approximately 70 percent of the petroleum mass was removed during the initial seven months of system operation, from May to November 2002. The highest mass removal rates were measured from July to September in both 2002 and 2003.

3.4   Microbial Assessment

Soil samples were collected for microbial analysis from the saturated and unsaturated zone in locations up gradient, within, and down gradient of the treatment systems. Following start-up of the treatment systems in May 2002, the populations of diesel-degrading bacteria in the saturated zone increased by as much as two orders of magnitude by October 2002. Diesel-degrading bacteria were less responsive to system start-up in the unsaturated zone down-gradient of the treatment system; however, in the treatment area, the population of diesel-degrading bacteria in the unsaturated zone increased by over three orders of magnitude by October 2002. From October 2002 to June 2004, bacteria populations have generally declined or remained at previous levels; however, they are still present in appreciable numbers and at numbers in excess of pre-treatment estimates.

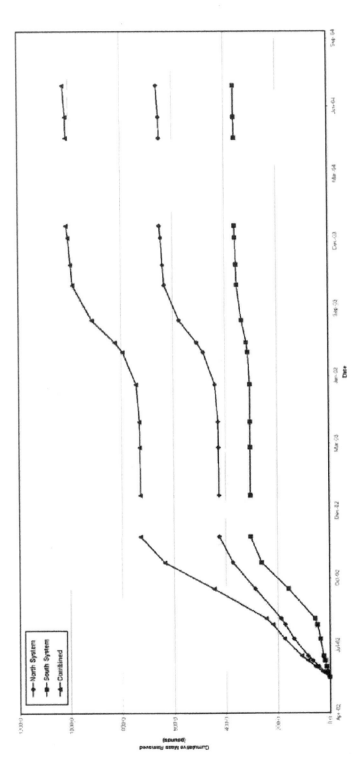

Figure 2. Cumulative Mass Removal

4. DISCUSSION AND CONCLUSIONS

The use of *in situ* air sparging and soil vapor extraction allowed for the successful remediation of petroleum-impacted soil at the site with no disruption of service on main line railroad tracks through the middle of the treatment area. The multi-faceted approach to treatment of the soil has decreased petroleum mass at the site by both volatilization and capture with activated carbon and microbial degradation. Concentrations of regulated VOC and SVOC compounds in soil and groundwater have been reduced to levels below applicable New York State standards in almost all monitored portions of the site.

Cost analysis for site remediation was performed following completion of the subsurface site investigation. The estimated cost for excavation and off-site disposal of petroleum-affected soil was approximately $3.7 million, which does not include costs associated with the temporary relocation of active railroad track. To date, approximately $235,000 have been spent on the SVE-AS remedial program, $65,000 of which was spent on the treatment buildings which may be reused at other sites. It is estimated that the remedial program will cost less than $350,000 to complete.

The efficiency of the treatment system was observed to be strongly correlated with subsurface soil and groundwater temperatures. Historic regional seasonal groundwater temperatures recorded during a study in the Albany, New York area, which fluctuate by as much as eight degrees Celsius, are shown on Figure 3 (Heath, 1964). Seasonal fluctuations in mass removal rates for the treatment systems are shown on the figure as well. As shown on the figure, mass removal rates and groundwater temperatures increase and decrease through the season in an almost identical pattern. Furthermore, according to the USEPA (1994), microbial growth rate is a function of temperature. When groundwater temperatures are within the range of 10 to 45 degrees Celsius, microbial activity typically doubles for every 10 degree Celsius rise in temperature. Microbial activity decreases substantially at temperatures less than 10 degrees Celsius and almost ceases completely at temperatures less than 5 degrees Celsius. Therefore, providing oxygen to diesel degrading microorganisms during winter months does not facilitate enhanced biodegradation. These data, and the support of the regulatory agency, allowed the remedial system to be shut down over the months when subsurface temperatures did not support mass removal, typically from December to March. As a result, operation and maintenance costs for the site were reduced with negligible effects on petroleum mass removal and overall site remediation.

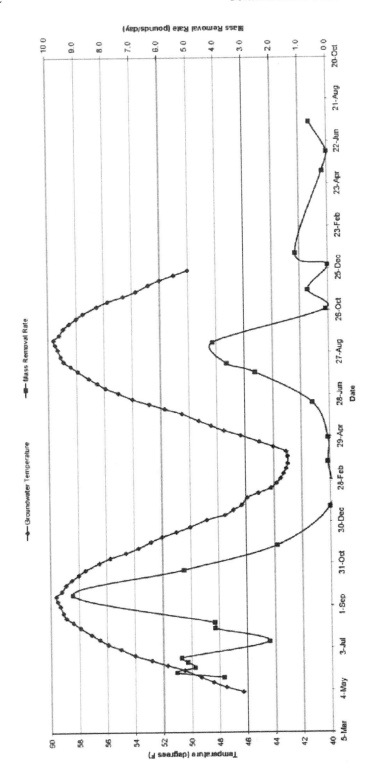

Figure 3. Seasonal Fluctuations in Groundwater Temperature and Mass Removal Rates

The dynamics of the diesel-degrading microbial populations generally conform to the anticipated growth curve for microorganisms. When microorganisms in a limiting environment are supplied with the limiting factor, microbial populations respond by growing exponentially for a period of time until another limiting factor is encountered. At this site, it was likely that an inadequate supply of oxygen limited microbial metabolism of diesel-compounds prior to treatment. Oxygen supplied by the treatment system allowed for the growth of diesel-degrading microbial populations, as observed in population estimates in samples collected within the treatment area. The current decreasing populations from samples within and down gradient from the treatment area likely result from a reduction in diesel compounds available for microbial metabolism, since other nutrients are present at sufficient concentrations and ratios. Regardless of diesel degrading bacteria population size, the aerobic environment allows for the more efficient metabolism of diesel compounds as compared to anaerobic metabolism.

Another indication that microbial activity has slowed within and down gradient of the treatment area is that dissolved oxygen concentrations in the groundwater have increased considerably from June 2003 to June 2004, indicating that the treatment system is introducing more oxygen than can be used by microorganisms. This trend is likely the result of the decreasing amount of petroleum compounds available for metabolism, which would decrease the oxygen requirements of the microbial community.

As petroleum compounds are removed from the site, it is expected that the efficiency and effectiveness of the soil vapor and air sparging systems will decline. Based on the concentrations of regulated VOCs and SVOCs in soil and groundwater at the site, it is anticipated that closure of the site by the state regulatory agency will be obtained during 2005. While current data indicates significant reduction in petroleum compounds at the site and that regulated VOC and SVOC compounds have been removed to acceptable concentrations, continued monitoring at the site after shut-down of the remedial system will determine the overall success of the remedial strategy.

REFERENCES

Caldwell, D.H. and R.J. Dineen. 1987. Surficial Geological Map of New York, Hudson-Mohawk Sheet, New York State Museum -Geological Survey, Map and Chart Series No. 40, Scale 1:250,000.

Heath, R.C., 1964. Seasonal Temperature Fluctuations in Surficial Sand Near Albany, New York. Article 168 in United States Geological Survey Short Papers in Geology and Hydrology, USGS Professional Paper 475 -D, pp. D204-D208

Naval Facilities Engineering Service Center (NFESC), 1998. Technical Memorandum TM - 2301-ENV: Application Guide for Bioslurping Volume II – Principles and Practices of Bioslurping.

United States Environmental Protection Agency (USEPA) – Office of Solid Waste and Emergency Response, 1995. How to Evaluate Alternative Cleanup Technologies for Underground Storage Tank Site – A Guide for Corrective Action Plan Reviewers.

CHAPTER 16

PHYTO-EXTRACTION OF FIELD-WEATHERED DDE BY SUBSPECIES OF *CUCURBITA* AND EXUDATION OF CITRIC ACID FROM ROOTS

Martin P.N. Gent*[1], Zakia D. Parrish[2], and Jason C. White[2]

[1]*Connecticut Agricultural Experiment Station (CAES), Dept of For estry and Horticulture, 123 Huntington Street, New Haven, CT 06504,* [2]*Department of Soil and Water, Connecticut Agricultural Experiment Station, POB 1106, New Haven CT 06504 *Corresponding author - Tel:203-974-8489, Fax:203-974-8502. Email:Martin.Gent@po.st ate.ct.us*

Abstract: Two subspecies of *Cucurbita pepo* L. (summer squash) differ in phyto-extraction of weathered DDE when grown in the field. Three cu ltivars were selected from each of the two subspecies; *Cucurbita pepo* ssp *pepo* (zucchini) with a greater ability to take up DDE, and *Cucurbita pepo* ssp *ovifera* (summer squash) with a lesser ability to take up DDE. When grown in the field, ssp *pepo* extracted 0.4 to 1.0 milligrams of DDE per plant, while ssp *ovifera* removed from the soil only 0.02 to 0.1 milligram per plant. These cultivars were grown in hydroponics to evaluate whether exudation of organic acids from the roots was involved in uptake of weathered DDE. Phosphorus nutrition played a significant role in exudation of organic acids into the hydroponics solution. For both subspecies, the bette r the phosphorus nutrition, the more tartaric and less citric acid was exuded. Subspecies *pepo* showed a greater increase in citric acid exuded under phosphorus depletion than *ovifera*. However, when solutions of root exudates w ere used to batch extract DDE from soil, effect of subspecies was opposite to that seen in the field. This suggests there are factors other than disruption of the soil matrix that play a role in phyto-extraction by plants. Nevertheless, among subspecies of *Cucurbita pepo*, the exudation of citric acid was related to phyto -extraction of more weathered organic contaminants in soil.

Key words: organic acid ; phosphorus depletion ; zucchini ; summer squash ; genotype.

1.   INTRODUCTION

Phyto-remediation is a process by which vegetation removes inorganic and organic contaminants from soil. Phyto-remediation is simple and cheap. However it is effective only if the combination of plant species and agronomic inputs result in significant removal of pollutant from the soil. This choice requires an understanding of the mechanism underlying phyto-remediation.

One mechanism by which plants remediate contaminated soil is by direct uptake into vegetation. This process of *phyto-extraction* can be viewed as a biological pump-and-treat system. Water flow through soil to the plant moves some contaminants from the soil and delivers them to the root surface (Matso, 1995; Schnoor, 2002). Because plants are able to take up and translocate to the shoot many cations that are similar in behavior to plant nutrients, phyto-remediation of soil through phyto-extraction has been reported for many heavy metals (Crews and Davies, 1985; Lombi et al., 2001; Weis and Wies, 2004). Phyto-extraction can also be used to remove organic contaminants from soil (Cunningham et al., 1996; Schnoor et al., 1995). To a large extent, the physical properties of organic compounds will determine whether plants can take up such contaminants. It is assumed that only moderately hydrophilic pollutants (log K_{ow} 0.5-3.0 for octanol-water partition coefficient) will move with water in the soil and in plants. These include pesticides such as alachlor and atrazine (Patterson and Schnoor, 1992), chlorinated solvents (Anderson and Walton, 1995), and nitroaromatic explosives (Rivera et al., 1998).

Persistent organic pollutants (POPs) are a group of compounds that include chemicals of environmental concern such as PCBs, dioxins and furans, and chlorinated pesticides such as chlordane, aldrin, DDT, and its degradation product DDE (Chaudry et al., 2002; Cunningham et al., 1996). These pollutants are of special concern as they accumulate in the tissues of animals, particularly those such as humans that are at the apex of the food chain. There is irreversible binding or sequestration in soil associated with weathering or aging of POPs, which leads to low bio-availability, and poor recovery with *in situ* remediation technologies such as pump and treat, air sparging, and bioremediation (Matso, 1995; Palmer and Nyer, 1997). These very hydrophobic compounds (log K_{ow} >6.0) are assumed to be too strongly bound to soil particles to be removed by direct uptake. However, we have found that certain plant species, including *Cucurbita pepo*, can phyto-extract very hydrophobic compounds such as DDE and chlordane (Mattina et al., 2002; White, 2002). Zucchini also accumulates dioxins via soil-to-plant uptake (Hülster et al., 1994). Given that most plant species cannot take up highly weathered POPs from soil, we hypothesize that the phyto-extraction of these contaminants by *C. pepo* is the result of root exudation of compounds that disrupt the soil matrix and facilitate release of sequestered organic pollutants (White et al., 2003a).

Low molecular weight organic acids, citric acid in particular, have been implicated in disruption of soil structure. *Lupinus albus* produces specialized root clusters known as proteoid roots which under phosphorus-depleted conditions release a large quantity of citric acid and markedly affect soil structure (Dinkelaker et al., 1989; Gerke et al., 1994). We believe that organic acids chelate polyvalent cations in soil, such as Fe, Al, Mg, Ca, and Mn, which disrupt the soil matrix, and promote the release of POPs that are sequestered among soil particles. Adding organic acids to contaminated soil in which plants are growing results in greater quantities of DDE taken up by the vegetation (White and Kottler, 2002; White et al., 2003a). Abiotic batch elution of soil with solutions of organic acids or other chelates, releases organic carbon and organic contaminants bound to mineral particles in soil (Yang et al., 2001). Thus, organic acid exudation from roots should bind cations in soil, and release organic compounds bound to the soil matrix.

Field studies show that subspecies of *Cucurbita pepo* differ in phyto-extraction of POPs. Among domesticated *Cucurbita pepo*, the subspecies *pepo* includes the "true" zucchini and many pumpkins, while ssp. *ovifera* includes summer squash known as straight necks, crooknecks, patty pans, and winter squash (Decker, 1988; Paris, 2001). A comparison among 21 cultivars within these two subspecies showed that uptake of DDE was correlated to the uptake of various nutrients and heavy metals available in trace quantities in the same soil (White et al., 2003b). These subspecies appear to differ in ability to forage for nutrients in soil, and perhaps in their ability to exude organic acids to disrupt soil structure and chelate metals. Other plant species exude a variety of organic acids in response to environment perturbation (Bhattacharyya et al., 2003; Luo et al., 1999; Ohwaki and Hirata, 1992; Rengel, 1997), and differences in organic acid exudation have been noted among genotypes within plant species (Gahoonia et al., 2000; Ishikawa et al., 2000; Keller and Romer, 2001).

We examined three cultivars in each of the two subspecies of *Cucurbita pepo*, namely ssp *ovifera* and *pepo* that appeared to differ in uptake of DDE from soil. Our objectives were to relate the uptake of DDE from plants grown in soil in the field to the rate of exudation of organic acid when the same genotypes were grown in hydroponics under phosphorus-sufficient or phosphorus-depleted conditions.

2.   MATERIALS AND METHODS

Six cultivars of summer squash (*Cucurbita pepo* L.) were used in these experiments. Early Prolific, Hybrid Crescent, and Zephyr belong to the subspecies *ovifera*, while Black Beauty, Gold Rush, and Raven belong to the subspecies *pepo*. Seeds sources were: Gold Rush, Raven, Zephyr (Johnny's Selected Seeds, Albion, ME); Black Beauty (Seedway, Hall, NY); Early Prolific, Hybrid Crescent (Gurney's Seed Co., Yankton, SD).

2.1   Field Plot

These cultivars were grown at Lockwood Farm (Hamden, CT, Lat. 42 N Long. 73 W 50m ESL) in a fine sandy loam (56% sand, 36% silt, 8% clay; pH 6.7, 1.4% organic carbon). Analysis after the first growing season indicated the soil contained the following concentrations of nutrients in mg kg^{-1} dry soil: K 1032, Ca 1144, Mg 2718, P 755, Fe 9585, Mn 312, Zn 47, and Cu 10. No fertilizer was added to the soil in the two years of this experiment. The plot was covered with black polyethylene mulch. Seeds were germinated for 3 days at room temperature before hills of several seedlings were planted into 0.1-m^2 holes cut at 2-m intervals in the mulch. Hills were later thinned to 3 plants. The plot was weeded and watered as necessary. Fruit were harvested and weighed when they reached a size of at least 200 g fresh weight. Seedlings were planted in early June, fruit was harvested beginning in early July, and plants were destructively harvested in early August to obtain tissue samples from vegetation. (White et al., 2003b)

2.2   Analysis of Tissue

The fresh weight of fruit was summed over the period of harvest. The fresh weights of roots, stems, and leaves were measured for each hill at the destructive harvest, and sub-samples of each tissue were saved for chemical analysis. Vegetation was thoroughly rinsed with tap water to remove soil particulates, finely chopped, and stored in a freezer. To determine the dry matter content of each plant part, sub-samples were dried at 70°C until constant weight was achieved. DDE (2,2-bis(p-chlorophenyl)-1,1-dichloroethylene) was extracted from fresh-frozen tissue using organic solvent and quantified by gas chromatography with trans-nonachlor as an internal standard (White et al., 2003b). All tissue concentrations were calculated on a dry weight basis, in micrograms per kg dry weight of plant tissue. Soil samples were also taken from each hill of plants to determine any variation in DDE concentration across the field plot. These samples were extracted and analyzed for DDE as described previously (White et al., 2003b).

2.3 Hydroponics Studies

A series of three hydroponics experiments was conducted to compare root exudates of these six cultivars. Seeds were germinated on filter paper, then transferred to a complete nutrient solution with adequate phosphorus and grown until the seedlings had three true leaves. The seedlings were transferred to a flowing solution without phosphorus, and allowed to grow for one week to induce phosphorus deficiency. All other nutrients were maintained at an adequate concentration. Nutrient solutions were as described in (Gent et al., 2005). After the plants ceased growing due to phosphorous depletion, phosphorus was re-supplied to some plants for three to five days. This withdrawal and re-supply procedure minimized differences in size and vigor of plants at the time of collection of root exudates. All experiments commenced in a growth room maintained at 25°C with a 12-hr photoperiod of 200 μmol m^{-2} s^{-1} photosynthetic flux density. Experiments in July and August were completed in this condition. In June, the plants were transferred to a greenhouse and grown under ambient light during the collection of exudates.

2.4 Collection and Analysis of Root Exudates

Collection of root exudates started in the middle of the photoperiod three to five days after phosphorus replenishment of half the plants, and after disinfection of the roots. Roots were immersed for one hour in a solution of streptomycin, 50 μg L^{-1}, and chloramphenicol, 25 μg L^{-1}, in 0.5 mM CaCl$_2$, and then immersed for another hour in 0.5 mM CaCl$_2$ without antibiotic. The flasks containing the plant roots were filled with fresh 0.5 mM CaCl$_2$. This solution was poured off and replaced after 24 hours. In order to concentrate the organic acids, the solution containing exudates was poured into 250-mL plastic screw-cap centrifuge tubes and agitated with two 5.0 x 0.8 cm pieces of anion exchange membrane in the NaHCO$_3$ form (Excellion I-200, Electropure Corp. Laguna Hills CA). A second batch of solution containing exudates was agitated with the anion exchange membranes after another 24-hour collection period. The anion exchange membranes were transferred to small screw-cap vials to which 3 mL of hydrochloric acid was added. This was agitated for two hours to release the organic acids. This method of concentration and extraction of organic acids was tested with known concentrations of organic acids standards to determine the efficiency of retention on the anion exchange membranes and any background due to the membranes.

Organic acids were detected by liquid chromatography on an Agilent model 1100 LC (Agilent Corp, Palo Alto CA). A 10 to 50 uL sub-sample was injected onto an anion-exclusion column (Supelcogel 610H, 7.4 x 300 mm, Supelco, Belefonte PA) operated at 40°C and eluted with 0.1% v/v H_3PO_4 at 1.0 mL min^{-1}. Organic acids were detected by UV absorption at 210 nm and quantified by comparison to standards prepared from the pure substances (Sigma-Aldrich, St. Louis, MO).

2.5   Batch Extraction of POPS with Root Exudates Solutions

Samples from the upper 15 cm of soil in a plot contaminated with DDE and chlordane was mixed by hand and passed through a number 10 sieve. Five replicate 10-g soil samples were weighed into 60-mL vials with 100 uL of 10 ppm pentachloronitrobenzene as an internal standard. Root exudates solution was collected as described above, except it was not concentrated using anion exchange membranes. A 60-mL subsample of root exudates was added to each vial, which were sealed and agitated at room temperature for two days. The vials were centrifuged at 1000 rpm for 12 min, and the supernatant was collected. A C-18 (octadecyl) disk was added to the supernatant and these solutions were shaken for three days. The C-18 disk was transferred to an 8-mL vial containing anhydrous sodium sulfate, 4 mL of hexanes was added, and the vial was heated at 70°C for two hours. A sub-sample was analyzed for DDE as described previously (White et al., 2003b). A chemical extraction determined the amount of DDE in soil. Six replicate 3-g soil samples were weighed into 35-mL amber vials, to which 15-mL of hexanes and 100-uL of 10 ppm internal standard was added. These vials were heated to 70°C for 5 hr. A 1-mL sub-sample was filtered and analyzed for DDE.

2.6   Experimental Design and Statistical Analysis

There were duplicate hills for each cultivar, placed randomly in the field plots. A different randomization was use in 2002 and 2003. Replicate measurements of DDE in soil or plant tissue consisted of repeated sub-samples of a single combined tissue sample for each cultivar. Year and sub-species were used as main effects in analysis of variance.

Three hydroponics experiments were combined for analysis. Within each experiment there were three replicate plants for each cultivar and treatment condition. Plants were placed randomly in blocks corresponding to plus and minus phosphorus treatments. Experiment and subspecies were the main factors. Phosphorus depletion or re-supply was considered a treatment factor within each experiment, as the days of re-supply, and reversal of phosphorus depletion varied between experiments.

3. RESULTS

*Table 1.*Shoot biomass, concentration and amount of DDE in six cultivars of two subspecies of *Cucurbita pepo* L. when grown in field soil in 2002 and 2003.

Year Cultivar	Sub-species	Biomass	Whole plant tissue			Phyto-extracted
			Specific	Total	Soil	
		g	µg kg^{-1} dwt	µg plant^{-1}	µg kg^{-1}	%
2002						
Early prolific		319	79	25	352	0.025
Hybrid crescent	*ovifera*	379	62	23	382	0.023
Zephyr		855	83	70	363	0.070
Average		**518**	**75**	**39**	**366**	**0.039**
Black beauty		496	404	176	856	0.176
Gold rush	*pepo*	1158	917	1063	515	1.063
Raven		761	579	426	356	0.426
Average		**805**	**633**	**555**	**576**	**0.555**
2003						
Early prolific		753	95	71	251	0.071
Hybrid crescent	*ovifera*	647	143	93	136	0.093
Zephyr		233	101	24	317	0.024
Average		**545**	**113**	**63**	**235**	**0.063**
Black beauty		1406	264	371	131	0.371
Gold rush	*pepo*	1022	533	544	667	0.544
Raven		589	787	464	280	0.464
Average		**1006**	**528**	**460**	**359**	**0.460**
Significance						
	Year	ns	ns	ns	ns	
	Subspecies	0.001	<0.001	<0.001	<0.001	
	Interaction	ns	ns	ns	ns	

3.1 DDE in Field-grown Plants

The *ovifera* and *pepo* subspecies of *Cucurbita pepo* differed in the tissue concentration of DDE, when averaged over the whole plant after a period of fruit production (Table 1). The concentration of DDE in plant tissues was eight and five times greater in ssp *pepo* than ssp *ovifera*, in 2002 and 2003, respectively. In part these differences may be due to variation in DDE in the soil. DDE concentrations in soil were assayed for each hill of plants in each year. On average, hills planted to ssp *pepo* had about 50% more DDE in soil, in both years, than did hills planted with ssp *ovifera*. However, the specific concentration of DDE in plant tissue varied far more between subspecies than did the respective concentrations in soil. Even when soil DDE concentration was used as a covariate in analysis, the difference between cultivars in tissue specific DDE was very significant, $P < 0.001$. Above-ground biomass, including dry weight of the picked fruit, was also greater for ssp *pepo* than for *ovifera*. Thus, when stated on a total uptake basis, in micrograms of DDE per plant, cultivars of ssp *pepo* extracted 18 and seven times more DDE from the soil than did *ovifera*, in 2002 and 2003, respectively.

We estimated mass of the soil compartment for each plant or hill in the field was about 285 kg (White, 2002). Given the concentration of contaminant for each species, subspecies *pepo* was able to phyto-extract 0.34 and 0.45 percent of the weathered DDE in the soil, in 2002 and 2003, respectively, while ssp *ovifera* extracted only 0.04 and 0.09 percent. This uptake of DDE into plant tissues implied a mechanism for release of these sequestered hydrophobic compounds from the soil structure. The greater uptake of DDE by ssp *pepo* when grown in the field may be due in part to greater ability to release the weathered DDE from the soil matrix. Organic acids exuded from roots may perturb soil structure and release chemicals bound to the soil. These subspecies were grown in hydroponics, in order to collect and quantify the exudation of organic acids.

3.2   Exudation in Hydroponics

Table 2. Exudation of organic acids into hydroponics solution by cultivars and subspecies of *Cucurbita pepo* L. in August.

Cultivar	Treatment Subspecies	Exudation rate, μmol day^{-1} kg^{-1} plant dry weight				
		Oxalic	Citric	Tartaric	Malic	Succinic
Early prolific	Minus P	nd	4.0	0.2	0.2	3.2
Hybrid crescent	*ovifera*	nd	4.3	0.0	0.0	1.5
Zephyr		nd	2.9	0.2	0.0	2.4
Black beauty	Minus P	nd	7.0	0.0	0.0	0.7
Gold rush	*pepo*	nd	11.9	0.0	1.8	3.9
Raven		nd	3.1	0.2	0.0	1.6
Early prolific	Plus P	17.8	5.1	2.6	0.1	4.6
Hybrid crescent	*ovifera*	19.1	1.6	6.7	0.0	2.9

Cultivar	Treatment Subspecies	Exudation rate, μmol day^{-1} kg^{-1} plant dry weight				
		Oxalic	Citric	Tartaric	Malic	Succinic
Zephyr		15.7	5.2	0.8	1.0	3.3
Black beauty	Plus P	13.0	2.2	0.8	0.0	3.7
Gold rush	*pepo*	15.6	4.3	2.3	3.0	0.0
Raven		15.8	1.7	2.4	0.3	1.0
Significance						
	Subspecies	ns	ns	ns	ns	0.079
	Treatment		0.015	0.003	ns	ns
	Interaction		0.044	ns	ns	ns

nd-not detected due to interference with other peaks in chromatogram

When plants were grown in hydroponics, oxalic, citric, tartaric, pyruvic, malic, succinic, formic, lactic, and acetic acids were detected in solutions of root exudates. The amount of pyruvic acid was an order of magnitude less than the other acids detected (data not shown). Only the concentrations of di- and tri-carboxylic acids are shown in the tables, as they chelate metals far more effectively than mono-carboxylic acids.

The rates of exudation of organic acids from roots was calculated and normalized to plant biomass. Results for each cultivar are shown for the August trial in Table 2. Averaged over trials and cultivars within subspecies, several of the organic acids that were detected in root exudates varied with phosphorus treatment (Table 3). Exudation of citric and succinic acids increased under phosphorus depletion, while that of tartaric acid declined. Citric acid was unique in that the change in exudation in response phosphorus deficiency differed among subspecies. Subspecies *pepo* exuded more citric acid than *ovifera* when under phosphorus deficiency (Table 3). There was no difference for plants re-supplied with phosphorus, except in August when *pepo* exuded less citric acid. Thus, ssp *pepo* appears to have a greater ability to exude citric acid from roots under phosphorus deficiency. An interesting observation is that *Cucurbita pepo* Gold Rush exuded more citric acid than any other cultivar under phosphorus deficiency (Table 2). Gold Rush was also most effective at extracting weathered DDE from the soil (Table 1).

Table 3. Exudation of organic acids into hydroponics solution by subspecies of *Cucurbita pepo* as affected by subspecies and phosphorus depletion treatment averaged over three trials.

Subspecies	Treatment	Exudation rate, μmol day^{-1} kg^{-1} plant dry weight				
		Oxalic	Citric	Tartaric	Malic	Succinic
Ovifera	Minus P	10.1	5.6	0.1	1.9	7.6
pepo	Minus P	12.7	7.6	0.1	2.4	7.7
Ovifera	Plus P	13.4	4.4	2.9	2.3	6.8

| Subspecies | Treatment | Exudation rate, μmol day^{-1} kg^{-1} plant dry weight | | | | |
		Oxalic	Citric	Tartaric	Malic	Succinic
pepo	Plus P	11.8	4.2	4.1	2.0	2.9
Significance						
	Subspecies	ns	ns	ns	ns	ns
	Treatment	ns	0.001	0.001	ns	0.076
	Interaction	ns	0.041	ns	ns	ns

3.3   Batch Extraction of POPS with Root Exudates Solutions

Batch extraction of DDE from soil by solutions of root exudates resulted in removal of about one half the DDE available by chemical extraction, perhaps due to profound disturbance of the soil structure. The solutions containing root exudates extracted more DDE than did control solution without exudates, but the controls extracted nearly half the contaminant. The data for Zephyr were not included as collection of exudates from this cultivar was delayed, and could not be compared to the other cultivars. Although exudates from ssp *pepo* and *ovifera* differed in elution of DDE from soil (Table 4), exudates of *ovifera* released more DDE from soil than did those of *pepo*, 697 compared to 601 ug kg^{-1} soil, respectively, averaged over treatments. This contrasts with extraction of metals from the same soil by the same root exudates solutions. Subspecies *pepo* was better than subspecies *ovifera* in extracting most nutrients (Gent *et al.*, 2005) and the plus P treatment resulted in extraction of more nutrient than the minus P treatment, perhaps because the former were larger plants. It should be noted that the method we used to collect exudates resulted in solution concentrations of organic acids that were very low, on the order of 10^{-7} M. Other studies required additions of 10^{-2} to 10^{-1} M of organic acids to release of organic contaminants from soil (White et al., 2002; Yang et al., 2001).

Table 4. Batch elution of DDE from field soil by solutions of root exudates of *Cucurbita pepo* L. grown in hydroponics.

Cultivar	Treatment Subspecies	DDE μg kg^{-1} soil
Early prolific	Minus P	650
Hybrid crescent	*ovifera*	685
Black beauty	Minus P	668
Gold rush	*pepo*	500
Raven		614

Cultivar	Treatment Subspecies	DDE µg kg^{-1} soil
Early prolific	Plus P	684
Hybrid crescent	*ovifera*	769
Black beauty	Plus P	601
Gold rush	*pepo*	612
Raven		609
Control	Minus P	507
Control	Plus P	564
Chemical extraction		1457
Significance		
	Subspecies	0.001
	Treatment	0.176
	Interaction	0.390

4. DISCUSSION

We hypothesized that the ability of *Cucurbita pepo* ssp. *pepo* to phyto-extract significant quantities of weathered POPs may be due to exudation of organic acids from roots. Under hydroponics conditions with sufficient P, these subspecies did not differ in exudation of organic acids. However, under P-deficient conditions, ssp *pepo* increased the exudation of citric acid more than did ssp *ovifera*. Thus, there was relation between the amount of citric acid exuded from roots in hydroponics under phosphorus deficiency and the amount of weathered DDE extracted from lightly-contaminated soil when the same cultivars were grown in the field. This supports the hypothesis that phyto-extraction of POPs requires organic acids to free weathered DDE from soil. Citric acid, in particular, has been implicated in disturbing soil structure (Dinkelaker et al., 1989; Gerke et al., 1994). However, this relation may be confounded by the fact that phosphorus depletion plays an important role in determining the amount of organic acid exuded. When grown in the field, these subspecies differed in phosphorus concentrations in the stem (White et al., 2003b). However, the phosphorus in the stem of field-grown plants of both subspecies was substantially less than that seen in phosphorus-sufficient plants grown in hydroponics. Thus, we expect that field-grown plants were partially phosphorus deficient and that their citric acid exudation systems were likely induced. Under this condition, the greater ability of ssp *pepo* to exude citric acid likely played a role in its greater uptake of DDE.

The relation between plant phosphorus status and citric acid exudation has been studied most thoroughly in *Lupinus*. Various subspecies differ in tolerance for P depleted soils, and also in exudation of organic acids in response to phosphorus (Egle et al., 2003). The formation of proteoid roots that release a large amount of citric acid is a well-characterized response of *Lupinus albus* to P deficiency (Gardner et al., 1982). There is a report of proteoid root formation in zucchini in hydroponics in response to Fe deficiency (Waters and Blevins, 2000). However, we did not see any formation of proteoid roots in *Cucurbita pepo* in hydroponics under P deficiency.

Exudates other than organic acids may also play a role in the phyto-extraction of weathered POPs. Surfactants present in the mucilage of maize, lupin, and wheat physically alter the rhizosphere soil structure, and increase nutrient availability to plants (Read et al., 2003). *Lupinus albus* exuded acid phosphatase that led to an appreciable depletion of organic P in the rhizosphere (Li et al., 1997). We have found ssp *pepo* released four times more peroxidase than ssp *ovifera* when grown in hydroponics or on solid media (unpublished data). Thus, release of POPs from soil by *C. pepo* may be a result of exudation of several compounds.

The results from batch extraction of soil with solutions of root exudates did not agree with the phyto-extraction seen in field grown plants. Exudates from ssp *ovifera* released more DDE from soil than did those of ssp *pepo*. These results also contrast with extraction of metals from the same soil by the same root exudates solutions, in which ssp *pepo* was better than ssp *ovifera* in extracting most nutrients (Gent et al., 2005). The hydroponics method tends to dilute the organic acids substantially, and this may explain why batch extraction of DDE did correlate with results in the field. The concentration of organic acids in the exudates solution was very low, $10^{-7}\,M$, whereas the concentration of organic acids at the root surface of plants growing in soil may be as high as $1\,M$. Abiotic batch extraction using solutions of organic acids to release metals or organic contaminants from soil only showed an effect at concentrations of 10^{-1} to $10^{-2}\,M$ (White et al., 2002; Yang et al., 2001). It is hard to explain why batch extraction released so much of the DDE from the soil. Presumably, processes other than chelating of metals in soil disrupted soil structure and the batch extraction conditions.

It should be noted that exudation of organic acids achieves only one step in the phyto-remediation process, namely disruption of soil structure and release of sequestered organic pollutants from tight binding in the soil matrix. After this step, POPs are likely to associate with organic matter in soil. Release of organic matter such as lignin or humic substances from bulk soil in water solution appears to require titration of the various chemical groups that bind to clay particles (Vadachari et al., 1994). The concentration of organic acid in solutions of root exudates was not high enough to achieve such a titration. Indeed this batch elution process did not release colored substances from the soil, a simple assay for extraction of humic substances. In this respect, elution of POPs by batch desorption using a dilute solution of root exudates was not an accurate simulation of phyto-extraction by plants. Under natural conditions, the reversal the sequestration of weathered POPs by root exudation of organic acids will make them more available for uptake at the root surface. Whether and to what extent plants do take up these organic constituents depends on other aspect of their physiology. These aspects include the permeability of the root cortex, binding to root tissues, and properties of the translocation system that enhance movement of relatively hydrophobic compounds from the root to the shoot.

ACKNOWLEDGEMENTS

This work was funded partially through EPA STAR Grant R829405. We thank Lydia T. Wagner, Craig Musante, Michael Short and Terri Arsenault for technical assistance.

REFERENCES

Anderson, T.A., and Walton, B.T. 1995. Comparative fate of 14C -trichloroethylene in the root zone of plants from a former solvent disposal site. *Environ. Toxicol. Chem.* 14, 2041-2045.

Bhattacharyya, P., Datta, S.C., and Dureja, P. 2003. Interrelationship of pH, organic acids, and phosphorus concentration in soil solution of rhizosphere and non -rhizosphere of wheat and rice crops. *Commun. Soil Science & Plant Analysis* 34, 231-245.

Chaudry, Q., Schroder, P., Werck-Reichhart, D., Grajek, W., and Marecik, R. 2002. Prospects and limitations in phytoremediation for the removal of persistent pesticides in the environment. *Environ. Sci. Poll.* Res. 9, 4-17.

Crews, H.M., and Davies, B.E. 1985. Heavy metal uptake from contaminated soil by six varieties of lettuce (Lactuca sativa L.). *J. Agric. Sci. Camb.* 105, 591-595.

Cunningham, S.D., Anderson, T.A., Schwab, A.P., and Hsu, F.C. 1996. Phytoremediation of soils contaminated with organic pollutant s. *Adv. Agronomy* 56, 55-114.

Decker, D.S. 1988. Origin, evolution, and systematics of Cucurbita pepo (Cucurbitaceae). *Econ Bot.* 42, 4-15.

Dinkelaker, W., Romheld, V., and Marschner, H. 1989. Citric acid secreti on and precipitation of calcium citrate in the rhizosphere of white lupin (Lupinus alba). *Plant Cell Envir.* 12, 285-292.

Egle, K., Romer, W., and Keller, H. 2003. Exudation of low molecular weight organic acids by Lupinus albus L., Lupinus angustifolius L. and Lupinus luteus L. as affected by phosphorus supply. *Agronomie* 23, 511-518.

Gahoonia, T., S, Asmar, F., Giese, H., Gissel-Nielsen, G., and Nielsen, N.E. 2000. Root-released organic acids and phosphorus uptake of two barley cultivars in laboratory and field experiments. *European J. Agronomy* 12, 281-289.

Gardner, W., K, Parberry, D., G, and Barber, D.A. 1982. The acquisition of phosphorus by Lupinus albus L. *Plant and Soil* 68, 33-41.

Gent, M.P.N., Parrish, Z.D., and White, J.C. 2005. Exudation of citric a cid and nutrient uptake among subspecies of Cucurbita. *J. Amer. Soc. Hort. Sci.* submitted.

Gerke, J., Römer, W., and Jungk, A. 1994. The excretion of citric and malic acid by proteoid roots of Lupinus albus L.; effects on soil solution concentrations of ph osphate, iron, and aluminum in the proteoid rhizosphere in samples of an oxisol and a luvisol. *Zeitschrift für Pflanzenernährung und Bodenkunde* 157, 289-294.

Hülster, A., Muller, J.F., and Marschner, H. 1994. Soil-plant transfer of polychlorinated dibenzo-p-dioxins and dibenzofurans to vegetables of the cucumber family (Curcurbitae). *Environ. Sci. Technol.* 28, 1110-1115.

Ishikawa, S., Wagatsuma, T., Sasaki, R., and Ofei-Manu, P. 2000. Comparison of the amount of citric and malic acids in Al media of seven p lant species and two cultivars each in five plant species. *Soil Science Plant Nutrition* 46, 751-758.

Keller, H., and Romer, W. 2001. Cu, Zn, and Cd acquisition by two spinach cultivars depending on P nutrition and root exudation. *Zeitschrift fur Pflanzener nahrung und Bodenkunde* 164, 335-342.

Li, M., Shinano, T., and Tadano, T. 1997. Distribution of exudates of Lupin roots in the rhizosphere under phosphorus deficient conditions. *Soil Science Plant Nutr.* 43, 237-245.

Lombi, E., Zhao, F.J., Dunham, S.J., and McGrath, S.P. 2001. Phytoremediation of heavy metal- contaminated soils: natural hyperaccumulation versus chemically enhanced phytoextraction. *J. Environ. Qual.* 30, 1919-1926.

Luo, H., M, Watanabe, T., Shinano, T., and Tadano, T. 1999. Comparison of alumi num tolerance and phosphate absorption between rape (Brassica napus L.) and tomato (Lycopersicon esculentum Mill.) in relation to organic acid exudation. *Soil Science and Plant Nutrition* 45, 897-907.

Matso, K. 1995. Mother nature's pum p and treat. *Civil Engineering* 65, 46-49.

Mattina, M.I., White, J.C., Eitzer, B.D., and Iannucci-Berger, W. 2002. Cycling of weathered chlordane residues in the environment: compositional and chiral profiles in contiguous soil, vegetation, and air compartm ents. *Environ. Toxicol. Chem.* 21, 281-288.

Ohwaki, Y., and Hirata, H. 1992. Differences in carbolic acid exudation among P -starved leguminous crops in relation to carboxylic acid contents in plant tissues and phospholipid levels in roots. *Soil Science Plan t Anal.* 38, 235-243.

Palmer, P.L., and Nyer, E.K. 1997. *In situ* treatments using water as the carrier. *Environmental Technology.* 7, 21-22.

Paris, H.S. 2001. History of the cultivar-groups of Cucurbita pepo. *Hortic. Reviews* 25:71-170.

Patterson, K.G., and Schnoor, J.L. 1992. Fate of alachlor and atrazine in a riparian zone field site. *Water Environ. Res.* 64, 274-283.

Read, D.B., Bengough, A.G., Gregory, P.J., Crawford, J.W., Robinson, D., Scrimgeour, C.M., Young, I.M., Zhang, K., and Zhang, X. 2003. Plant roots release phospholipid surfactants that modify the physical and chemical properties of soil. *New Phytol.* 157, 315-326.

Rengel, Z. 1997. Root exudation and microflora populations in rhizosphere of crop genoty pes differing in tolerance to micronutrient deficiency. *Plant and Soil* 196, 255-260.

Rivera, R., Medina, V.F., Larson, S.L., and McCutcheon, S.C. 1998. Phytotreatment of TNT -contaminated groundwater. *J. Soil Contam.* 7, 511-529.

Schnoor, J.L. 2002. Phytoremediation of soil and groundwater. Groundwater Remediation Technology Analysis Center Evaluation Report TE -02-01.

Schnoor, J.L., Licht, L.A., McCutcheon, S.C., Wolfe, N.L., and Carreira, L.H. 1995. Phytoremediation o f organic and nutrient contaminants. *Environ. Sci. Technol.* 29, 318A-323A.

Vadachari, C., Mondal, A.H., and Nayah, D.C. 1994. Clay -humic complexation: effect of pH and the nature of bonding. *Soil. Biol. Biochem.* 26, 1145-1150.

Waters, B.M., and Blevins, D.G. 2000. Ethylene production, cluster root formation, and localization of iron III reducing capacity in Fe deficient squash roots. *Plant and Soil* 225, 21-31.

Weis, J.S., and Wies, P. 2004. Metal uptake, transport, and release by wetland plants: implications for phytoremediation and restoration. *Environ. Intern.* 30, 685-700.

White, J.C. 2002. Differential bioavailability of field -weathered p,p'-DDE to plants of the Cucurbita and Cucumis genera. *Chemosphere* 49, 143-152.

White, J.C., and Kottler, B., D. 2002. Citrate mediated increase in uptake of weathered 2,2 -bis(p-chlorophenyl)1,1 -dichloro ethylene residues by plants. *Environ. Toxicol. Chem.* 21, 550-556.

White, J.C., Mattina, M.J.I., Lee, W.Y., Eitzer, B.D., and Iannucci -Berger, W. 2003a. Role of organic acids in enhancing the uptake of weathered DDE by plants. *Environ. Pollut.* 124, 71-80.

White, J.C., Wang, X., Gent, M.P.N., Iannucci -Berger, W., Eitzer, B.D., Schultes, N.P., Arienzo, M., and Mattina, M.J.I. 2003b. Subspecies -level variation in phytoextraction of weathered p,p' -DDE by Cucurbita pepo. *Environ. Science Technol.* 37, 4368-4373.

Yang, Y., Ratte, D., Smets, B., Pignatello, J., and Grasso, D. 2001. Mobilization of soil organic matter by complexing agents and implications for polycyclic aromatic hydrocarbon desorbtion. *Chemosphere* 43, 1013-1021.

CHAPTER 17

PHYTOREMEDIATION OF LEAD-CONTAMINATED SOIL IN THE URBAN RESIDENTIAL ENVIRONMENT USING SEED MUSTARD

Ilana S. Goldowitz[1] and Joshua Goldowitz[2]

[1]Cornell University Plant Science, 228 Plant Science Building, Ithaca, NY 14853 [2]Rochester Institute of Technology, Civil Engineering Technology Environmental Management & Safety Department. 78 Lomb Memorial Drive, Rochester, NY 14623

Abstract: Lead contamination in the urban environment is a continuing serious public health concern. Historically lead entered the urban residential area though paint pigment and gasoline additives. This legacy persists as the two most important lead sources that affect children in the urban environment: contaminated paint residue and contaminated soil. One technique for remediation of lead in urban soils is phytoremediation . Previous research has shown Brassica juncea, Indian mustard, to be a promising phytoremediator of lead in soil. Researchers commonly use broad leafed mustard such as Southern Giant Curly Leaf, because it is an accumulator of lead, has extremely high production of mustard green biomass in a short growing season, and is adaptable to poor soil conditions. The authors believe that use of broadleaf mustard in the urban environment may be problematic. The greens are enjoyed as a food, are easily recognized, and may be pilfered and eaten. This research tested the hypothesis that seed mustard , which produces abundant flowers but few greens, would be more suitable. Two 64 m^2 plots were prepared in a Greater Rochester Urban Bounty garden, located at a busy intersection in Rochester New York's low-income northeast neighborhood. One plot was sown thickly with Southern Giant, and the other with seed mustard. As the plants approached maturity the entire crop of Southern Giant mustard was pilfered, but the seed mustard remained untouched. At maturity the seed mustard produced 550g dry biomass per m^2. Subsequent testing included germination of seed mustard in lead -contaminated soil, and trials to determine maximum biomass production using seed mustard by varying planting density. Results indicate no detriment to germination rate, and a maximum biomass production capacity approaching 7 Kg/m^2

Key words: phytoremediation; lead; Brassica juncea; urban phytoremediation; residential
              phytoremediation; sowing density.

1.   INTRODUCTION

Elevated blood lead levels continue to be public health concern. Elevated
blood lead level is defined as a concentration over 10 micrograms per
deciliter (ug/dcl). It is recognized that elevated blood lead levels contribute
to decreased IQ and decreased attention span among school age children.
Recent studies have shown that concentrations less than 10ug/dcl also have
an effect on IQ, and that there may not be a lower threshold for effects
(Canfield et al., 2003). The highest incidence of elevated blood lead levels is
found in low-income areas associated with older or poorly maintained
housing stock. These conditions are most often found in the urban residential
environment. Sources of ingestible lead in the urban residential environment
include lead oxide paint pigment on interior and external painted surfaces, as
well as fallout residue from lead-based gasoline additives.

Recent evidence has implicated soil in the house drip line and runoff area
as a source of indoor lead (Lanphear et al., 1998). This study indicates that
soil tracked in on shoes can account for as much as one third of lead in
indoor dust. The source is lead leached from exterior walls by rainfall. The
current US Environmental Protection Agency standard for lead in residential
soil is 400 mg/Kg for bare soil, and 1,200 mg/Kg for vegetated soil (USEPA,
2001).

When a child is found to have elevated blood lead levels a remediation is
usually performed on the house. Current acceptable practices address only
interior surfaces, and may include encapsulation or removal. The above
mentioned research indicates that the remediation should include soil
adjacent to the dwelling. Brassica juncea has been used to phytoremediate
lead-contaminated soil (Salido et al., 2003). This paper attempts to
investigate a low cost remediation option for the house drip line of an
occupied urban residential dwelling using a mustard variety that would not
be viewed as a food.

2.   MATERIAL AND METHODS

2.1   Southern Giant Vs. Seed Mustard Mass Comparison

A side-by-side plant mass study was conducted to compare the mass
produced by southern giant leaf mustard and seed mustard. The study was

conducted at a Greater Rochester Urban Bounty garden located in an urban residential neighborhood. Two adjacent 8m by 8m plots were established within the fenced garden, tilled in two directions and raked smooth. Mustard was over sown in each plot and kept uniformly moist until germination. The plots were subsequently watered weekly to supplement rainfall and ensure abundant moisture. The plots were maintained until maturity (approximately 50 days). At this point the entire giant leaf mustard crop was pilfered. Two 1m^2 areas of seed mustard were harvested, dried in a muffle oven at 100°C for 24 hours, and weighed in a Mettler balance.

2.2   Seed Mustard Germination Trials

Standard 400-seed germination tests were performed. Nine-inch-square aluminum trays were filled to a depth of 1 cm with Miracle Gro seed starting mix. A 1 cm on-center planting grid was incised into the soil surface. Planting hole locations were gridded with a metric ruler and t-square, and planting holes were created by pushing a 3mm diameter dowel approximately ½ cm into the soil. A single seed was placed in each hole in the 20 by 20 grid. Soil was pushed over the seeds, the tray was misted to field capacity, and kept evenly moist until germination. The trays were not heated, so temperature fluctuated with ambient weather. After germination the germination rate was determined as a percent.

2.3   Seed Mustard Biomass Maximization

Planting density tests were carried out in a field receiving full sun. The subsoil was weeding, tilled to approximately 10 cm, and raked smooth. The subsoil was covered with approximately 8 cm of Miracle Gro topsoil mix and then raked smooth. Six subplots were gridded for plant spacing of 15 cm, 10 cm, 8 cm, 4 cm, 2 cm, and 1 cm. Planting holes were gridded, sowed and watered using the same methods described for germination. The plot received abundant rainfall and weekly applications of Miracle Gro fertilizer solution. Holes in the grid from non-germinating seeds were replanted after germination. The plots were harvested at maturity. The outer row of each subplot was not collected to avoid edge effects. Plants from each subplot were oven dried at 100°C for 24 hours. Mass/area was calculated as average mass per plant times number of plants per square meter.

3.   RESULTS

3.1   Southern Giant Vs. Seed Mustard Mass Comparison

Our intent was to determine the relative mass/area that could be produced with seed mustard as compared to broadleaf mustard. Growing conditions were similar to what might be found in an urban residential area that is the target of this study. As the entire broadleaf crop was pilfered at maturity, this comparison is not possible. The seed mustard produced 550 g dry biomass/m^2. Although we can not compare the relative mass production of the two mustard types, this result indicates that broadleaf mustard or any other "mustard green" type should not be used for phytoremediation in the urban environment.

3.2   Seed Mustard Germination

This phase of the research was performed to determine if seed mustard germination would be adversely effected by the presence of soil lead. Standard 400 seed germination tests were performed in five increasingly contaminated soils. The lead concentration ranged from 0 to 2,000 mg/kg. Results are shown in Table 1.

Germination rates for the 0, 1000, 1500, and 2000 mg/kg soils ranged from 96.5% to 93.5. One soil, the 500 mg/kg appears to be an outlier, with a significantly lower germination rate. The data indicates that there is little if any effect on germination rates from the lead levels likely to be found in urban residential soils.

Table 1. Seed mustard germination in lead contaminated soil

Lead Conc. (mg/Kg)	Number Germinating per 400	Germination Rate
0	386	96.5
500	354	88.5
1000	382	95.5
1500	375	93.8
2000	382	95.5

3.3   Seed Mustard Biomass Maximization

The side-by-side mass study clearly indicated that seed mustard is preferable to broadleaf mustard for urban phytoremediation. The question still remained as to the planting density for seed mustard to yield the highest biomass per planted area. In that a percentage of the biomass will be made up of lead, it is necessary to optimize the mass per area to facilitate phytoremediation. At the onset of this study the authors believed that some intermediate planting density would yield the most biomass. It was assumed that a wide spacing would yield fewer, larger plants, close spacing would yield many stunted plants, and that some intermediate spacing yielding many full size plants would produce the highest biomass. A plot of dry biomass/m^2 Vs planting density indicates that this was not the case (Figure 1). Not only did the closest spacing yield the most mass, there was a decline towards roughly 500 mg/m^2 at spacing of 4 cm. and above.

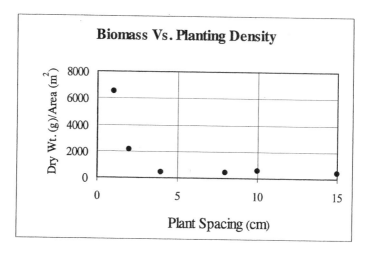

Figure 1. Calculated seed mustard biomass production per square meter

4. DISCUSSION

Lead contaminated soil is a continuing source of health concerns for children living in the urban residential environment. The relative low cost pf phytoremediation compared to other remediation strategies makes it a viable alternative. This is especially true of low-income residential urban areas where funding and resources are likely to be low.

This research has shown that there is a potential to phytoremediate lead-contaminated soil in the urban residential environment without subjecting residents to the possibility of eating contaminated greens. This can be accomplished with Brassica juncea provided that a high plant biomass can be produced. With seed mustard this requires over-seeding, with a planting density of roughly 10,000 seeds per square meter. Seed mustard's short growing season allows for up to three crops per season in the northeastern US.

5.   CONCLUSIONS

Results of the mass comparison study indicate that use of any broadleaf mustard variety is unwise in the urban residential setting. This result can be generalized; no recognizable food crop should be used for phytoremediation in the urban residential setting.

Germination of seed mustard does not seem to be decreased by soil lead concentrations up to 2,000 mg/Kg.

The seed mustard biomass study indicates that the greatest mass is produced with 1 cm spacing. This spacing provides for 10,000 plants per m^2 and approximately 7 kg biomass per m^2.

REFERENCES

Salido A.L., Hasty K.L., Lim J.M., Butcher D.J. Phytoremediation of arsenic and lead in contaminated soil using Chinese brake ferns (Pteris vittata) and Indian mustard (Brassica juncea). 2003. *Int. J. Phytoremediation.* 5(2):89-103

Lanphear B.P., Dietrich, K.N., Berger O. Prevention of lead toxicity in US Children. 2003. *Ambulatory Pediatrics.* S: 27-36,

Canfield R.N., Henderson CR J.R., Cory-Selecta D.A., Cox C, Jusko T.A., Lanphear BP, Intellectual Impairment of Children with blood lead concentrations below 10 mcg per deciliter. 2003 *New England Journal of Medicine.* 348(16):1517-26

Lanphear B.P., Matte T.D., Rogers J, et al. The contribution of lead -contaminated house dust and residential soil to children's blood lead levels. 1998. *Environ Res* 79:51-68

USEPA (U. S. Environmental Protection Agency). 2001 Lead : identification of dangerous levels of lead; Final rule. 40 CFR Part 745, Federal Regis ter 2001:66(4)

CHAPTER 18

CVOC SOURCE IDENTIFICATION THROUGH *IN SITU* CHEMICAL OXIDATION IN FRACTURED BEDROCK

Mark D. Kauffman, P.E.[1], Andrea M. Traviglia[1], James H. Vernon, Ph.D., P.G.[1], John C. LaChance[2]

[1]ENSR International, 2 Technology Park Drive, Westford, MA 01886; [2]Terra Therm, 356 Broad St., Fitchburg, MA 01420, Tel: .978-343-0300, Fax: 978-343-2727, Email: lachance@terratherm.com

Abstract: An *in situ* chemical oxidation (ISCO) pilot program, using Fenton's Reagent (hydrogen peroxide and a ferrous sulfate catalyst), was performed to assess its effectiveness in destroying chlorinated volatile organic compounds (CVOCs) in a fractured-bedrock aquifer. This case study is unique because it was one of the first applications of ISCO in fractured bedrock. In addition, the targeted CVOC reduction from 1,500 to 100 micrograms per liter ($\mu g/L$) was relatively aggressive compared to most ISCO applications. This pilot program also provided the opportunity for an independent, third party evaluation of ISCO in a fractured-bedrock environment. The site geology consists of approximately 6 meters (m) of unconsolidated glacial deposits overlying fractured bedrock, with a groundwater depth of approximately 2 m. Initial characterization activities, including injection testing and multi-level packer sampling, identified a pre-ISCO CVOC plume extending approximately 90 m long by 45 m wide and spanning a vertical depth between 3 and 35 m. Packer sampling results indicated the pre-ISCO plume had an asymmetric configuration that was consistent with the injection-test results. The ISCO pilot program involved the injection of 14,237 liters of 50% hydrogen peroxide, combined with a ferrous sulfate and pH-buffering catalyst. Two injection events were performed, with overlapping performance sampling. Samples collected 30 to 45 days after each injection event showed CVOC concentrations below the treatment objective in many areas of the plume. However, samples collected 60 to 100 days after each event revealed significant rebound in most areas, at concentrations that approached initial pre-ISCO aquifer conditions. An assessment of the results suggests that the injected oxidants primarily influenced the more transmissive fractures in the treatment zone, whereas the less transmissive fractures were less influenced. Geochemical data and calculations indicate that the peroxide and catalyst may persist in the subsurface for prolonged periods (>200 days), thus complicating the assessment of rebound and the actual effectiveness of the technology. Although the success of treatment was limited, i t proved to be successful in enhancing the conceptual site model of the subsurface, better defining the applications and limitations of ISCO treatment in fractured bedrock, and most importantly, clearly identifying the source of residual C VOCs at the site.

Key words: ISCO, Fenton's, Peroxide, Fractured Bedrock

1.   INTRODUCTION

The subject site is relatively small in area, encompassing 2,800 square meters (m^2), located within a former Navy training facility in southeastern Massachusetts (Figure 1).   Historically, the site was used for vehicle maintenance activities from the 1940s through the 1990s, during which time incidental shallow subsurface releases of CVOCs mixed with waste oil occurred from a leaking underground storage tank (UST), immediately west of the former maintenance building (Figure 1).   The site is underlain by approximately 6 m of poorly sorted silty sands. A pink, equigranular and slightly porphyritic, fractured granite lies beneath the silty sands, and the water table is approximately 3 m below the ground surface. The bedrock unit is in direct hydraulic connection with the overburden, and groundwater flow is predominately horizontal with the flow direction east-northeast to west-southwest across the site in both the overburden and bedrock. Although no direct evidence of dense non-aqueous phase liquid (DNAPL) has been seen to date, it is suspected that there is a potential for a small amount of DNAPL to exist below the water table in both the overburden and bedrock.   Prior to ISCO treatment, the main axis of the CVOC plume (primarily tetrachloroethene [PCE] and vinyl chloride [VC]) trended approximately 90 m, east-to-west, from MW-20D to MW-22D2 (Figure 1), and extended to a depth of at least 30 m below the top of the fractured bedrock (i.e., 35-40 m below the ground surface). Laterally, the plume is between 20 and 45 m wide and is bounded by MW-10D and MW-8D (Figure 1). Although the treatment of total CVOCs in fractured bedrock was the focus of the pilot program, the CVOCs and BTEX contaminants present in the overburden aquifer were also targeted.

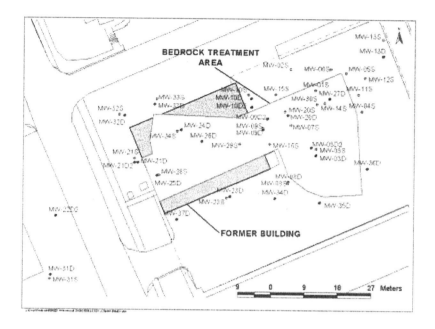

Figure 1. Site Plan and Monitoring Well Network

The U.S. Navy elected to perform the ISCO pilot program to evaluate its potential effectiveness in reducing CVOC concentrations in fractured bedrock, to gather specific data on the interconnectivity of bedrock fractures, and to aid in the selection and design of an appropriate remedial strategy for the site. Geo-Cleanse International, Inc. (GCI) was selected to implement the pilot program, while ENSR was selected to monitor and evaluate its performance (U.S. Navy, 2002).

2. MATERIALS AND METHODS

2.1 Bedrock Fracture Characterization

Because a significant portion of the CVOC plume was in fractured bedrock, extensive work was performed to characterize the physical and hydraulic properties of the bedrock system, including fracture frequency and orientation, effective hydraulic conductivity, interconnectivity, and fracture aperture/porosity. This included performing outcrop surveys, bedrock core extraction, borehole geophysics, packer sampling and pressure testing, and bedrock interconnectivity testing (U.S. Navy, 1999). This characterization effort was conducted prior to ISCO implementation. In general, there was good correlation among geophysical measurements, core observations, and influence testing.

Estimates of effective hydraulic conductivity in the bedrock treatment zone, based on slug tests and packer pressure tests, ranged between 4.3×10^{-5} to 2.6×10^{-4} centimeters per second (cm/sec). Data from the bedrock coring and packer pressure tests were also used to develop estimates of effective fracture aperture [156 to 349 micrometers (μm)] and bedrock porosity (1.03×10^{-3} to 2.86×10^{-3}). Based on these calculations, the volume of water within the bedrock treatment zone was estimated to be approximately 36,000 liters. Given the observed hydraulic gradient across the bedrock treatment zone, the estimated time to flush one fracture-volume through that zone (assuming all of the fractures are connected) was between 88 and 243 days.

2.2   ISCO Implementation.

A comprehensive monitoring well network of 44 wells (20 overburden and 24 bedrock), spanned the dimensions of the CVOC plume and were used as performance monitoring points (Figure 1). The ISCO process tested at the site was based on the *in situ* application of Fenton's Reagent. This process involves the simultaneous injection of hydrogen peroxide and a ferrous sulfate catalyst into the subsurface (overburden and fractured bedrock zones). GCI uses a patented process for the delivery of H_2O_2 and catalyst. The subsurface reaction results in the formation of hydroxyl free-radicals (OH$^-$), which lead to the destruction of CVOCs and the creation of reaction byproducts (H$^+$, Cl$^-$, H_2O and CO_2) as shown in the following equations for one of the target compounds, PCE:

(1) $Cl_2C=CCl_2 + OH^- + H_2O \rightarrow CHCl_2COOH + H^+ + Cl^- + Cl^-$

(2) $CHCl_2COOH + 4OH^- \rightarrow 2H^+ + 2Cl^- + 2H_2O + 2CO_2$

Two phases of injection were performed, an initial phase in October 2000 and a final phase in March 2001. For the initial injection, 48 injectors were installed (Figures 2 and 3). Of those injectors, 20 were screened across the saturated portion of the overburden, and 28 were screened within the fractured bedrock (up to 36 m deep). The spacing between both the overburden and bedrock injectors was approximately 3 m (Figures 2 and 3). Prior to the second and final injection, three additional bedrock injectors were installed in the vicinity of MW-9D in an attempt to better access the fractures intercepted by MW-9D (Figures 1 and 3).

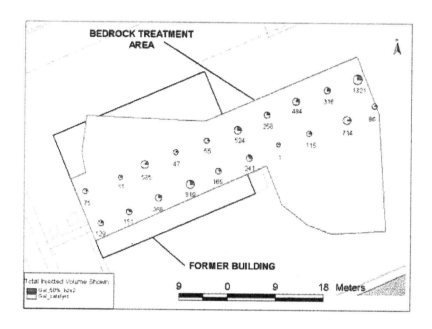

Figure 2. Overburden Injectors

The total volume of peroxide applied to the bedrock aquifer over both injection phases was 9,233 liters, and the total volume of catalyst (ferrous sulfate, acid, and water) injected was 28,174 liters (Table 1). The total amount of peroxide and catalyst injected into the bedrock aquifer over the two phases of ISCO treatment was 37,407 liters.

Table 1. Summary of ISCO Injection Volumes (Liters)

Target Zone	Initial Injection		Final Injection		Total Volume
	Peroxide	Catalyst	Peroxide	Catalyst	
Overburden	4,758	19,210	246	295	24,509
Bedrock	5,841	21,292	3,392	6,882	37,407
Total	10,599	40,502	3,638	7,177	61,916

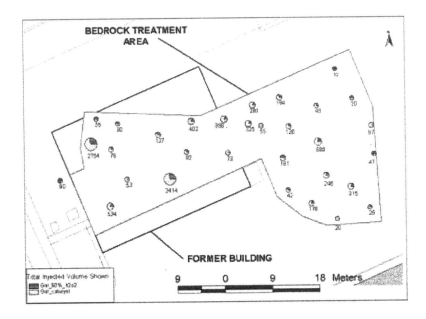

Figure 3. Bedrock Injectors

Performance sampling consisted of groundwater quality parameters, including VOCs, dissolved gasses and chloride, hydrogen peroxide, iron, dissolved oxygen, and pH. Samples were collected on six occasions, prior to implementation, during initial injection, 30 and 60 days after the initial injection, and 45 and 100 days after the final injection.

3.   RESULTS AND DISCUSSION

Prior to the pilot program, baseline groundwater sampling revealed that two wells exhibited total CVOCs in excess of 1,000 μg/L (MW-09D and 21D2), both of which are screened within the bedrock zone. An additional 14 wells were in the 100-1,000 μg/L range, and the remaining 29 wells were less than 100 μg/L. The apparent concentration of CVOCs was depressed during and immediately following each phase of injection, but then rebounded over time. For example, the concentration of total CVOCs in bedrock monitoring well MW-09D (the well closest to the suspected release/source area) was reduced from 1,007 μg/L to 85 μg/L during the initial injection. The concentration rebounded to 300 μg/L, 30 days after initial injection, and to 1,020 μg/L, 60 days after initial injection. After the final injection event and a 45-day recovery time, the concentration at that particular location remained depressed, at 218 μg/L, but then rebounded to 868 μg/L, 100 days following the final injection.

An evaluation of the representativeness of the performance monitoring data was conducted by considering: 1) the impacts of dilution due to injection of the peroxide and catalyst and 2) evidence for the persistence of the reactants and reaction by-products. A comparison of the estimated total volume of water in the fractures within the bedrock treatment zone (36,000 liters) with the total volume of peroxide and catalyst added to the bedrock over the two phases of ISCO treatment (37,407) indicates that a significant amount of dilution occurred. Using the upper-bound of the calculated flushing time (243 days) as a conservative measure for the amount of time necessary to flush one fracture-volume through the bedrock treatment zone, and considering the impacts of mass-transfer limitations from poorly connected and low-permeability fractures, a theoretical flushing time can be derived. Applying this approach, it is reasonable to expect that it would take 9 to 12 months for the bedrock system to flush the injected fluids and for CVOC concentrations within the system to re-equilibrate.

Figures 4 and 5 present time-series trend-plots of VOCs and geochemical conditions from MW-09D and MW-21D2 during the six sampling events conducted during the ISCO pilot test (baseline, during ISCO implementation, and post treatment). The geochemical data presented in the plots indicate the impacts of ISCO treatment and the persistence of the reactants and reaction by-products (i.e., the slow rate of flushing and re-equilibration). Immediately following ISCO treatment, dissolved oxygen (DO) concentrations increased (maximum ~45 mg/L) and pH decreased (minimum ~1). Data from MW-09D from the last sampling event following final injection (post 100 days), indicated that DO concentrations had returned to normal and pH had partially recovered. As discussed previously, CVOC concentrations for MW-09D had rebounded to approach pre-ISCO treatment levels by that time. These results indicate that residual CVOCs present near MW-09D were not effectively treated by ISCO. This zone appears to be at the low-permeability overburden/bedrock interface that could be resistant to penetration by the injection fluids.

Data from MW-21D2 presents a different pattern, in that DO levels remained elevated (~40 mg/L) and pH reduced (~1), 100 days after final injection. CVOC concentrations remained reduced in MW-21D2 at that time. Similar trends were observed in 34 (77%) of the 44 performance-monitoring wells. These data indicate that the injected fluids significantly affected the ability to accurately assess treatment effectiveness during and immediately after treatment (e.g., 30-day post-injection sampling). Because

MW-21D2 is west (downgradient) of MW-09D, it is possible that the decline of DO and rebound of downgradient CVOC concentrations could still be observed after more time has elapsed.

4.   CONCLUSIONS

In summary, the primary mechanisms contributing to the contaminant reductions observed in the fractured bedrock aquifer include chemical destruction via Fenton's reagent, and dilution.   An accurate assessment of ISCO treatment effectiveness can only be accomplished after allowing sufficient time to flush reactants and reaction by-products, and for the system to re-equilibrate. For this particular pilot program, a minimum of 12 months is estimated to be required before representative aquifer conditions can be expected.

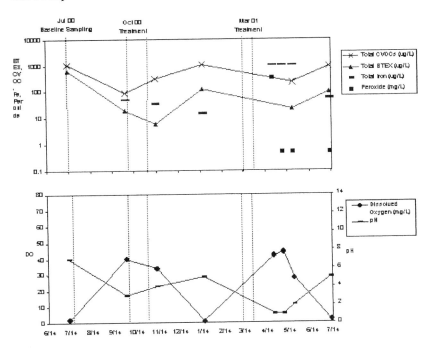

Figure 4. VOC and Geochemical Conditions at MW -09D

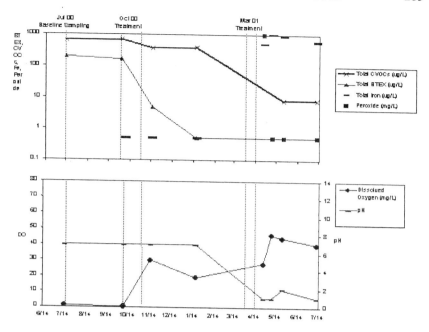

Figure 5. VOC and Geochemical Conditions at MW -21D2

Although the success of treatment may have been limited, it proved to be successful in refining the delineation of the residual source, supporting the characterization of bedrock fractures, and better defining the applications and limitations of ISCO treatment in a fractured-bedrock setting. The injection results also suggest that residual CVOCs are present at the overburden/bedrock interface, at a depth of ~5 m near MW-09D. This interpretation warrants direct field confirmation, and may focus further site investigation and remediation efforts.

REFERENCES

Geo-Cleanse International, Inc, 2000, Injection Work Plan Geo -Cleanse Pilot Test Program, Kenilworth, NJ.

Geo-Cleanse International, Inc., 2002, Effectiveness Evaluation Report, Geo -Cleanse Treatment Program, Kenilworth, NJ.

U.S. Navy (ENSR), 2002, *In situ* Chemical Oxidation (ISCO) Pilot Test Performance Assessment, Environmental Field Activity Northeast (EFANE), Philadelph ia, PA.

U.S. Navy (ENSR), 1999, September 1999 Bedrock Characterization, Engineering Field Activity Northeast (EFANE), Philadelphia, PA.

CHAPTER 19

ISCO TECHNOLOGY OVERVIEW: DO YOU REALLY UNDERSTAND THE CHEMISTRY?

Ian T. Osgerby, Ph.D., PE
US Army Corp of Engineers New England District, 696 Virginia Road, Concord, MA 01742

Abstract: The reaction chemistry of ISCO is presented for the common oxidant systems employed in ISCO: catalyzed peroxide propagations (Modified Fenton's), persulfate, ozone/ozone -peroxide (peroxone), and permanganate. All of these oxidant systems, with the exception of permanganate are described by reaction schemes employing free radical generation, and all are dependen t to some degree on local conditions such as water chemistry and pH. A less familiar reactant condition may be the influence of inorganic and organic compounds in the soil matrix, which can have a strong influence over the intended outcome of the ISCO app lication. Thus, naturally occurring organic compounds may overwhelm the contaminant demand for oxidant or prevent the transition of the adsorbed contaminant to the aqueous phase where ISCO reactions occur. Naturally occurring inorganic compounds may actu ally cause destruction of the oxidant or modify the catalytic component. Some experience with soils having markedly different matrix properties are discussed to provide an illustration of some of the difficulties which may be faced in the practice of ISCO .

1.   INTRODUCTION

A concise collation of the reaction chemistries and a more accurate portrayal of the nomenclature have been prepared to provide some clarification of the errors in the literature for this industry. The formulations demonstrate the synergism of the radical-based chemistries (peroxide, persulfate, ozone/peroxone) and the susceptibilities to common components of the subsurface environment (anionic such as chlorides, carbonates; cationic such as inorganic metal ions and natural (humic) organic compounds, excess primary oxidant, etc.). The formulation appropriate to the so-called (modified) Fenton's Reaction system is discussed, illustrating that the conditions actually preclude the original Fenton's Reaction mechanism, but that the fundamental free-radical reaction is an integral part of the "new" modified system. Some discussion is provided concerning the necessity for the presence of an iron catalyst and whether it makes sense to add the oxidized form (ferric) or the reduced form (ferrous) if it is lacking in the subsurface soils (unlikely), and as an inorganic or in chelated form. The discussion also includes reference to the hydroxyl free-radical being almost ubiquitous to the reaction chemistries including peroxide, ozone/peroxone and basic persulfate systems. The exception obviously being permanganate since it is not a radical based chemistry. Each oxidant system is described as a stand alone system with some brief remarks on materials compatibility since each is a potent oxidizer, and some limitations are described or referenced to indicate applicabilities of the oxidants to common contaminants. The literature cited is not inclusive but provides some relevant background and notes some of the primary contributors to the development of the description of the oxidant systems.

2. HYDROGEN PEROXIDE SYSTEMS INCLUDING CATALYZED PEROXIDE PROPAGATION (CPP)

Although hydrogen peroxide is an oxidant, it is not strong enough to degrade most hazardous organic contaminants, for example toluene, phenol, etc. Fenton (1894) discovered that adding a ferrous salt (Iron II) dramatically increased the oxidation of tartaric acid. It was not until 1934 (Haber & Weiss) that this increase was attributed to the production of hydroxyl radicals (OH·) thus, henceforth the reaction of iron catalyzed oxidation was called a Fenton's reaction and the hydroxyl radical Fenton's reagent. Fenton's reaction occurs efficiently at peroxide concentrations of about 300 ppm, oxidizing the Iron II (Fe^{2+}) to soluble Iron III (Fe^{3+}). If the pH is maintained between pH 2.5 to 3.5 the Iron III is reconverted to Iron II, and the iron remains in solution to continue the catalyzed production of hydroxyl radicals.

A classical Fenton's system cannot be readily created *in situ* as it is generally too difficult to maintain the well-mixed low peroxide concentration via injection of the chemicals. In practice, more concentrated solutions of hydrogen peroxide are injected, ranging from 5% to 50%, with iron in acid solution either co-injected (separate injection strings or nozzles) or injected separately, sequentially. Also, the well-mixed setup in a laboratory beaker is not realizable in the subsurface; rather, displacement of water in the pore spaces occurs, with mixing/diffusion occurring at the interfaces.

3.   MODIFIED FENTON'S SYSTEMS

The simple one-step Fenton's reaction with dilute peroxide concentrations which occurs in the well-stirred, laboratory beaker becomes significantly more complicated when strong peroxide solutions are injected. Thus, if ferrous iron is injected, the ferrous iron is immediately oxidized to ferric iron and the hydroxyl radicals (an electrophile and strong oxidant) react with the excess peroxide to form perhydroxyl radical ($HO_2^{\cdot}$), an electrophile and a relatively weak oxidant:

$$Fe^{2+} + H_2O_2 \rightarrow Fe^{3+} + OH^{\cdot} + OH^{-} \qquad \text{(Fenton's reaction)} \quad (1)$$

$$OH^{\cdot} + H_2O_2 \rightarrow HO_2^{\cdot} + H_2O \qquad \text{(Peroxide > 2 to 3\%)} \quad (2)$$

$$HO_2^{\cdot} + Fe^{3+} \rightarrow O_2 + H^{+} + Fe^{2+} \qquad (3)$$

However a catalyzed peroxide propagation sequence [Watts (1a,b)] generally takes place which can generate perhydroxyl radical ($HO_2^{\cdot}$), superoxide ($O2^{\cdot-}$), hydroperoxide anion (HO_2^{-}) and organic radicals ($R^{\cdot}$); – super oxide anion is a nucleophile and weak reductant and hydroperoxide is a nucleophile and a strong reductant:

$$HO_2^{\cdot} + Fe^{2+} \rightarrow HO_2^{-} + Fe^{3+} \qquad (4)$$

$$OH^{\cdot} + H_2O_2 \rightarrow HO_2^{\cdot} + H_2O \qquad (5)$$

$$HO_2^{\cdot} \rightarrow O_2^{\cdot-} + H^{+} \qquad pKa = 4.8 \qquad (6)$$

$$OH^{\cdot} + RH \rightarrow R^{\cdot} + OH^{-} \qquad (7)$$

$$R^{\cdot} + H_2O_2 \rightarrow ROH + OH^{\cdot} \qquad (8)$$

According to Watts (1a), the catalyzed peroxide propagation sequence of reactions may produce many more radicals than are consumed. This is in marked contrast to Fenton's reaction which terminates at the first step in reaction 1. Additional reactions occur between the oxidant and the hydroxylated organic intermediate which may oxidize or hydrolyze it to more intermediate compounds which bacteria will consume readily or in some (rare) cases proceed to CO_2 and H_2O. In almost all cases, the intermediates are relatively innocuous compared to the parent compound. Acid compounds such as HCl produced in the oxidation of chlorinated solvents will react with carbonates in the aquifer matrix resulting in a return to a "normal" pH, sooner or later.

Two important side reactions also occur:

$$Fe^{3+} + O_2{}^{\cdot -} \rightarrow Fe^{2+} + O_2 \tag{9}$$

$$Fe^{3+} + n\ OH^- \rightarrow \text{Amorphous iron oxides (precipitate)} \tag{10}$$

The importance of selecting the right pH to use with the oxidant system cannot be stressed enough. Inorganic metal compounds can cause decomposition of peroxide to oxygen as well as provide the conditions (in principle) for a modified Fenton propagation sequence. Inorganic metal compounds in the subsurface, for example manganese, can prevent the chain reactions from occurring. In recent tests [Osgerby (2a)] PCB-contaminated (Aragonite) coral-based soils, hydrogen peroxide solutions at pH > 13 consistently achieved conversions in the mid 90% range. In this latter case the reaction was a reducing reaction, hydroperoxide anion stripping chlorine off the bi-phenyl molecule.

$$H_2O_2 \rightarrow H^+ + HO_2{}^- \quad \text{Conjugate Base in NaOH solution} \tag{11}$$

The excess [OH⁻] ions in the strong base solution removes the [H⁺] ions, driving the reaction to the right. In another test with a volcanic soil from Maui, in the presence of abundant, naturally occurring iron compounds, manganese limited the oxidation of PCBs in a soil slurry to < 10%. Presoaking the soils in acid at a pH 2 – 3 followed by a modified Fenton's system generated in excess of 90 % conversion in a single pass [Osgerby (2b)]. These same soils, treated with iron-EDTA chelant at pH 9 generated 90% conversion without the acid pre-soak. Again, an alternative oxidant system [Lundy (3)] achieved 60% conversion at a pH of about 8 in a single pass. This alternative oxidant system incorporated a chelant catalyst, pH buffered, metal peroxide. The buffered peroxide slowly dissolves, releasing H_2O_2 at a rate determined by the pH. Selecting a chelant also may depend on the reactant system and pH as the required active form may be strongly dependant on its solubility at varying pH conditions.

Hydrogen peroxide is an extremely versatile reagent, in which both oxidizing and reducing reactions, electrophyllic and nucleophyllic can be brought into play as needed. The skill of the applicator/vendor can be very important to the success of an application in the field, and bench testing by a competent organization is the best way the response/behavior and findings can be diagnosed. This is particularly important when difficulties are experienced in any field application which should (almost) never be attempted to be solved via pilot or full-scale activities.

Certain physical properties are very important to *in situ* oxidation with hydrogen peroxide and solid peroxides. For example calcium peroxide [www.solvayinterox.com/ www.fmcchemicals.com] will rapidly decompose to O2 at low pH and take 6 months to decompose to hydrogen peroxide and oxygen at its natural (unbuffered) solution pH (10 to 12) – a fact that can be exploited in a combined oxidation/enhanced bioremediation approach. At pH 8 the controlled rate of production of hydrogen peroxide from calcium peroxide is nearly optimal and will support the modified Fenton's system of reactions. As practiced [Lundy/ Cool-Ox^{TM} (3)], controlled insitu oxidation (CISCO) is feasible even where substantial free product is present. This fact has been exploited in a proprietary technique as a means of locating the source and plume boundaries [Nuttal (4a,b)]. This may be very important when it becomes apparent that the prior remedial investigation has been inadequate in defining the site contaminant distribution. One of the common difficulties in oxidative site remediation with strong oxidizers is the natural elimination of readily oxidizable components of residual fuels from the original highly paraffinic aviation and jet fuels. Other contaminants such as chlorinated solvents may have become entrapped in the residual waxy residues left after the light ends have been transported away with the groundwater. The waxy residues are the compounds which were originally co-solved in the light ends but now form the majority with only a portion of the original light ends and even chlorinated solvents bound into them. It requires an understanding of the chemistry of the contaminants and oxidizers and the site history (where known) to effect efficient and economic remediation by ISCO.

3.1   Stoichiometry

Estimating the ratio of peroxide to contaminant plus other oxidant demands is an art practiced by the successful vendor. One can balance the equation readily for simple compounds such as benzene:

$$C_6H_6 + 20 H_2O_2 \rightarrow 6 CO_2 + 23 H_2O + 2.5 O_2 \tag{12}$$

It has little meaning in terms of the overall oxidant demand including naturally occurring minerals and organic (humic) compounds. Chemical reactions do not occur by stoichiometric reactions since even a simple molecule such as benzene would have to be surrounded by 20 molecules of hydrogen peroxide – a highly unlikely circumstance. Reactions proceed in many small steps, where hydroxylation would be the first step followed by many other steps producing hydroxycyclohexadienyl intermediates, to phenol, to dihydroxybenzene, to orthobenzoquinone, to muconic acid, to maleic acid, to oxalic acid, to formic acid to carbon dioxide and water. In addition, it is not hydrogen peroxide which rapidly oxidizes the benzene but hydroxyl and other radicals, and it is by no means a one-to-one correspondence between hydrogen peroxide and hydroxyl radical production. Fenton's reaction is efficient with one hydroxyl radical produced per peroxide in the presence of the ferrous ion, but again it is by no means certain that this reaction proceeds without chemical losses as shown above. Ferrous ions are initially oxidized to ferric in strong peroxide solutions, simply because peroxide is there in excess, needlessly consuming peroxide. Inject soluble ferric salts and cut the injection losses.

Even if we write the stoichiometry of oxidation of PCE in terms of hydroxyl radicals, the presence of 4 free electrons are required to balance the charges, and this supply is not normally available in the manner shown. They can however be obtained (but not limited to) in the presence of iron:

$$C_2Cl_4 + 4\,OH^{\cdot} + 4\,e^- \rightarrow 2\,CO_2 + 4\,HCl \qquad (13)$$

$$C_2Cl_4 + 4\,OH^{\cdot} + 4\,Fe^{2+\,-} \rightarrow 2\,CO_2 + 4\,HCl + 4\,Fe^{3+} \qquad (14)$$

Reactions of the type shown cannot possibly take place as written stoichiometrically. Although vendor experience indicates what is a reasonable supply ratio for the reactants, frequently in excess of 10 to 20 times the stoichiometric demand, it is the role of the bench test to determine the most suitable combination of reactants. The reactant destruction path in strong solutions of hydrogen peroxide involves a multiplicity of radicals as shown in the above equations.

4.   PERSULFATE SYSTEMS

As with hydrogen peroxide, potassium/sodium/ammonium persulfate salts dissolved in water dissociate to persulfate anions $(S_2O_8{}^{2-})$ are not a strong enough oxidant to degrade most hazardous organic contaminants, and the persulfates are kinetically slower than peroxide. As with permanganates, the potassium salt has the lowest solubility: ammonium 46%; sodium 40% (73grams/100 grams H_2O) and potassium 6% @ 25°C. The ammonium salt is somewhat unstable due to oxidation of ammonium and ammonia by persulfate; thus the sodium salt is the typically preferred form. They are widely used in the chemical process industry for a variety of tasks including initiators in polymerization, polymeric concrete, coatings, oxidants in metals cleaning/plating, cleaning/activation of adsorbents such as graphite and carbon, and in organic synthesis [www.fmcchemicals.com]. Boiling persulfates have been used for many years for routine laboratory analysis, including the measurement of total and dissolved organic carbon (TOC & DOC). It is this latter capability which is currently being formulated for the environmental remediation industry, where the very active sulfate free radical ($SO_4^{-\cdot}$) is generated, roughly equivalent to the hydroxyl free radical in reactivity. Coupled with group II metal (ferrous iron Fe^{2+}; -also copper, silver and manganese ions) catalysis, the sulfate free radical is also generated but not as efficiently as with heat.

The sulfate free radical is most commonly generated with the ferrous iron (Fe^{2+}; -also copper, silver and manganese ions) activated system, and though not as reactive as the heated persulfate system, is a very potent oxidizer, roughly equivalent to the catalyzed peroxide propagations (CPP) system. A generalized organics chain reaction process is described comprising three steps, initiation, propagation and termination as follows [Kishenko (5), Kolthoff (6) and Liang 7a,b,c)]:

Initiation Reactions (Me denotes a metal ion; R denotes an organic compound):

$$S_2O_8^{2-} \rightarrow 2\,SO_4^{-\cdot} \tag{15}$$

$$Me^{n+} + S_2O_8^{2-} \rightarrow 2\,SO_4^{-\cdot} + Me^{(n+1)+} + SO4^{2-} \tag{16}$$

$$S_2O_8^{2-} + RH \rightarrow 2\,SO_4^{-\cdot} + R^{\cdot} + HSO_4^{-} \tag{17}$$

Propagation Reactions:

$$Me^{(n+1)+} + RH \rightarrow R^{\cdot} + Me^{n+} + H^{+} \tag{18}$$

$$SO_4^{-\cdot} + RH \rightarrow R^{\cdot} + HSO_4^{-} \tag{19}$$

$$SO_4^{-\cdot} + H_2O \rightarrow OH^{\cdot} + HSO_4^{-} \tag{20}$$

$$OH^{\cdot} + RH \rightarrow R^{\cdot} + H_2O \tag{21}$$

$$R^{\cdot} + S_2O_8^{2-} \rightarrow SO_4^{-\cdot} + HSO_4^{-} + R \tag{22}$$

$$SO_4{}^{-\bullet} + Me^{n+} \rightarrow \quad Me^{(n+1)+} \quad + \quad SO_4{}^{2-} \tag{23}$$

$$OH^{\bullet} + Me^{n+} \quad \rightarrow \quad Me^{(n+1)+} \quad + \quad OH^{-} \tag{24}$$

Termination Reactions:

$$R^{\bullet} + Me^{(n+1)+} \quad \rightarrow \quad Me^{n+} \quad + \quad R \quad + \quad H \tag{25}$$

$$2\,R^{\bullet} \rightarrow \quad \text{Chain termination} \tag{26}$$

Ferrous Iron activation (Me = Fe) :

$$Fe^{2+} + S_2O_8{}^{2-} \quad \rightarrow \quad Fe^{3+} + SO_4{}^{-\bullet} + SO_4{}^{2-} \tag{27}$$

$$SO_4{}^{-\bullet} + Fe^{2+} \rightarrow \quad Fe^{3+} + SO_4{}^{2-} \tag{28}$$

$$2\,Fe^{2+} + S_2O_8{}^{2-} \rightarrow 2\,Fe^{3+} + 2\,SO_4{}^{2-} \tag{29}$$

Equation (29) decribes the stoichiometric representation of equations (27) and (28). Note that as with hydroxyl radicals (excess iron and peroxide), excess reactants can deplete the pool of free radicals as shown in equations (27) and (28). The excess ferrous iron may deplete the free radicals essential for chain propagation. As with Fenton's reaction, persulfate activation is not catalysis because the iron molecule is changed, ferrous to ferric. Sulfate free radical chain propagation is reactive over the range of pH acidic to alkaline as is catalyzed peroxide propagations; however, it has not yet been established what specific changes in the chain propagation sequence occur under different pH conditions. Sulfate free radicals have a reported half-life of about 4 seconds under ambient conditions [Banerjee (8)]. Although not reported it may be expected that hydroxyl radicals have a similar or shorter half-life.

Since iron is a natural constituent, and one of the most abundant elements, one might anticipate that iron injection is not necessary when injecting persulfate system chemicals in *in situ* oxidative remediation. This may be similar to the case of iron-activated peroxide systems, where the iron mineral in nature may not be accessible or available in the reduced ionic form which serves as a catalytic agent. Ferrous ions require highly reducing conditions and an acidic pH to remain in solution. This means an excess of H^{+} ions in solution and may be as problematical as with activated peroxide chemistry. A more promising approach may be with the use of chelating agents to eliminate the competition of excess ferrous/ ferric ions with sulfate free radical as shown in equations (27) and (28). Ferrous ions control the rate of production of sulfate free radicals (much like the activated peroxide chemistry); however, 2 moles of iron per persulfate anion (2.8 pounds ferrous sulfate heptahydrate to 1 pound of persulfate) are required as shown in equation (29). Complexing the ferrous ions in solution is readily accomplished using carboxyl groups of inorganic acids (oxalic; citric), EDTA (also NTA, STPP, HEDPA, NTA) as routinely used in activated peroxide chemistry, and has been demonstrated in reactions with TCE by Liang (7a,b,c). Chelation, though not fully described in these oxidation systems (peroxide and persulfate) may be more representative of a catalyzed reaction system [Bruell (9)].

Hydrogen ions (H$^+$) will react via the water equilibrium under low pH conditions to remove hydroxyl ions which also contribute to the radical pool, combining persulfate and hydroxyl free radical chemistry to the oxidation processes:

$$SO_4^{-\cdot} + OH^- \rightarrow OH^\cdot + SO_4^{2-} \tag{30}$$

$$H^+ + OH^- \rightarrow H_2O \tag{31}$$

In a basic solution the participation of hydroxyl free radicals should be even more influential in the oxidation reactions via activated propagation reactions (CPP). Hoag (10) has utilized the dual advantage of adding hydrated lime to generate a surfeit of hydroxyl ions and thermal activation of persulfate to sulfate free radicals (from the heat of hydration of the lime) as an effective means of dechlorinating PCBs.

As with hydroxyl free radicals, chloride ions, carbonate and bicarbonate ions act as radical sinks for sulfate free radicals. Although the reaction rates with excess reactants and radical sinks is generally quite low compared with the free radical reactions with the intended contaminants, if their concentrations are relatively high (they may be several orders of magnitude greater) the net effect is to reduce the oxidant effectiveness or even eliminate it. Enough chloride ions can result in the production of chlorine and since this is aqueous phase chemistry, may either react to form hypochlorous acid or even chlorinated compounds such as trihalomethanes in the presence of natural organics compounds.

In terms of stoichiometry for chlorinated solvents, a useful calculation in estimating a starting point for quantity of persulfate required (to be multiplied by a factors of ten or a hundred) depending on whether the natural organic/inorganic demand is relatively low (not typical):

$$6\,SO_4^{-\cdot} + C_2HCl_3 + 4\,H_2O = 2\,CO_2 + 9\,H^+ + 3\,Cl^- + 6\,SO_4^{2-} \tag{32}$$

As discussed with peroxide chemistry, reactions do not take place in groups such as that shown above, where 6 sulfate radicals are required per mole of TCE (plus 4 water molecules). Charged species in aqueous solution are typically surrounded by hordes of water molecules which has it's own asymmetrical distribution of electrical charges. The quantity of persulfate required for $SO_4^{-\bullet}$ oxidation of contaminants will be strongly dependant on the method of radical producing reactions (metal activated or thermal) and efficiency of chemical utilization. At this stage of development of the ISCO persulfate reaction chemistry, little is known concerning the effective ratio of chemicals to contaminant concentration, or the natural organic/inorganic demand, thus bench scale and pilot scale tests are essential tools for site remediation. Although the number of bench scale tests [Dahmani (11)] with a wide range of organic compounds and pilot scale persulfate tests for ISCO projects is growing, there have been no well defined applications reported at full scale, and testing has been carried out by a very limited number of vendors. An interesting variant is currently being investigated in which an insitu thermal technology is being applied on a site where the contaminants are chlorinated ethanes (temperature accelerated hydrolysis) to be followed by a persulfate application, utilizing the synergistic mechanism of thermal activation of persulfate to sulfate free radicals [Lewis (12)]. An item of importance to persulfate chemistry applications is the extreme corrosivity to unprotected iron hardware such that stainless steels or plastics (PVC) are essential materials for injection hardware and wells (injection or monitoring) [Pac (13)].

5.   OZONE/OZONE-PEROXIDE (PEROXONE) SYSTEMS

Ozone-based processes are unique to most other ISCO processes in that they involve application of a gas (ozone) posing very different design and operational issues than those faced with the application of the peroxide and permanganate systems. For ozone-based ISCO systems, ozonation (application of ozone alone) and peroxone (application of ozone and hydrogen peroxide) are most often considered. However, ISCO ozonation can be also applied using adjusted elevated pH conditions. This is done as a strategy for increasing the degradation rates of some pollutants (such as phenolics) via increased hydroxyl radical formation and potential dissociation into ionic forms that have been found to be much more oxidizable [Qiu (14)].

5.1   Chemistry

Chemical oxidation processes may be broken down into two categories; primary oxidation and advanced oxidation [Zappi (15)]. Primary oxidation involves oxidation of the targeted chemical by the parent oxidizers such as ozone. Primary oxidation does not rely heavily on the hydroxyl radical (OH) for achieving targeted results. This process has found significant usage in water treatment [Langlais (16)]. The second form of oxidation is advanced oxidation processes which are processes that rely on the hydroxyl radical (OH) as the main reactant for contaminant oxidation [Glaze (17); Langlais (16) and Hong (18)]. Oxidation products of most organic compounds are usually hydroxylated products, aldehydes, ketones, carboxylic acids, carbon dioxide, and water [Adams (19); Trapido (20); Zappi (15 and Yao (21)].

5.2   Ozonation

Ozone, O_3, has found significant usage for oxidation of complex organic compounds (Staehelin (22); Hoigne (23) and Yao (21). Ozone is an allotrope of oxygen which is approximately 20 times more soluble than oxygen and is typically generated on-site using ozone generators. Commercial generators produce air or oxygen streams containing ozone typically within the 2 - 10 percent (weight [w]/weight [w]) range. Since ozone is introduced via carrier gases, then the solubility of ozone using these streams is controlled by the phase equilibrium between stripping and absorption. When using typical ozonated feed gases sparged into contactors containing clean water, the equilibrium ozone concentrations in these aqueous solutions generally range from 5 mg/l to 30 mg/l [Langlais (16)].

Ozonation is a very common municipal water treatment technology. Over the past 20 years, more and more literature has been published that supports the concept of using ozonation for treating complex organic pollutants [Trapido (20)]. The mechanisms of organic contamination destruction during ozonation is primarily twofold. Firstly, the organic compounds can be directly oxidized by ozone via primary oxidation. Typical modes of attack involve the insertion of the ozone molecule into unsaturated carbon-carbon bonds which results in the formation of an ozonide [Qui (14)]. The second mode of organics oxidation is through the reaction with hydroxyl radicals, which are usually formed during ozonation due to reaction with the hydroxide ion at neutral to basic pH ranges as summarized below:

$$O_3 + OH^- \rightarrow O_2^{\cdot -} + HO_2^{\cdot} \text{ (intiation step)} \tag{33}$$

$$HO_2^{\cdot} \leftrightarrow O_2^{\cdot -} + H^+ \tag{34}$$

$$O_3 + HO_2^- \rightarrow OH^{\cdot} + O_2^{\cdot -} + O_2 \text{ (radical production step)} \tag{35}$$

Note that the initiator for the ultimate production of the hydroxyl free radical as shown above is the hydroxyl anion; however, other initiators are possible including the hydroperoxide anion (HO_2^-). However, within soil systems, the hydroxyl anion and cationic initiation steps are the most likely initiators. Promotion of hydroxyl radical formation may occur in the presence of key organic compounds such as alcohols, carboxylic acids, and humics (all fairly common constituents of most soil matrices and providers of natural chelants). Finally, radical inhibitors are also common within soil matrices including carbonates, humic acids, and tertiary alcohols (note that only carbonates are a common inorganic soil matrix component and others may be produced in the course of chemical oxidation reactions).

Use of ozone during ISCO has been documented with numerous successes noted [Nelson (24); Masten (25)]. Minimal mechanistic research detailing the reaction of ozone with soil constituents has been performed; however, effective transport within soil matrices with subsequent pollutant oxidation has been documented [Masten (25)].

5.3   Peroxone Oxidation

The reactions between ozone and hydrogen peroxide which result in the formation of the hydroxyl radical have been under investigation since the early 1940's when chemists first described potential radical formation reactions [Taube (26)]. The ozone-hydrogen peroxide reactions resulting in hydroxyl free radical formation were later defined by the engineering community as peroxone [Glaze (17)]. The mechanism for the formation of the hydroxyl free radical during peroxone treatment has been defined as [Langlais (16)]:

$$H_2O_2 + H_2O \leftrightarrow HO_2^- + H_3O^+ \tag{36}$$
$$O_3 + HO_2^- \rightarrow OH^{\cdot} + O_2^{\cdot -} + O_2 \tag{37}$$
$$O_2^{\cdot -} + H^+ \leftrightarrow HO_2^{\cdot} \tag{38}$$
$$O_3 + O_2^{\cdot -} \rightarrow O_3^{\cdot -} + O_2 \tag{39}$$
$$O_3^{\cdot -} + H^+ \leftrightarrow HO_3^{\cdot} \tag{40}$$
$$HO_3^{\cdot} \rightarrow OH^{\cdot} + O_2 \tag{41}$$

Peroxone has been used for treating waters contaminated with 2,4,6-trinitrotoluene (TNT), aminodinitrotoluenes, and pesticides [Zappi (15); Yao

(21)]. Of all the dark AOPs considered good candidates for ISCO, peroxone is considered to be the most aggressive due to the high yields of hydroxyl free radicals obtainable [Hoigne (23); Hong (18) and Kuo (27)]. Both Wang (28) and Tiang (29) found that peroxone was far more aggressive for the removal of petroleum-based pollutants than ozonation and Fenton's Reagent within the soil systems tested.

6. OXIDIZER SCAVENGERS

In terms of ozonated ISCO processes, the key aquifer soil constituents of concern are high levels of bacteria, total organic carbon, iron, manganese, hydrogen sulfide, and carbonates. The actual level of each that pose potential problems cannot be simply listed because of unique matrix effects that the combination of different matrix chemistries may impart. However, soils containing higher levels of any one constituent listed above than an average soil [Dragun (30)] should offer some level of concern, and thus it is suggested when there is doubt that some laboratory or pilot testing be initiated to ensure site compatibility. One particular scavenger that is not soil-derived and yet can be very problematic is the overdosing of hydrogen peroxide (a common practice with Fenton's Reagent applications) resulting in the scavenging reaction of the hydrogen peroxide with the generated hydroxyl free radical. This counter-productive reaction is well described by Hong et al. (18).

7. PROCESS EQUIPMENT

Since ozone is corrosive with many materials, special consideration is required for selection of process equipment that will contact ozonated media. Stainless steel and teflon are excellent materials for this purpose. Additionally, safety issues concerning the explosiveness of gases evolving from ozonation activities must also be considered.

8. PERMANGANATE SYSTEMS

In situ chemical oxidation (ISCO) of soil and groundwater using permanganate has been examined quite broadly [Vella (31); Leung (32); Vella (33); Gates (34); Yan (35); Schnarr (36); West (37a,b); Siegrist (38 a,b); Lowe (39); Siegrist (40); Yan (41); Struse (42); Urynowicz (43); Gates-

Anderson (44); Crimi (45)]. Potassium permanganate ($KMnO_4$) is a crystalline solid from which aqueous solutions of a desired concentration (mg/L up to wt%) can be prepared on site using groundwater or tapwater, avoiding the cost of transporting large quantities of dilute aqueous chemicals. Several grades of potassium permanganate (e.g., technical, free-flowing, etc.) having different particle sizes and physical properties are commercially available. Selection of the potassium permanganate grade will depend on system design and site characteristics. For example, the free-flowing (not technical) grade contains an additive that makes it more suitable "under high humidity conditions and where the material is to be dry fed through a chemical feeder or stored in a bin or hopper," (Carus Chemical Company, Free-Flowing Grade fact sheet). Sodium permanganate ($NaMnO_4$) is supplied as a concentrated liquid (min. 40 wt% as $NaMnO_4$). In this form, permanganate anion is provided at much higher concentrations and without the potassium (for sites where ^{40}K is a concern), and without dusting hazards associated with dry $KMnO_4$ solids.

The stoichiometry and kinetics of permanganate oxidation at contaminated sites can be quite complex as there are numerous reactions in which Mn can participate due to its multiple valence states and mineral forms. The primary redox reactions for permanganate are giving in equations 42 to 44. These half-cell reactions are useful for two purposes; (1) to evaluate stoichiometric requirements of the oxidant for complete mineralization of contaminants via electron transfer balances (e.g., see example for TCE, equation. 47), and (2) to determine potential environmentally significant reaction products based on reaction conditions. For example, the half-cell reaction for permanganate under acidic conditions involves a 5-electron transfer as shown in equation 42, with Mn^{2+} resulting (regulated in drinking water). In the pH range of 3.5 to 12, the half-cell reaction involves a 3-electron transfer as shown in equation 43 with MnO_2 as the primary reaction product. At high pH (> 12), a single electron transfer occurs as given in equation 44 resulting in MnO_4^{2-}. In these three reactions, manganese is reduced from Mn^{+7} to either Mn^{+2} (eqn. 42), Mn^{+4} (eqn. 43), or Mn^{+6} (eqn. 44). Eqn. 43, which represents the typical half-cell reaction under common environmental conditions, leads to the formation of a manganese dioxide solid.

$$MnO_4^- + 8H^+ + 5e^- \rightarrow Mn^{+2} + 4H_2O \qquad (42)$$

$$MnO_4^- + 2H_2O + 3e^- \rightarrow MnO_2(s) + 4OH^- \qquad (43)$$

$$MnO_4^- + e^- \rightarrow MnO_4^{-2} \qquad (44)$$

The Mn^{+2} cations formed under highly acidic pH conditions (equation 42) can be oxidized subsequently by excess (unreacted) permanganate (equation 45):

$$3MnO_2 + 2MnO_4^- + 2H_2O \rightarrow 5MnO_2(s) + 4H^+ \tag{45}$$

In acidic solutions, the Mn^{+4} in MnO_2 can be reduced slowly to yield Mn^{+2} as shown in equation 46:

$$MnO_2(s) + 4H^+ + 2e^- \rightarrow Mn^{+2} + 2H_2O \tag{46}$$

This is significant from a regulatory perspective since highly acidic conditions may result in long-term elevated concentrations of Mn^{2+}, possibly above regulatory limits, in drinking water. The initial pH and system buffering capacity will influence the potential for this to occur.

Permanganate can also react with water, but at very slow rates, resulting in nonproductive depletion of permanganate and further generation of MnO_2 solids. When reduced species (contaminant or natural) are no longer available to react with permanganate, this slow decomposition process will eventually result in depletion of excess permanganate that may remain in the subsurface after treatment.

Permanganate decomposition and disproportionation reactions can also occur, but only at appreciable rates under extremely high pH. The half-cell reactions of permanganate as a function of pH (equations 42-45) yield the potential for oxidation of reductants in the system. These reductants can include reduced inorganics (e.g., Fe^{+2}, Mn^{+2}), natural organic matter (NOM), or target organic chemicals of concern (e.g., trichloroethylene (TCE), tetrachloroethylene (PCE)).

For degradation of organic chemicals such as TCE, the oxidation involves direct electron transfer (rather than free radical processes that characterize oxidation by Fenton's reagent or ozone). The stoichiometric reaction for TCE destruction occurring at common groundwater pH's of about 3.5 to 12, as an example, is provided in equation 47:

$$2MnO_4^- + C_2HCl_3 \rightarrow 2CO_2 + 2MnO_2 + H^+ + 3Cl^- \tag{47}$$

The rate of organic chemical degradation by permanganate in the absence of substantial natural organic matter (NOM) or other reductants depends on the concentration of both the TCE and the permanganate and can be described by a second-order kinetic expression as given in equation 48.

$$d[R] / dt = -k_2[R][MnO_4^-] \tag{48}$$

where [R] = concentration of the organic being oxidized (mol/L), $[MnO_4^-]$ = the concentration of permanganate in water (mol/L), and k_2 = second-order reaction rate constant $(mol^{-1}Ls^{-1})$. The second-order rate constants reported for TCE degradation, for example, are in the range of 0.6 to 0.9 $mol^{-1}Ls^{-1}$ and the reaction rate appears to be independent of pH (at least over the range of 4-8) and ionic strength (up to 1.57 M Cl^-) [Yan (41); Huang (46); Urynowicz (47)]. It is important to acknowledge that these degradation rates are readily affected by the presence of competing species, such as naturally occurring organic matter or reduced mineral species. Currently it is most common for degradation rates to be determined experimentally on a site-specific basis. The rate of reaction is temperature dependent as described by the Arrhenius eqn. with an activation energy on the order of 35 to 70 kJ/mol [Case (48); Huang (46); Yan (41)], providing a means of extrapolating laboratory temperature-derived rates to field conditions.

Many studies have been conducted examining the reaction of permanganate with chlorinated ethenes [e.g., Schnarr (36); Gates (34); Yan (35); Case (48); Tratnyek (49); Huang (46); Siegrist (50); Struse (51); Yan (41); Urynowicz (43); and others]. Reaction stoichiometries and kinetics have been determined for common contaminants such as TCE, PCE, DCE, and vinyl chloride. In general, halocarbons with higher chlorine substitution consume less oxidant (per the stoichiometric requirement) and produce less MnO_2 solids.

Permanganate oxidation of other contaminants has also been examined to some extent. Under conditions common for ISCO applications, permanganate is not an effective oxidant for degradation of chlorinated alkanes (saturated compounds containing single-bonded carbons), such as 1,1,1-trichloroethane [Gates (34); Tratnyek (49); Gates-Anderson (44)]. Experimental evidence regarding the oxidation of BTEX compounds by permanganate during ISCO suggests that substituted aromatics, containing available electron pairs, could be degraded; however, benzene is stable in the presence of permanganate [Verschueren (52)].

Saturated aliphatic compounds have no readily available electron pairs and are thus not easy to chemically oxidize. It appears that permanganate oxidation is favored in unsaturated compounds containing a carbon-carbon double bond (formed by a single bond coupled with a pi bond of electrons shared between the two carbon atoms forming an "electron cloud" surrounding the single bond) or other electrophilically favorable chemical structures, where the oxidant can readily react with the electron pair provided by the carbon-carbon double bond. Electrons of the pi bonds of aromatic compounds are more tightly shared, and thus more stable, than in aliphatic compounds. However, the available electron pairs constituting the pi bond of both aromatics and aliphatics are more reactive when associated with substituted carbon atoms (i.e., chlorine present or an organic group such as methyl – toluene) due to the longer, less stable bond (here the shared electrons are oriented more "loosely").

Permanganate has been used for chemical oxidation of phenolic compounds during wastewater treatment. According to Vella (31), the oxidation of phenol proceeds through organic acid intermediates before yielding carbon dioxide. Mineralization of phenol consumes a relatively large amount of permanganate (15.7 g of $KMnO_4$ per g of phenol). Polyaromatic hydrocarbons such as naphthalene, phenanthrene, and pyrene can be oxidized by permanganate. Cleavage of one of the aromatic rings can occur. These compounds also exert a high demand for the oxidant. Limited studies with PCBs suggest permanganate is not an effective oxidant for these contaminants. Finally, permanganate oxidation of high explosives (e.g., HMX, RDX, TNT, etc.) has been demonstrated in laboratory studies [IT Corp. and SM Stollar Corp. (53)]. The viability of applying permanganate in the case of high demand contaminants is determined on a case-by-case basis, and will depend on the extent of contamination (i.e., cost), the presence of competing naturally reduced materials (i.e., overall oxidant demand), and treatment goals (including treatment time frames and final contaminant mass/concentration).

9.   CONCLUSION/RECOMMENDATION

The choice of oxidant has been generally quite arbitrary between preferences of the client and/or the vendor (which may have more to do with personal experiences) than sophisticated selection/evaluation methods. A more systematic approach would be to carry out bench scale tests with more than one of the oxidants, looking for reactivity with the soil matrix as well as the contaminants. Some charts typical of established comparisons are included as tables 1 through 4, and are offered more for interest than veracity. The relative oxidizing power of fluorine has been ignored since it is not a typical oxidant of choice for remediation. The comparative oxidizing power is an excellent starting point for those not so familiar with the limitations of each oxidant if it helps orient thinking towards appropriate exclusions (selecting permanganate for saturated alkanes such as the chlorinated paraffins - TCA, etc., or selecting a fast radical chemistry for remediation of a dilute plume of TCE).

Table 1. Impact of Soil Oxidant Demand [54]

Oxidant	Soil Oxidant Demand (Kg/m^3) Average
Permanganate	1.0
Hydroxyl Radical	0.5
Hydrogen Peroxide	0.2
Persulfate	0.1
Ozone	0.1

Table 2. Oxidation Potentials and Costs [54]

Oxidant	Potential (volts)	Potential (Chlorine = 1)	Equivalent Weight (grams)	Cost $/lb	Cost $/1000 Equiv	Weight (Lb) 1000 Equiv
Hydroxyl Radical	2.76	2.06	17			
Persulfate Radical	2.5	1.84	96			
Ozone	2.07	1.52	24			
Sodium Persulfate	2.01	1.48	119	1.08	283	262
Hydrogen Peroxide	1.78	1.31	34	0.75	56	75
Perhydroxyl Radical	1.7	1.25	33			
Permanganate (K/Na)	1.68	1.24	52.6/47.3	1.4/ 5.95	162/620	115.8/ 104.2
Chlorine Dioxide	1.57	1.15	67			
Hypochlorous Acid	1.49	1.1	51			
Chlorine	1.36	1	35			
Calcium Peroxide[1]	0.9	0.66	36	3.	237	105.7
HydroPeroxide Anion[2]	0.88	0.65	33			

Notes: 1. Without pH control (pH = 12), chelant/catalyst, etc.; 2. Aggressive reducing agent

Table 3a. Reactivity of Oxidants

Oxidant	Amenable CVOCs	Reluctant CVOCs	Recalcitrant CVOCs
Hydroxyl Radical	PCE, TCE, DCE, VC, CB	DCA, CH_2Cl_2	TCA, CT, $CHCl_3$
CPP[1]	PCE, TCE, DCE, VC, CB, DCA, TCA	CH_2Cl_2, CT, $CHCl_3$	
Calcium Peroxide*	PCE, TCE, DCE, VC, CB, DCA, TCA	CT, $CHCl_3$	
Permanganate (K/Na)	PCE, TCE, DCE, VC	NR	NR

Oxidant	Amenable CVOCs	Reluctant CVOCs	Recalcitrant CVOCs
Persulfate/Fe[11]	PCE, TCE, DCE, VC, CB	NR	NR
Persulfate Radical (Heat: 60+ $^\circ$C)	Chlorinated Alkenes	Common Chlorinated Alkanes[11]	Uncommon Chlorinated Alkanes[11]

Note: * pH Modified to Obtain Desired Reaction System (W. Lundy/ Cool -Ox ^{TM})

Table 3b. Reactivity of Oxidants [54]

Oxidant	B	TEX	PAHs	Phenols	Explosives	PCBs	Pesticides
Hydroxyl Radical	H	H	M	H	M	L	L
CPP[3]	H	H	H	H	H	H	H
Permanganate (K/Na)	NR	H	H	H	H	L	M
Persulfate/Fe[11]	H	H	M	H	M	L	M
Persulfate Radical (Heat)	H	H	H	H	H	H	H
Ozone	M	M	H	H	H	H	H
Peroxone	H	H	H	H	M	L	L

ACKNOWLEDGEMENTS

The chemistries were developed with contributions from Prof. R. Watts, WSU, Pullman, WA (Peroxide); Prof. C. Bruell, Umass Lowell, MA (Persulfate); Prof. M. Zappi, MSU, MO; and Dr. M. Crimi, Colo. School of Mines, CO (Permanganate).

REFERENCES

1. R. J. Watts; Private communication, 2003a; Watts, R. J., Bottenburg, B. C., Hess, T. F., Jensen, M. D., Teel, A. L., "Role of Reductants in the Enhanced Desorption and Transformation of Chloroaliphatic Compounds by Modified Fenton's Reactions", *Environ. Sci. Technol.,* 33, 3432-3437, 1999b.
2. Osgerby, I.T., Takemoto, H.Y., Watts, R.W. "PCB Destruction in Pacific Soils Using Modified & Fenton Like Reactions." Battelle Conference on Remediation of Recalcitrant Chlorinated Compounds, May, 2002a; Osgerby, I.T., Takemoto, H.Y., Watts, R.J., Lundy,W. "Approaches to treating contaminated soils in the Pacific Basin", May 2004b.
3. Lundy, W., Private communication, 2003
4. Nuttal, H. E.; Bowden, J; Lundy, W.L. "Remediation of a PAH Release at a Railroad Switchyard"; Nuttal, H. E.; Alin, S.; Lundy, W. L., "Remediation of a VOC Plume at an Industrial Park".

5. Kislenko, V.N., Berlin, A.A., Litovchenko, N.V. 1995. "Kinetics of glucose oxidation with persulfate ions, catalyz ed by iron salts". *Russian Journal of General Chemistry* Part 2 65(7): pp1092-1096

6. Kolthoff, I.M., Medalia, A.I., Raaen, H.P. 1951. "The reaction between ferrous iron and peroxides. IV. Reaction with potassium persulfate". *Journal of American Chemical Society* 73: 1733-1739

7. Liang, C., Bruell, C. J. (UMass, Lowell), Marley, M. C., Sperry, K.L. (XDD); Persulfate Oxidation for *In situ* Remediation of TCE: I. Activated by Ferrous Ion with and without a Persulfate-Thiosulfate Redox Couple: *Chemosphere* v55 pp 1213-1223 2004a; Persulfate Oxidation for *In situ* Remediation of TCE: II. Activated by Chelated Ferrous Ion. Chemosphere v 155 pp 1225-1233 2004b; Thermally Activated Persulfate Oxidation of TCE and 1,1,1-TCA in Aqueous Systems and Soil Slurries. *Soil & Sediment Contamination* 12(2): 2-7 – 228, 2003c.

8. Banerjee, M.; Konar, R.S. 1984. Comment on the paper " Polymerization of acrylonitrile initiated by $K_2S_2O_8$-Fe(II) redox system, *Journal of Polymer Science: Polymer Chemistry Edition* 22: 1193-1195

9. Private communication; Dr. Clifford Bruell; Dr. Chenju Liang, UMass, Lowell.

10. Hoag, G. E.; Private communication.

11. Dahmani, A., Huang, K., Hoag, G., Zhao, Zhiqiang, Block, P: Degradation of Volatile Organic Compounds with Activated Pe rsulfate, 19th. Int. Soils and Sediments Conf., UMass, MA, October 2003

12. Lewis, R., ERM/New England: Private communication

13. Pac, T., ERM/New England: Private communication

14. Qui, Y., Zappi, M., Kuo, C., and Fleming, E., 1999, "A Kinetic and Me chanistic Study of the Ozonation of Dichlorophenols in Aqueous Solutions", *Journal of Environmental Engineering,* V125, N5, pp. 441-450

15. Zappi, M., 1995, "Peroxone Oxidation of 2,4,6,-Trinitrotoluene With and Without Sonolytic Catalyza tion", Dissertation presented to the Chemical Engineering Faculty of Mississippi State University.

16. Langlais, B., Reckhow, D., and Brink, D., 1991, "O zone in Water Treatment", Lewis Publishers Inc., Chelsea, MI

17. Glaze, W. and Kang, 1988, "Advanced Oxidation Processes for Treating Groundwater Contaminated With TCE and PCE" Laboratory Studies@, *Journal of the American Water Works Association*, V80.

18. Hong, A., Zappi, M., Kuo, C., and Hill, D., 1996, "Modeling the Kinetics of Illuminated and Dark Advanced Oxidation Processes", *Environmental Engineering,* V122, N1.

19. Adams, C. and Randtke, S., 1992, "Ozonation Byproducts of Atrazine in Synthetic and Natural Waters", *Environmental Science and Technology*, V26, N11.

20. Trapido, M., Veressinina, J., and Munter, R., 1994, "Ozonation and AOP Treatment of Phenanthrene in Aqueous Solutions", *Ozone Science and Engineering,* v16.

21. Yao, C., Spanggord, R., and Mill, T., 1996, "Peroxone Oxidation of Aminodinitrotoluenes", Proceedings of the Second Advanced Oxidation Conference, San Francisco, CA.

22. Staehlin, J. and Hoigne, J. 1982, "Decomposition of Ozone in Water: Rate of Initiation by Hydroxide Ions an d Hydrogen Peroxide ", *Environ. Sci. Technol*., v16, pp 2100-2106.

23. Hoigne, J., and Bader, H., 1983, "Rate Constants of Reactions of Ozone With Organic and Inorganic Compounds In Water -II", *Water Research*, v17.

24. Nelson, C. and Brown, R., 1994, "Adapting Ozonation for Soil and Groundwater Cleanup", *Chemical Engineering,* Nov. Issue.

25. Masten, S. and Davies, S., 1998, "Efficacy of *In situ* Oxidation for the Remediation of PAH Contaminated Soils", *Journal of Contaminant Hydrology,* v28, pp. 327 - 335.

26. Taube, H. and Bray, W., 1940, "Chain Reactions in Aqueous Solutions Containing Ozone, Hydrogen Peroxide , and Acid",. *Journal of the American Chemical Society* , V62.

27. Kuo, C. and Chen, S., 1996, "Ozonation and Peroxone Oxidation of Toluene in Aqueous Solutions", *Industrial and Engineering Chemistry Research,* v35, N11.

28. Wang, W., Zappi, M., Albritton, G., Crawley, A., Singletary, J., Hall, N., Karr , L., 2001, "Using Chemical Priming as a Means of Enhancing the Performance of Biocells for Treating Petroleum Products Containing Recalcitrant Species", MSU E-TECH Report 2001, Mississippi State University.

29. Tiang, I., and Zappi, M., 2003, "Chemical Oxidation Enhanced Bioremediation of Polycyclic Aromatic Hydrocarbon Contaminated Sediment", MSU E-TECH Report, Mississippi State University.

30. Dragun, J. and Chiasson , A., 1991, " Elements in North American Soils", Hazardous Materials Control Resources Institute, Greenbelt, MD

31. Vella P.A., G. Deshinsky, J.E. Boll, J. Munder, and W.M. Joyce (1990). "Treatment of low level phenols with potassium permanganate". *Res. Jour. Water Pollution Con t. Fed.,* 62(7):907-914.

32. Leung S.W., R.J. Watts, G.C. Miller (1992). "Degradation of perchloroethylene by Fenton's reagent: speciation and pathway". *J. Environ. Qual*. 21:377-381.

33. Vella P.A., and B. Veronda (1994). "Oxidation of trichloroethene: a comparison of potassium permanganate and Fenton's reagent". 3rd Intern. Symposium on Chemical Oxidation. In: *In situ* Chemical Oxidation of the Nineties. Vol. 3. Technomic Publishing Co., Inc. Lancaster, PA. pp. 62 -73.

34. Gates D.D., R.L. Siegrist, and S.R. Cline (1995). "Chemical oxidation of contaminants in clay or sandy soil". Proceedings of ASCE National Conference on Environmental Engineering. Am. Soc. of Civil Eng., Pittsburgh, PA.

35. Yan Y. and F.W. Schw artz (1996). "*In situ* oxidative dechlorination of trichloroethene by potassium permanganate". Proc. Third International Conference on AOTs. October 26-29, Cincinnati, OH.

36. Schnarr M.J. and G.J. Farquhar (1992). "An *in situ* oxidation technique to destroy residual DNAPL from soil". Subsurface Restoration Conference, Third International Conference on Groundwater Quality, Dallas, TX. June 21 -24, 1992.

37. West O.R., S.R. Cline, W.L. Holden, F.G. Gardner, B.M. Schlosser, J.E. Thate, D.A. Pickering, and T.C. Houck (1998a). "A full -scale field demonstration of *in situ* chemical oxidation through recirculation at the X -701B site". Oak Ridge National Laboratory Report, ORNL/TM-13556; West O.R., S.R. Cline, R.L. Siegrist, T.C. Houck, W.L. Holden, F.G. Gardner, and R.M. Schlosser (1998b). "A field -scale test of *in situ* chemical oxidation through recirculation". Proc. Spectrum '98 International Conference on Nuclear and Hazardous Waste Management. Denver, Colorado, Sept. 13-18, pp. 1051-1057.

38 Siegrist R.L., K.S. Lowe, L.D. Murdoch, W.W. Slack, and T.C. Houck (1998a). "X -231A demonstration of *in situ* remediation of DNAPL compounds in low permeability media by soil fracturing with the rmally enhanced mass recovery or reactive barrier destruction". Oak Ridge National Laboratory Report. ORNL/TM -13534; Siegrist R.L., K.S. Lowe, L.D. Murdoch, T.L. Case, D.A. Pickering, and T.C. Houck (1998b). "Horizontal treatment barriers of fracture-emplaced iron and permanganate particles". NATO/CCMS Pilot Study Special Session on Treatment Walls and Permeable Reactive Barriers. EPA 542-R-98-003. May 1998. pp. 77 -82.

39. Lowe K.S., F.G. Gardner, R.L. Siegrist, and T.C. Houck (1999). "Field pilot test of *in situ* chemical oxidation through recirculation using vertical wells at the Portsmouth Gaseous Diffusion Plant". EPA/625/R-99/012. US EPA ORD, Washington, DC. pp. 42-49.

40. Siegrist R.L., K.S. Lowe, D.R. Smuin, O.R. West, J.S. Gunderson, N.E. Korte, D.A. Pickering, and T.C. Houck (1998c). "Permeation dispersal of reactive fluids for *in situ* remediation: field studies". Project report prepared by Oak Ridge National Laboratory for US DOE Office of Science & Technology. ORNL/TM-13596.

41. Yan Y. and F.W. Schwartz (1999). "Oxidative degradation and kinetics of chlorinated ethylenes by potassium permanga nate". *Journal of Contaminant Hydrology,* 37:343-365

42. Struse A.M. and R.L. Siegrist (2000). "Permanganate transport and matrix interactions in silty clay soils". In: Wickramanayake, G.B., A.R. Gavaskar, and A.S.C. Chen (ed.) Chemical Oxidation and Re active Barriers. Battelle Press, Columbus, OH. pp. 67-74.

43. Urynowicz M.A., and R.L. Siegrist (2000). "Chemical degradation of TCE DNAPL by permanganate". In: Wickramanayake, G.B., A.R. Gavaskar, and A.S.C. Chen (ed.). Chemical Oxidation and Reactive Barriers. Battelle Press, Columbus, OH. pp. 75-82.

44. Gates-Anderson D.D., R.L. Siegrist, and S.R. Cline (2001). "Comparison of potassium permanganate and hydrogen peroxide as chemical oxidants for organic ally contaminanted soils". *J. Environmental Engineering,* 127(4):337-347.

45. Crimi M.L. and R.L. Siegrist (2003). "Geochemical Effects on Metals Following Permanganate Oxidation of DNAPLs". *Groundwater*, 41(4):458-469.

46. Huang K., G.E. Hoag, P. Chhed a, B.A. Woody, and G.M. Dobbs (1999). "Kinetic study of oxidation of trichlorethene by potassium permanganate". *Environmental Engineering Science*, 16(4):265-274.

47. Urynowicz M.A. (2000). "Reaction kinetics and mass transfer during *in situ* oxidation of dissolved and DNAPL trichloroethene and permanganate". Ph.D. dissertation, Environmental Science and Engineering Division, Colorado School of Mines. Golden, CO. May 2000.

48. Case T.L. (1997). "Reactive permanganate grouts for horizon tal permeable barriers and *in situ* treatment of groundwater ". M.S. Thesis, Colorado School of Mines, Golden, CO.

49. Tratnyek P.G., T.L. Johnson, S.D. Warner, H.S. Clarke, and J.A. Baker (1998). " *In situ* treatment of organics by sequential reduction and oxidation". Proc. First Intern. Conf. on Remediation of Chlorinated and Recalcitrant Compounds. Monterey, CA. pp. 371 -376.

50. Siegrist R.L., K.S. Lowe, L.C. Murdoch, T.L. Case, and D.A. Pickering (1999). " *In situ* oxidation by fracture emplaced reactive solids". *J. Environmental Engineering,* 125(5):429-440.

51. Struse A.M. (1999). "Mass transport of potassium permanganate in low permeability media and matrix interactions". M.S. thesis, Colorado School of Mines. Golden, CO.

52. Verschueren, K. (1983). "Handbook of Environmental Data on Organic Chemicals". Van Nostrand Reinhold Company, New York. p. 239, p. 1105, p. 1194.

53. IT Corporation and SM Stollar Corporation (2000). "Implementation report of remediation technology screening and treatability testing of possible remediation techniques for the Pantex perched aquifer". October, 2000. DOE Pantex Plant, Amarillo, TX.

54. Brown, R., 2003. AFCEE Technology Transfer Workshop

CHAPTER 20

BROWNFIELD SITE ASSESSMENT AND REMEDIATION
City of Lawrence Gateway Bridge Project

Ronald Richards[1], Christen Sardano[1], Lester Tyrala[1], and John Zupkus[2]
[1]Stone & Webster Massachusetts, Inc., 100 Technology Center Drive, Stoughton, MA 02072; [2]Massachusetts Department of Environmental Protection, 1 Winter Street, Boston, MA 02108

Abstract: This paper describes how the Massachusetts Brownfields process was implemented at the Oxford Paper Mill Site in Lawrence, Massachusetts. The City employed a dynamic process where the community, City, and regulators developed a detailed plan to address integration of stakeholder schedules and provide the necessary monies to remediate a three acre Brownfield site. Successful implementation of this complex process is allowing the City to abate contamination, develop a large-scale park, and construct the Gateway Bridge. This is considered integral to future development of Lawrence.

Significant stakeholders include the City of Lawrence, Massachusetts Department of Environmental Protection (MDEP), Massachusetts Highway Department (MHD), the Environmental Protection Agency (EPA), Mass Development, and Community Action Groups.

Contaminants of concern at the site include asbestos, PCBs and PAHs. On site concentrations of PCBs significantly exceed the EPA and Massachusetts Contingency Plan (MCP) cleanup goals.

The challenges for this project include assisting the City in obtaining sufficient funding, coordinating with the abutting property owner cleanup of the raceway (man-made waterway) that runs through the site, and coordinating areas of remediation with MHD and the City.

Key words: asbestos, Massachusetts, PAH, PCB, remediate

1.   INTRODUCTION

Since 1999, the City of Lawrence, Massachusetts has been focused on remediation and site closure at the Oxford Paper Mill (OPM) Site. The path forward has been a highly participatory process with the City, the US Environmental Protection Agency (EPA), the Massachusetts Department of Environmental Protection (DEP), the US Army Corps of Engineers (USACE), citizen groups, the Massachusetts Highway Department (MHD), MassDevelopment, abutting property owners, and regional planning groups all involved. The unique synergy of these groups has provided the means and impetus to move the project forward. However, the need to move forward with appropriate agreement by all parties has resulted in a slower pace. This paper provides details regarding the framework that has been developed to move the project forward and the funding sources that have been identified.

2.   BACKGROUND

Among the Northeast's oldest and most historic mill cities, Lawrence sits along the banks of the Merrimack River (Figure 1). It came to prominence in the mid-nineteenth century with the production of textiles as its primary industry. Using the transportation power of the Merrimack River, the city's industrial roots were planted on an extensive system of canals, penstocks, and turbines that powered and transported raw materials to the city's huge mills. By the 1970s, the textile industry began to collapse due to the increase in foreign imports. Resulting in thousands of workers being laid off. Dozens of factories shut down operations. Many of these factories remained abandoned or underused for a generation.

Figure 1. Site Locus Map, Former Oxford Paper Mill, Lawrence, MA

Today the city is among the poorest in New England, facing one of the area's greatest poverty and unemployment rates. The average household income in 1999 was only $24,500. Over 21% of the population lived below the poverty line in 2000. However, the city encompasses a large and community-oriented Latino population. People of Latino heritage make up over 54% of the city's population and 22% of Latino residents are of Puerto Rican heritage.

Through a coordinated effort between public agencies, businesses, and community groups, an all-encompassing plan for economic, environmental, and quality of life renewal and development called the Lawrence Gateway Project (LGP) was created.

Named after the historic downtown canal district that comprises the redevelopment target area, the LGP is funded through private and public sources – including over $1 million in EPA Brownfields funds provided to the city since 1996. The multimillion dollar project focuses on transportation and infrastructure redevelopment, assessment and cleanup of the many mill complexes located in the target area, transformation of landfills, creation of affordable housing, and the implementation of numerous quality-of-life and community organized programs.

The LGP will use private and public investments to revitalize the city's downtown residential, commercial, and industrial centers that have been most affected by economic and environmental hardship. The Gateway Project promotes a practical and balanced leveraging of projects that, taken collectively, make up a new and stronger foundation for the community. The well-balanced plan encompasses all levels of city, community, state, and federal assistance and will create lasting change in Lawrence through various focused plans of action and remediation.

Transportation improvements center around the Interstate-495 Interchange project made possible by the Massachusetts Highway Department (MHD). The new interstate interchange will create a defined gateway into the downtown corridor and become a catalyst of change. As more traffic flows into the area, so does more potential business and the need to improve the area's appearance for passersby. The link to I-495 would also improve traffic flow patterns into the Gateway area and link the interstate to the new Spicket River Bridge. Additionally, ramps will be realigned and roads and bridges (like the Spicket River Bridge) will be constructed or reconstructed.

The program will also address the problem of the vast number of abandoned or underused mills. The project targets priority sites and calls for environmental site assessments at each site using portions of the grant monies offered through the EPA Brownfield Assessment Grant Program. To date, the Assessment Grant Program has awarded the city of Lawrence $400,000 to help assess several key properties within the Gateway area, including the former OPM Site.

The OPM Site, formerly used to manufacture printing paper, is now entering its final stage of remediation following the extensive efforts of the city and local contractors. A portion of the property is set to become a park that will offer the urban population a chance to enjoy open greenspace everyday.

The remediation of the nearby GenCorp property is also a major cornerstone of the Gateway project. The GenCorp site, an 8.6-acre Brownfield property within the Gateway area, has an industrial history dating back to 1848 and has an extensive history of contamination. GenCorp, a Fortune 500 company, purchased the property in 1955 and manufactured rigid and soft plastic products until the plant closed down in 1981 due to poor economic conditions and excess capacity. The estimated $100 million investigation and remediation of the GenCorp site is in its final stages and has been conducted voluntarily and entirely at the expense of GenCorp under the approval authority of EPA. The property will eventually be redeveloped into needed parking and landscaped community spaces.

The all-encompassing LGP will bring needed and permanent change to the urban community and has already inspired the very culture-oriented community to come together in the spirit of community involvement and education. Three major community groups have developed together alongside the LGP.

Lawrence Community Works, Inc. is a non-profit organization that focuses on job training, improved housing, and the creation of economic opportunities during Gateway remediation. The organization has successfully implemented the Summer Street Home Ownership Project and the Our House Family Learning Center.

Groundwork Lawrence was established by the city of Lawrence in partnership with the EPA and the National Park Service. The goal of Groundwork is to bring improvement to the environment through community partnerships. Groundwork supports park improvement, cleanup and planting days, Adopt-a-Space programs, and environmental education initiatives.

The Groundwork team also partnered with the Reviviendo Gateway Initiative, a 38-member steering committee that formed a guide to development that keeps the safety and happiness of the community in mind through all aspects of the Gateway remediation project.

3. OPM SITE CONDITIONS

The OPM Site has been divided into three areas (Figure 2) that are being remediated through both the DEP Massachusetts Contingency Plan (MCP) and federal EPA processes. Area 1 includes areas south of the raceway that transverses the property, including what is commonly referred to as the south raceway wall, the raceway itself, the arches and bridge that span the raceway, and the area east of the Spicket River that includes Building 28 (also referred to as the South Area). RAM Plans for Area 1 are currently being conducted for work done by the MHD and the City of Lawrence. Area 2 consists of an approximate 30-foot strip adjacent to and along the raceway to the north, including what is commonly referred to as the north raceway wall (also referred to as the wedge). Area 3 includes the remaining portion of the site north of the raceway (also referred to as the North Area).

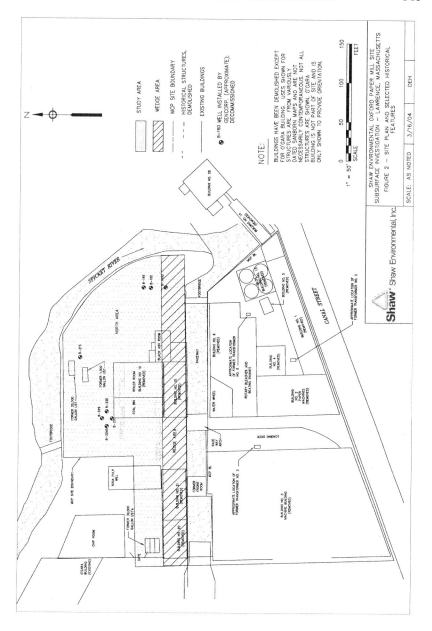

Figure 2. Site Plan and Selected Historical Features

GenCorp, which owns the abutting property, has taken responsibility for the remediation of polychlorinated biphenyl (PCB) contaminated sediments (after MHD removes the debris from the raceway) that are found in the existing raceway, which transects the OPM Site.

At present, it is anticipated that the entire area north of the raceway will be used for parkland with a passive- or low-intensity nature. The use would be restricted via an Activity and Use Limitation that would require either grass or pavement to be maintained on the site, and no facilities or activities would be proposed that would result in high-intensity use.

Within the OPM Site, there are no buildings on the area north of the raceway due to demolition activities that occurred during the 1970s. Metcalf & Eddy (M&E) investigated an area immediately north of the raceway, referred to as the wedge area, as part of an EPA Targeted Brownfields Assessment (TBA) in May of 2002. This assessment consisted of a geophysical survey to assess the presence of buried structures and a test pit investigation to sample and analyze the soils. The results of this investigation concluded that the primary contamination found in the soils of the wedge area were PCBs and asbestos-containing materials (ACM), although certain metals and polycyclic aromatic hydrocarbons (PAHs) were also detected. Figure 3 shows the complete area north of the raceway.

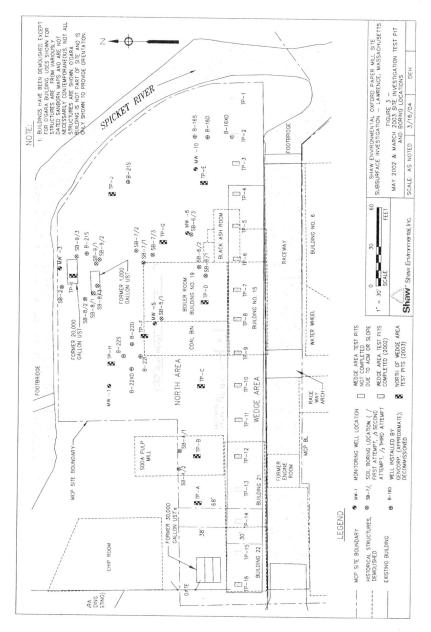

Figure 3. May 2003 & March 2003 Site Investigation Test Pit and Boring Locations

The area north of the wedge, referred to as the North Area, was investigated in March of 2003 through a joint effort between Stone &

Webster Massachusetts, Inc. (S&W) and M&E. Test pits, soil borings, and monitoring-well installations were conducted to evaluate the nature of contamination in the North Area. The results of this investigation concluded that the primary contamination found in the soils of the North Area were ACM, metals, and PAHs. PCBs were not detected in the soil or groundwater of the North Area.

4.   S&W AND M&E STUDIES

Environmental investigations for the area north of the raceway at the OPM Site were conducted by M&E and S&W from 2000 to 2003. The following section presents information regarding the contaminant conditions north of the raceway (wedge and North Area) at the OPM Site.

4.1   S&W, 2003

In August of 2003, S&W collected 12 surface samples (0 to 1 foot) along the perimeter boundaries of the North Area and analyzed the samples for ACM. The majority of the surface samples were collected along the banks of the Spicket River (see Figure 4). The primary purpose of this sampling effort was to provide supporting data for use in determining the placement of a geotextile cap extending from the wedge area to the boundaries of the North Area. All samples were tested for the presence of asbestos (PLM Qualitative) and, subsequently, each sample that was found positive from the qualitative analyses was then tested quantitatively using the EPA Region 1 protocol.

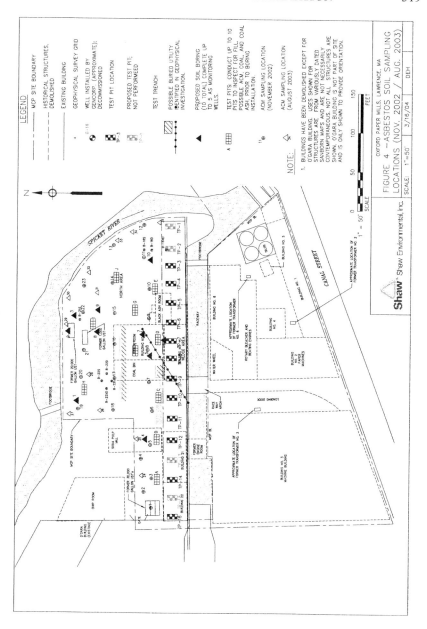

Figure 4. Asbestos Soil Sampling Locations (Nov. 2002 / Aug. 2003)

4.2   M&E, 2003 – "Technical Memorandum and Method 3 Risk Characterization Oxford Paper Mill Site – Area North of Raceway and North of Bridge Construction Area"

In March 2003, through a joint effort between M&E and S&W, investigations to the north of the wedge and raceway areas (North Area) were conducted to support the performance of a MCP Method 3 Risk Characterization and to prepare a planning-level cost estimate for remediation of the North Area. This investigation consisted of test pits, soil borings, and well installations located throughout the North Area (see Figure 3). The results of this investigation concluded that the primary contamination found in the soils of the North Area were asbestos, PAHs and certain metals, while PCBs were all below MCP reportable concentrations. Groundwater sampled in the North Area indicated that selenium and vanadium were present at concentrations above MCP reportable concentrations. It is believed that these compounds are present due to the presence of coal ash. PCBs were not detected in any of the groundwater samples. GenCorp will conduct remedial actions in the raceway area sediments after MHD has removed the raceway debris.

The investigation of the North Area included test pits to evaluate and characterize the type of fill and obstructions present. A total of 10 test pits were excavated throughout the North Area. The objectives of the test pits were divided into four general categories:

1. Evaluate geophysical anomalies that were discovered during the TBA investigation in 2002 (M&E).
2. Determine the extent and depth of fill (up to 15 feet deep).
3. Characterize the type of fill across the site with particular attention to the amount of asbestos.
4. Investigate former UST graves.

The possible explanation for the geophysical anomaly (pipeline that appears to run from the engine room to a stack) identified in the electromagnetic (EM) survey and ground penetrating radar (GPR) survey during the TBA investigation in 2002 (M&E) was not discovered during the March 2003 M&E/S&W investigation. This anomaly was not reached due to two-feet-thick brick and concrete obstructions. The lateral extent of fill was undetermined due to the inability to access the slopes to the Spicket River. All 10 test pits contained fill and building debris. Most of the test pits contained large quantities of debris, building walls, foundations, and floors. No native soil was found in any test pit. Since the test pits cover a significant portion of the site (test pits from March 2003 and May 2002), it is reasonable to conclude that the entire Site has been filled to a depth of 15 feet or more. During test pit activities in the former UST graves, field observations did not indicate the presence of petroleum contamination in any of the soils. Also during test pit investigations, bulk samples of asbestos-looking material were taken and later confirmed to be positive.

Within this program, there were 10 soil borings, of which five were completed as monitoring wells. The soil borings were divided into five groups:

- Upgradient of the former Site buildings (SB-1)
- Downgradient of the former Site buildings (SB-10)
- In the area of the former USTs (SBs 2, 3, 8, and 9)
- In the vicinity of the former soda pulp mill (SB-4)
- In the vicinity of the wedge area TP-7 (SBs 5, 6, and 7)

During the TBA in May of 2002, M&E observed a sheen on the water found in test pit 7. Due to the difficult nature of drilling at the site, because of subsurface obstructions, attempts to reach the target depth were generally limited to a maximum of three attempts at each location. Monitoring wells were installed at locations 1, 3, 5, 6, and 10. The results of this investigation concluded that the primary contamination found in the soils of the North Area were asbestos, PAHs, and certain metals; while PCBs were below MCP reportable concentrations. The soil borings were designed to collect data for a Method 3 Risk Characterization and, in particular, to determine the extent of PCB contamination.

The groundwater was not sampled by M&E as part of their TBA in May 2002. Monitoring wells MW-1 and MW-3, and MW-10, were typical overburden monitoring-well installations. MW-5 and MW-6 were both situated so that the bentonite plug straddled a 12-inch concrete/brick floor that was penetrated during drilling, effectively isolating the groundwater from water running off the former foundation floors of the buildings. Groundwater sampled in the North Area concluded that selenium (188 µg/L) and vanadium (3,710 µg/L) were present at concentrations above MCP reportable concentrations. These concentrations are related to the selenium and vanadium found in coal ash that is abundant throughout the Site, which perhaps leached into the groundwater. PCBs were not detected in any of the groundwater samples.

4.3   M&E, November 2002 – "Targeted Brownfield Assessment Final Report: Test Pit Investigation of Area North of Raceway to be Graded for Bridge Construction"

In May of 2002, M&E, under contract to the EPA, completed a test pit investigation of an area north of the raceway known as the wedge area, which was conducted as part of a TBA. The test pit investigation concluded that soils were primarily contaminated with PCBs, asbestos, PAHs, and certain metals.

PCB analysis was done by mobile and fixed laboratories. PCB analysis was conducted by the mobile laboratory for samples for every one-to-two feet of soil collected from each of the 12 test pits. The mobile laboratory results indicated the presence of low levels of PCBs (1 to 15 mg/kg) throughout the wedge area soils. The fixed laboratory sample results did not confirm the mobile laboratory results in all cases. It was determined that the mobile laboratory sample results provided a conservative measurement of the PCBs present, because the mobile laboratory sample results were nearly always higher than the fixed laboratory sample results.   Since the PCB concentrations do not show a pattern with either depth or test pit location, segregation of PCB-contaminated wedge material from uncontaminated material is not believed to be feasible. The concentrations do not differ over a wide range, and are uniformly less than 50 mg/kg.

As noted in the TBA, together with the site knowledge of the Region 1 PCB coordinator and GenCorp representatives, initial releases of PCBs at the OPM Site occurred before 1978. Disposal of pre-1978 wastes that have PCB concentrations less than 50 mg/kg is not regulated under the Toxic Substances Control Act (TSCA). However, the PCB concentrations in the wedge area soils are greater than what can be accepted by Massachusetts landfills and/or recycling/reuse facilities (2 mg/kg or less).

Soil samples and bulk waste samples were collected for asbestos analysis, based on field team observations of suspect ACM and asbestos warning tape. The presence of asbestos in soils was confirmed by laboratory analysis at 9 of 12 test pit locations. Soil asbestos concentrations ranged from less than 1% (detectable) to several percent, and bulk ACM was also discovered. ACM was not visually observed at all locations, and it is considered likely that it is present throughout soils of the wedge area, even though some soils contained concentrations less than 1%.

Surface soil samples at each test pit location were screened for arsenic, lead, and chromium in a mobile laboratory using a x-ray fluorescence (XRF) instrument. Samples from each of the test pit locations were also sent to a fixed laboratory for analysis. For wedge area soils from test pits located furthest east, arsenic concentrations are elevated relative to samples from other areas. The elevated arsenic concentrations (>60 mg/kg) may be partially due to the presence of coal ash in that area. A "black ash room" and "boiler room" were formerly located in this area. Metals detected in the soil are believed to primarily originate from the ash. Toxic Characteristics Leaching Procedure (TCLP) extraction and analysis did not reveal concentrations of leachable metals in excess of toxicity characteristic regulatory levels.

The primary petroleum hydrocarbon fraction present in wedge area soils is the aromatic fraction, plus target PAH compounds. The presence of PAHs is consistent with field observations of coal and wood ash, coal chips, asphalt, and partially burned timbers. The PAHs found in wedge area soils are likely to be predominantly from asphalt, coal, coal ash, and wood ash, but petroleum and/or creosote releases cannot be definitely excluded as contributors to the PAH contamination. The highest total PAH concentrations (>100 mg/kg) were detected around the vicinity of the former "black ash" room.

As defined by the Massachusetts hazardous waste regulations (310 CMR 30.000), a waste exhibits hazardous characteristics if it is ignitable, corrosive, reactive, or toxic. Based on visual and laboratory evidence, it was confirmed that the wedge area soils that were sampled were non-corrosive and non-reactive.

4.4 S&W, 2002

In November of 2002, S&W collected and analyzed surface soil samples from the area north of the raceway for ACM. Twenty-three (23) samples were collected using a grid-sampling plan within the area north of the raceway (Figure 4) and north of the wedge area that was investigated in the TBA by M&E. The primary purpose of this sampling effort was to provide supporting data for use in determining the engineering controls, air monitoring requirements, and action levels that were emphasized during the investigations in March 2003. All samples were tested for the presence of asbestos (PLM Qualitative) and, subsequently, each sample that was found positive from the qualitative analyses was then tested quantitatively using the EPA Region 1 protocol.

Since asbestos has been identified at the Site in both surface soil and subsurface soil, an asbestos air monitoring program for workers and surrounding receptors was implemented for the investigations in March 2003.

5.   FUNDING SOURCES

Brownfields development monies have been responsible for turning idle, underused properties into beneficial use. Due to their location in urban industrial areas, most of these properties have pre-existing environmental impacts that require investigation, characterization, and often remediation before they can be put back to beneficial use. State and federal Brownfields programs have facilitated this process at OPM, providing several financial tools to pay for site investigation as summarized below.

5.1   EPA Demonstration Pilots

In New England, approximately $8 million in EPA grant money has been awarded primarily to municipalities who hire environmental consulting firms to perform investigations and evaluate remedial options. $150,000 from the EPA was used to assess the OPM Site.

5.2   EPA Targeted Brownfields Assessments

Approximately $1.6 million has been spent by the EPA to perform Brownfields site assessments and RI/FS-type work. This funding source was used to assess the area north of the raceway.

5.3   Massachusetts Brownfields Redevelopment Fund (BRF)

Monies from the MassDevelopment Brownfields fund are being used to cleanup remaining contamination on the OPM Site to allow passive recreation and the construction of the new bridge over the Spicket River.

5.4   EPA Revolving Loan Fund

The EPA also awards municipalities low interest loans on a competitive bid process. The RLF pilot funding will be combined with both Community Development Block Grant funds and US Housing and Urban Development Section 108 loans (BEDI) to complete the project remediation.

5.5   Other Sources

The OPM Site will move to completion through the use of additional funding including, monies and services provided by abutting property owners, and transportation/parking monies used for providing parking at an adjacent property.

CHAPTER 21

TREATMENT OF PCP-CONTAMINATED SOIL USING AN ENGINEERED *EX SITU* BIOPILE PROCESS ON A FORMER WOOD TREATMENT SUPERFUND SITE

Carl Rodzewich[1] , Christian Bélanger, Ph.D[2], Nicolas Moreau[1], Michel Pouliot[1], and Nile Fellows[3]

[1]*Biogenie Corporation, 2085 Quaker Pointe Drive, Quakertown, PA 18951;* [2]*Biogenie S.R.D.C. Inc., 350, rue Franquet, entrée 10, Sainte -Foy, Québec, Canada G1P 4P3;* [3]*Minnesota Pollution Agency, 520 Lafayette Rd, St. Paul, MN 55155*

Abstract: Releases of creosote and pentachlorophenol (PCP) at a New Brighton, Minnesota wood treatment facility, resulted in widespread soil and groundwater contamination. Site investigations lead to the facility's inclusion on the National Priorities List (NPL), and the Minnesota Pollution Control Agency was designated lead regulatory agency. Biological treatment was the preferred remedy for 18,000 yd³ of contaminated soil, and Biogenie was contracted to design, construct and operate an *ex situ* Biopile to achieve the remedial objectives.

Most project owners and remedial managers perceive the Biopile proc ess as a passive remediation technology, not fully appreciating the scientific and engineering expertise required to design and operate an effective Biopile. This paper highlights the "behind the scenes" efforts of the scientific and engineering team responsible for a Biopile project, and specifically how those efforts achieved difficult remedial objectives within a treatment performance guarantee contract.

The New Brighton project required that an *ex situ* Biopile be used to achieve site reuse criteria of 10 mg/kg PCP and 5 mg/kg cPAHs, with a schedule dictating operation during winter months reaching -22°C. Biopile operating parameters were developed during laboratory Treatability Studies designed to optimize the biodegradation capacity of indigenous microorganisms. The TS included the configuration of miniature Biopiles with various amendment and operational strategies to stimulate and sustain preferential bacteria.

Additionally, specific operational paramet ers were developed to biodegrade
soil containing a large percentage of highly impacted woodchips.

An engineering team applied the TS results during scale -up Biopile design and
the project team mobilized to the site for construction. Average PCP reduction
rates achieved 95% for soils (Phase I) and 76% for soils containing wood
chips (Phase II), and over 96% of the contaminated soil achieved treatment
criteria for reuse at the site.

1. INTRODUCTION

In the early 1920s the MacGillis and Gibbs Company and Bell Pole
Company began wood treatment plant operations on adjoining properties in
New Brighton, Minnesota. Wood products were treated with creosote and
pentachlorophenol (PCP) and placed in open-air staging areas throughout
both sites. Dripping and spillage from these operations resulted in
widespread soil, sediment and groundwater contamination. Regulatory
investigations of the site were initiated in the early 1990s and the 68-acre
site was included on EPA's National Priority List (Superfund).

A series of removal actions were implemented to address major source
areas and groundwater. In 1999 a Record of Decision (ROD) Amendment
was completed to address residual site contamination consisting of over
18,000 cubic yards (yd^3) PCP and carcinogenic polycyclic aromatic
hydrocarbon (cPAH) contaminated soils. The ROD Amendment specified
that these soils be treated using an on-site biopile prior to off-site disposal in
a RCRA Subtitle C landfill. The amendment also stipulated that treated soils
achieving site cleanup standards and RCRA Land Disposal requirements
could be placed at the site following treatment.

The Minnesota Pollution Control Agency (MPCA) was the lead
regulatory agency for the biological treatment phase of the remedial action.
MPCA conducted a search to identify a company with the necessary
scientific expertise to design an effective biological treatment program, and
the engineering capabilities to implement the program in the field. Based
on the results of this search Biogenie Corporation (Biogenie) was selected to
design and implement a biological treatment program for the site.

The key elements of an effective biopile are simple: isolate the
indigenous microorganisms capable of degrading the CoCs; determine the
optimum growing conditions for those microorganisms; design a cost
effective system capable of reproducing the optimum conditions in large
volumes of soil; and constructing and operating the system in the field. A
wide range of site-specific factors must also be considered including: soil
composition, site-wide contaminant profiles, and available space. Finally,

regulatory program and project management constraints must be addressed. These include, but are not limited to treatment criteria, budget, schedule, permitting, oversight, and public participation.

An effective project team must be capable of addressing these issues and others in a seamless and transparent manner. As a result the project owners and remedial managers are often not aware of the incredible level of technical expertise and coordination required to implement a successful biological treatment program. Compounding the problem is the fact that an effective biopile looks very similar to an in-effective biopile – a pile of dirt.

Those responsible for implementing this program quickly realized the importance of the work that occurs "behind the scenes" on a successful biological treatment program, and importance of an integrated team of scientists, engineers and construction professionals to accomplish the project.

This paper uses the biological treatment program implemented by Biogenie at the MacGillis and Gibbs Superfund Site to illustrate the "behind the scenes" science and engineering expertise necessary to perform a successful biological treatment program.

2. MATERIALS AND METHODS

2.1 Problem

Previous treatment experience with similar soils indicated that a well-designed treatment program could achieve site reuse criteria for the majority of the soils and the LDR Phase IV Treatment Standards for the remainder. Previous experience also indicated that these particular contaminants were difficult to biodegrade and that special care would be required both in the laboratory and on the design table to ensure an effective Biopile could be constructed in the field. To render this project even more difficult, the contract schedule stipulated that the biological soil treatment program be initiated, without additional treatability studies (TSs) and that the first phase of the treatment program be performed during the winter months, which in Minnesota can dip below -20°C.

2.2 Technical Approach

For all complex on-site biological treatment projects, experienced teams of scientists, engineers, and construction personnel are assembled with separate but equal responsibilities. Each component of the project team is responsible for not only the performance of their respective technical disciplines but also for communicating and interacting with members of the other disciplines to ensure the timely exchange of information and data. This project management philosophy is similar to EPA's Triad Approach.

For the New Brighton project the compressed project schedule combined with a large number of uncertainties required a high level of technical competence and effective communication from each team member. Each technical team is responsible for critical, but not always obvious, project tasks. The following highlights a partial list of the responsibilities of each component of the project team.

2.2.1   Scientific Team

The primary responsibility of the scientific team is performance of TSs and the evaluation of treatment performance data. TSs use site soils to identify the contaminant-specific indigenous microorganisms and the optimal soil conditions for the growth of those microorganisms. The scientists evaluate the degradation rates of the contaminants and provide recommendations to increase the rates through the modification of soil amendments and the control of environmental variables within the operating Biopile primarily pH, temperature and moisture content.

The TS is critical to the design of an effective Biopile, and seemingly minor variations in one or more of the Biopiles operational parameters can have a tremendous effect on unit's contaminant reduction efficiency. Additionally, understanding the dynamic nature of the Biopile as nutrients and contaminants are metabolized is critical to maintaining optimum efficiency and achieving the maximum possible degradation.

Data from the TS is provided to the engineering team to support the design of the treatment system and its components. Additionally, the scientific team evaluates critical parameters during the operational life of the Biopile to verify TS data and to anticipate and respond to fluctuations within the Biopile, to maintain optimal contaminant degradation rates. Recommendations are provided to the engineering and construction teams for implementation in the design table or in the field.

2.2.2   Engineering Team

The engineering team is responsible for the development of the overall site strategy and the design of the treatment system. The engineering team is typically responsible for the overall management of the project including interaction with the project owner and regulatory agencies. The most important responsibility of this team is the development of the technical approach. An effective technical approach combines the technical aspects of the project with contract requirements, regulatory requirements, site conditions and other factors into an implementation schedule.

The engineering team also is responsible for the detailed design of the treatment system and its components based on parameters provided by the scientific team, the project owner and others to ensure it functions as required. Additionally, this team is responsible for securing operational permits, identifying qualified subcontractors, and local materials and supplies. The team also works with the scientific and construction teams to coordinate the operation, maintenance and modification of the system. All project reporting is also coordinated through the engineering team.

2.2.3   Construction Team

The construction team is responsible for the implementation of the remedial project in the field, including the construction, operation and maintenance of the treatment system.   The construction team provides oversight of subcontractors, performs routine monitoring of the Biopile, and collects and manages field samples and data.   The construction team implements system design and operational changes provided by the scientific and engineering teams.

Because the largest percentage of the project cost is expended in the field, the construction team is integral to the planning process.   The construction manager is involved in the development of the technical approach, as well as project estimating, and scheduling.   Once the planning documents are in place, the construction team assumes full responsibility and accountability for project implementation. Many projects are performed on a guaranteed lump sum basis, making the construction teams incentive driven to complete, not sustain, the projects within budget and schedule. This situation requires that the construction team be supported with a scientific and engineering team that shares the same project closure orientation.

For the New Brighton project, two significant driving factors enforced the need to closely adhere to the project philosophy – an aggressive project schedule, and technical uncertainties with the implementation of a full-scale treatment program for PCPs. The following sections highlight specific critical tasks required by two of the three project teams, scientific and engineering, which lead to the success of the project.

2.3   Results

2.3.1   Behind the Scenes with the Scientific Team

A significant concern at the on-set of the New Brighton project was the fact that the construction of the biological treatment program was required to begin immediately upon contract award.   Because TSs replicate actual Biopiles, they often require up to 20 to 28 weeks to complete. As a result the project schedule did not provide the team the opportunity to systematically evaluate and develop the optimal design for the treatment unit. The project team made the determination to move forward with the engineering design and the field construction work without the benefit of advanced TS data.   The team opted to initiate a series of TSs and to apply the resultant data as a tool to optimize the treatment system after it was constructed and operational.

The engineering and construction teams determined, based on past project experience, that the system could be designed and constructed to accommodate the range of anticipated optimization modifications. It was determined that the most practical approach was to prepare the initial Biopile in accordance with a standard soil amendment recipe (soil mineral fertilizers) and configuration (treatment cells 10 feet wide and 8 feet high).

The project team planned that TS data would be available one to two months in advance of the actual soil treatment program and that this lead-time would provide adequate indication of the progression of the treatment program to support the optimization of the full-scale unit.

The scientific team used site soils to configure four distinct 15-liter mesocosms to simulate treatment conditions that would be effective for the New Brighton site. The mesocosms established for the New Brighton site include the following.

> Mesocosm #1   Mineral Fertilizer (Standard Mesocosm)

> Mesocosm #2   Mineral Fertilizer; pH Adjustment (from 7 to 8)

> Mesocosm #3   Organic Amendment and Fungal (white rot fungus) Inoculation

> Mesocosm #4   Organic Amendment and Fungal Inoculation

PCP concentrations were measured at the beginning of the TS and at 4-week intervals. In addition to PCP other operational parameters including pH, temperature, moisture content, soil fertility, and microbial counts, were monitored and evaluated.

Four key operational parameters were targeted for adjustment based on the trends observed in the TS including: the response to organic amendments, pH, temperature and moisture content.

2.3.1.1 Organic Amendment

At week 4, the two mesocosms that included organic amendments showed significant PCP degradation, while the other mesocosms showed minimal degradation. Based on this data the scientific team recommended that organic amendments be added to the Biopile to stimulate degradation. The construction team added organic amendments during the next mixing cycle and the Biopile responded as expected and significant PCP degradation was observed. To support the optimization objective of the TS, the scientific team then split Mesocosm #1 and added organic amendments to create a Mesocosm (#5) that duplicated the operating Biopile. Figure 1 shows the PCP degradation rates of the 5 mesocosms.

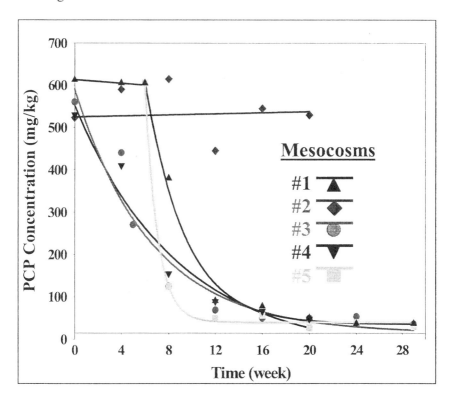

Figure 1

As a side note and as depicted on Figure 1, the Mesocosm #1 (mineral fertilizer) recipe was, in fact, a very effective treatment recipe. The TS graph for this mesocosm shows a 7-week lag time before significant PCP degradation is observed; however, after this period the rate of PCP degradation was excellent. Based on subsequent analysis it was determined that the lag was caused by the presence of total petroleum hydrocarbons (TPH) in the soil, upon which the microorganisms preferentially degrade before PCP.

Were the complete data set available prior to the field implementation, the project team would likely have continued with the standard amendment recipe; however, at the time the team was not fully confident when or if the PCP degradation would start, therefore the team opted to be proactive in consideration of the overall project schedule.

2.3.1.2   PH/Temperature/Moisture Contents

During the TS the impact of pH, temperature and moisture content on microbial activity was continuously monitored. Optimal ranges for each parameter were determined: pH ~ 7.0; temperature ~ 23° to 25° C; and moisture content ~ 8 and 12%. These optimal ranges were provided to the engineering and construction team who prepared contingency equipment specifications and operational adjustments for implementation in the event that field monitoring indicated deviations from these ranges.

In the field, operational modifications were required to maintain temperature and moisture content. The addition of organic amendment increased metabolic activity significantly requiring greater airflow through the pile and the addition of water on a more frequent basis. Once optimized, the metabolic activity sustained itself at a high level throughout the winter months that had recorded temperatures at -23° C. This metabolic activity required the aeration and water management measures to continue throughout the winter.

Throughout the project the scientific team analysed and interpreted data, providing the engineering and field teams the information necessary to make continuous adjustments to the Biopile. Without these adjustments, it is unlikely that the treatment objectives would have been achieved within the required time frame if at all.   Adjustments made during the first phase of treatment were incorporated into the second phase of the treatment program, rendering this phase more efficient and cost effective.

2.3.2   Behind the Scenes with the Engineering Team

Two major factors impacted the overall planning and design of the Biopile treatment system.

> Site soils contaminated with PCPs and cPAHs above certain concentrations would not achieve site reuse criteria based on maximum theoretical biological degradation rates; and

> Site-specific TS data would not be available to support the overall design.

In order to move forward with the project, the engineering team, in consultation with MPCA, developed a soil segregation strategy and a means to use the TS data to optimize the Biopile.

2.3.2.1   Soil Segregation Strategy

Treatment criteria were established for the site by the MPCA team based on risk-based determinations and applicable hazardous waste regulations. Site-specific risk values were used to develop site reuse criteria values for PCP concentrations of less than 10 mg/kg and cPAH concentrations (total of 7 specific cPAHs) of less than 5 mg/kg.

Based on experience with similar contaminants and discussions with MCPA's project team, a preliminary site strategy was developed.   The overall strategy was to involve two phases of the treatment: Phase I focused on approximately 14,500 yd³ of impacted soils, and Phase II focused on 3,500 yd³ of contaminated soil and woodchips.   The general strategy for each phase involved the segregation of site soils into two categories based on initial contaminant concentrations and predicted treatment efficiencies.

The scientific team conducted a review of past TSs and actual project performance data and an extensive literature review and determined that biodegradation rates for PCPs could not be expected to exceed 97% total reduction.   Based on the expected degradation and the known contaminant concentrations at the site, the engineering team developed a soil segregation strategy to effectively manage the treatment program.   The strategy involved establishing two soil categories for the site.

Category 1 soils were defined by PCP concentrations of less than 200 mg/kg and total cPAH less than 5 mg/kg. These soils were expected to meet site reuse criteria during the treatment program.   The engineering team determined that of the 18,000 yd³ of contaminated soil at the site, approximately 9,000 yd³ met the Category 1 criteria.

Category 2 soils had PCP concentrations exceeding 200 mg/kg and the cPAH values exceeding the LDR criteria.   These soils were expected to exceed site reuse criteria but achieve LDR criteria during treatment.   The engineering team estimated that another 9,000 yd³ of Category 2 soils existed at the site. It was also known that approximately 3,500 yd³ of the Category 2 soils contained significant percentage of contaminated wood chips.

With this knowledge the engineering team then evaluated the layout of the site to develop the most effective treatment system configuration and developed an overall implementation strategy for the site. The overall site strategy involved two treatment phases, utilizing two treatment pads for the first phase and one for the second.

Phase I focused on all Category 1 soil (9,000 yd³) and the Category 2 soils without wood chips (5,500 yd³) for a total of approximately 14,500 yd³ of soil. Anticipating operational variations for each category, separate treatment pads were designed to accommodate the volume of soil from each soil category. Treatment Pad 1 (designated for Category 1 soils) was constructed with 36 treatment cells, and Treatment Pad 2 (Category 2 soils) was constructed with 23 treatment cells. Site soils were then segregated and configured into Biopiles on the treatment pads to begin the biological treatment program.

Phase II of the treatment program focused on the remaining 3,500 yd³ of Category 2 soils containing wood chips. These soils were configured into treatment cells on a portion of Treatment Pad 1 and the footprint of the treatment system was reduced including the demolition of Treatment Pad 2.

Full-scale results (see below) showed that the strategy to segregate soils prior to treatment was an effective management decision developed by the engineering team that was implemented successfully on this project.

2.3.2.2   Missing TS Data

As discussed above, in order to address schedule issues the project team opted to move forward with the design and construction of the Biopile without the site-specific TS data that is critical to the design process. To compensate, the engineering team configured a treatment platform equipped to accommodate a wide range of design and operational modifications in response to the interpretation of the concurrent TS work. The operational components of the Biopile including the blowers, aeration pipes, and emission controls system were all slightly over-designed to ensure that the optimum operational parameters could be achieved by the system when the TS data was available.

The initial Biopile was then constructed using a standard treatment cell configuration and amendment recipe consisting of mineral fertilizers. This configuration and recipe was based on numerous successful past projects and provided the operational flexibility to adjust in response to the trends observed in the TSs.

After four weeks the TS data demonstrated that biological activity was greater for the mesocoms that included organic amendments. As a result, the engineering team provided the construction team specifications for the addition of organic amendments and the adjustment of the systems operational controls to compensate for the additional mass of the Biopile. As biological activities increased in response to the organic amendments, aeration and moisture control became increasingly critical; however, due to the conservative design of the Biopile controls, no change outs were required during the project.

2.4   Results of the Full Scale Program

The "behind the scenes" efforts of the scientific and engineering team resulted in the implementation of a biological treatment program for the MacGillis and Gibbs Superfund site that exceeded the project expectations for PCP degradation enabling over 96% of the contaminated soils to be reused at the site. The remaining soils were treated to below LDR and were placed in a RCRA Subtitle C landfill. The following summarizes the results of the two phases of the biological treatment program.

2.4.1   Phase I

➤ All of the 59 cells from the first phase achieved site reuse criteria (PCP < 10mg/kg). A volume of 14,672 yd^3 of treated soil was backfilled on site.

➤ The average treatment duration required to reach the remediation criteria for the 59 cells was 38 weeks.

➤ The average PCP reduction rate for Category 1 soil was 95% after an average of 9 months of treatment.

➤ The field PCP reduction rate in Phase I was similar to that obtained during the TS. However, additional time was necessary to reach project objectives. Low temperatures during the winter treatment may explain the additional time required to reach the objectives.

2.4.2   Phase II

➤ From the 15 cells treated during Phase II, 3 cells, a total 708 yd^3, met site reuse criteria and were backfilled on site. The 12 remaining cells, a total of 2,832 yd^3, were disposed of in a landfill after attaining LDR criteria (PCP < 74 mg/kg).

➤ The soil from the second phase was treated for an average of 18 weeks before reaching the remediation criteria.

➤ Category 2 soils exhibited an average reduction rate of 76% after 5 months of treatment.

> ➢ The field PCP reduction rate (76%) was less than that obtained during the treatability study. The reduction rate was attributed to a determination by the project team to end treatment when the LDR criteria were attained. Discussion and Conclusions

2.5   Discussion and Conclusions

The New Brighton project presented many technical challenges including the design and implementation of an effective biological treatment program for a difficult contaminant profile within the limits of a compressed project schedule.   To address these challenges, a dedicated project team was assembled, comprised of scientific, engineering and remedial construction specialists to develop and implement the technical approach.   The technical approach developed for this project was the direct result of the many critical "behind the scenes" project tasks performed by the scientific and engineering teams. These efforts were then implemented by the construction teams who provided ongoing feedback to the technical teams during the project that provided the basis for the optimization of the biological program.

The coordinated efforts of the team resulted in a successful biological treatment program far exceeding project expectations. The treatment objectives were achieved for all contaminated soil soils and over 96% of the soils were reused at the site.

PART VIII: RISK ASSESSMENT AND REMEDIAL APPROACHES TOWARDS RESTORATION AND MANAGEMENT OF CONTAMINATED RIVERS

CHAPTER 22

EXPLORING INNOVATIVE AND COST-EFFECTIVE SOLUTIONS TO CONTAMINATED SEDIMENTS TO ACHIEVE ECOLOGICAL RESTORATION OF THE LOWER NEPONSET RIVER

An Overview

Karen Pelto
River Restore Coordinator, Riverways Program, Massachusetts Department of Fish and Game Riverways Programs, Dept of Fish and Game, 251 Causeway Street, Suite 400, Boston, MA 02114

Abstract: This past decade has seen financial and volunteer investment in a cleaner, more accessible Neponset River. With these tangible improvements has also come a campaign to re-orient the public, and their perception of the river, from its previous reputation as an "open sewer" to a new role as a habitat for anadromous fish, venue for recreation on its waters, celebrations on its banks, and an inspiration for revitalizing the village of Lower Mills. From an ecological perspective, the Neponset is unique among urban rivers as its estuary remains largely intact and only two dams prevent anadromous fish from reaching fifteen miles of free-flowing habitat. However, the U.S. Geological Survey has documented elevated levels of polychlorinated biphenyls (PCBs) in the soft sediment impounded behind the two dams and the water column in the lower five miles of the Neponset River.

Decision-making about river restoration and sediment remediation must proceed concurrently when traditionally they would proceed on separate regulatory pathways. In addition, alternatives for fish passage and ecological restoration must not constrain remediation options and, conversely, remediation options must not limit the feasibility of alternatives for fish passage and ecological restoration. This workshop explores a range of potential remedial options, from traditional to innovative in the context of an ecological restoration effort that includes dam removal among the suite of alternatives.

Key words: Anadromous fish, river restoration, polychlorinated biphenyls, dam removal

1.   INTRODUCTION

Two decisions critical to the future ecological health and continued public commitment to the Neponset River must be made in the near future: 1) selection of an appropriate alternative to implement upstream fish passage for herring and shad; and 2) selection of an appropriate alternative to remediate riverine sediments found to be contaminated with polychlorinated biphenyls (PCBs), which are a persistent bioaccumulative toxin. Both of these decisions will require thoughtful input and knowledgeable understanding from the public and public agencies.

In 1995, as part of the Neponset River Pilot Project for the Massachusetts Watershed Initiative, the Massachusetts Division of Marine Fisheries surveyed the river's mainstem to evaluate spawning habitat for blueback herring and American shad. (Reback and Brady, 1996) These anadromous fish species were once abundant in the Neponset River, but populations dwindled and disappeared following the construction of mill dams, which obstructed upstream passage. Based on the quality and quantity of habitat observed during the 1995 survey, the Division recommended restoring the fishery and began stocking herring and shad on a yearly basis.

In order to obtain the technical information necessary to evaluate options for upstream fish passage for herring and shad, the Massachusetts Executive Office of Environmental Affairs (EOEA) requested that an ecological restoration study be conducted by the U.S. Army Corps of Engineers pursuant to Section 206 of the Water Resources Development Act. The Department of Fisheries, Wildlife & Environmental Law Enforcement's (now Department of Fish and Game) Riverways Program serves as liaison to the Corps, cooperating federal and state agencies, and the public.

A Preliminary Draft Environmental Restoration Report and Environmental Assessment for the Neponset River Fish Passage and Habitat Restoration Project, including the Tileston and Hollingsworth Dam and the Walter Baker Dam has been completed by the U.S. Army Corps of Engineers (USACE, 2002). This Preliminary Draft Report describes several fish passage alternatives, including fishways, dam removal, dam breaching, and sluiceway modification. The Preliminary Draft Report identifies dam removal as the preferred alternatives to restore anadromous fishery habitat, as it would make available 17 miles of river spawning habitat, 4.34 acres of restored aquatic habitat, and 11.46 acres of restored riparian wetlands.

One of the Preliminary Draft Report's findings was the presence of PCBs in the sediments impounded behind both the Baker (or Lower Mills) Dam and Tilestone-Hollingsworth (or Hyde Park) Dam. The sampling effort was limited, however, to one sample in each impoundment.

In 2002 and 2003, Riverways entered into an agreement with the U.S. Geological Survey to determine the thickness, grain-size characteristics, physical properties, and chemical quality of soft sediments impounded behind the two most downstream dams on the Lower Neponset River (at Lower Mills and Hyde Park), as well as conduct detailed bathymetry (water depths) of the river between Lower Mills and Hyde Park. In addition to the two most downstream dams, the braided channel area between the two dams was investigated; this area represents the sediments impounded behind the former Jenkins Dam that was breached in the 1950s. The USGS study consisted of the collection of 20 sediment grab samples (from the top 4 inches of sediment), 31 sediment core samples (composited from 5 to 50 inches, depending on the total depth of the sediment), and 12 measurements of PCB concentrations in the water column by using a passive water-sampling system (PISCES). Field and laboratory analyses to assess the level of PCB contamination in bottom-feeding fish in selected reaches of the Neponset River and evaluate potential sources are ongoing. Defining the monthly, seasonal, and annual load of PCBs from the Neponset River to Boston Harbor and the fate of those loads in the Neponset River estuary's water, sediment and fish, is the next phase of work (USGS, 2004).

2.   DISCUSSION

The Massachusetts Riverways Program entered into partnership with UMASS Amherst to engage a select group of national academic and private industry leaders in an exploration of available and emerging regulatory and technical approaches for managing and remediating contaminated sediment. Leaving contaminants in place while pursuing fish passage and river restoration is not a desired nor practical response in a dynamic river ecosystem as it would fall short of restoring the ecological function in the Neponset River.

Restoration alternatives identified in a Preliminary Draft *Environmental Restoration Report and Environmental Assessment for the Neponset River Fish Passage and Habitat Restoration Project, including the Tileston and Hollingsworth Dam and the Walter Baker Dam* by the U.S. Army Corps of Engineers (USACE, 2002) create a range of opportunities for up- and down-stream passage of anadromous fisheries and varying levels of restored riparian and riverine habitats. The technical and economic feasibility of potential restoration alternatives depends upon a thorough exploration of sediment management techniques, costs, and feasibility.

This workshop, "Exploring Innovative and Cost-Effective Solutions to Contaminated Sediments to Achieve Ecological Restoration of the Lower Neponset River," has been designed to provide insight on the scale and scope of appropriate remedial options that might exist for the Neponset River, recognizing the public investments in greenways and ecological restoration. Each workshop presenter was asked to consider the following questions when preparing their oral and written remarks:

- What are the ranges of "tried and true" and "innovative" options and technologies for managing and remediating PCB-contaminated sediment?
- What are the relative merits of the range of remedial options and technologies, particularly in the context of ecological restoration and fish passage?
- What remedial options and technologies have states allowed and what sediment cleanup standards were employed to determine the appropriate level of cleanup?
- What successful ecological restoration projects have included environmental cleanup of contaminated sediments?

REFERENCES

Reback, K. and P. Brady. 1996. *A Survey of the Neponset River to Determine its Potential for Anadromous Fish Development*. Massachusetts Division of Marine Fisheries, Sportfisheries Program.

USACE. 2002. *Environmental Restoration Report and Environmental Assessment, Neponset River Fish Passage and Habitat Restoration Project, including the Tilestone and Hollingsworth Dam and the Walter Baker Dam*. Preliminary Draft, February 2002, U.S. Army Corps of Engineers, New England District.

USGS. 2004. Sediment Quality and Polychlorinated Biphenyls in the Lower Neponset
    River, Massachusetts, and Implications for Urban River Restoration.. USGS SIR 2004-
    5109. U.S. Geological Survey.

CHAPTER 23

RESTORING AN URBAN RIVER
Polychlorinated Biphenyls and other Contaminants in Bottom Sediment of the Lower Neponset River, Massachusetts

Robert Breault and Matthew Cooke
U.S. Geological Survey 10 Bearfoot Road Northborough, MA 01532

Abstract: The U.S. Geological Survey studied sediment and water quality, with an emphasis on polychlorinated biphenyls (PCBs), in the bottom sediment and water of the Neponset River. The USGS completed this study in cooperation with the Riverways Program of the Massachusetts Executive Office of Environmental Affairs Department of Fish and Game, and the U.S. Environmental Protection Agency. The major findings of this study were: human activities have adversely affected sediment quality in the Neponset River; with the exception of polyaromatic hydrocarbons and PCBs, contaminant concentrations are similar to those of other urban rivers; sediment contaminant levels may adversely affect aquatic life and human health; PCBs continue to be released into the Neponset River; and PCBs in the water may have toxic effects.

Key words: Neponset River; sediment; elements; PCBs; PISCES.

1.   INTRODUCTION

In 1998, then Department of the Interior Secretary Bruce Babbitt reported that throughout its history America had constructed 75,000 dams. This number is, as Babbitt wrote, "the equivalent of one [dam being built] every day since Jefferson wrote the Declaration of Independence" (Babbitt, 1998). As these dams have aged, many have fallen into states of disrepair.

Dams commonly interfere with many natural processes in rivers. Blocking fish passage is perhaps the most widely recognized environmental effect of dams, but the reservoirs that dams create also inundate wetlands and terrestrial ecosystems; with dams, rivers become fragmented, and peak flows and other hydrologic characteristics are changed. Dams also change sediment regimes in a river by trapping most of the sediment moving downstream in impoundments behind the dams.

Dams have impounded the Neponset River, a tributary to Boston Harbor, for the past 350 years (fig. 1). The river historically supported abundant populations of American shad and river herring. Presently (2003), the estuary supports an important rainbow smelt fishery.

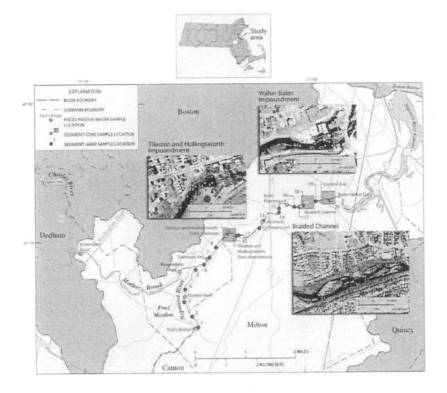

Figure 1. Sediment-grab (top 2–4 in. of sediment) and sediment-core (5–50 in., depending on the total depth of the sediment) sampling si tes and location of passive -water-sampler (PISCES) deployment, in the Neponset River of Massachusetts

Like most urban rivers in the Northeast, the Neponset River also has a long industrial history. Dam construction and settlement began as early as 1630. By the mid 1700s, the Neponset River drained one of the most heavily industrialized basins in the Nation. Industrial activity continued on the Neponset River until 1965, when the last major industrial facility relocated from the lower section of the river. This industrial past, combined with the urbanization that continues in the drainage basin, has likely contaminated water, biota, and bottom sediment throughout the river.

Much effort by State, local, and Federal agencies has already been put forth towards restoring the Neponset River and surrounding areas, including the abatement of both industrial and sanitary discharges, greenway expansion, economic revitalization, and improving access to the river. Although there has been much improvement, work remains to be done. To that end, environmental managers and local advocates have evaluated river-restoration efforts, such as channel restoration for aquatic-habitat improvements, and fish-passage alternatives, including the installation of engineered fishways, dam breaching, and removal of the most downstream dams on the lower Neponset River—the Walter Baker Dam and the Tileston and Hollingsworth Dam. Fish passage at these dams would provide access to more than 17 miles of habitat to shad and herring and help increase recreational use on the lower Neponset River, that section of the river from Fowl Meadow to the Walter Baker Dam in Milton, MA.

Increased public dialogue about restoration of the Neponset River, combined with river-channel alteration, the long history of industrialization and urbanization along the river, and some knowledge of the occurrence and geographic distribution of sediment contamination prompted this study of bottom-sediment quality and quantity. The U.S. Geological Survey (USGS) completed this study in cooperation with the Riverways Program of the Massachusetts Executive Office of Environmental Affairs (EOEA) Department of Fish and Game, and the U.S. Environmental Protection Agency (USEPA).

This report discusses the prevalence of trace elements and organic compounds in bottom sediment of the Neponset River. Because of the high concentrations of PCBs measured in the river and their high relative toxicities, this report focuses on the occurrence, source identification, and toxicity of PCBs in sediment and water of the Neponset River; other bottom-sediment contaminants are also briefly discussed.

2.   MATERIALS AND METHODS

The study described in this report consisted of the collection of 20 sediment-grab samples (from the top 2–4 in. of sediment), 31 sediment-core samples (composited from 5 to 50 in., depending on the total depth of the sediment), and 12 measurements of PCB concentrations in the water column by using a passive water-sampling system (PISCES) (fig. 1). Sample-collection design and techniques, laboratory analysis, data-analysis techniques, and quality-assurance and quality-control procedures are discussed in detail by Breault et al. (2004).

3.   RESULTS AND DISCUSSION

3.1   Human Activities Have Adversely Affected Sediment Quality in the Neponset River

Trace element concentrations measured in sediment samples from the lower Neponset River were compared to concentrations from selected New England streams, which for the purposes of this study represent "nonurban background" concentrations. This comparison shows that historical and recent human activities have affected many trace element concentrations, for example, concentrations of lead, a common contaminant resulting from human activities (fig. 2). High concentrations of some trace elements in sediment, such as lead, are not unexpected in the Neponset River because of the historical and continued use of these trace elements in industry and the use of automobiles in the area around the Neponset River.

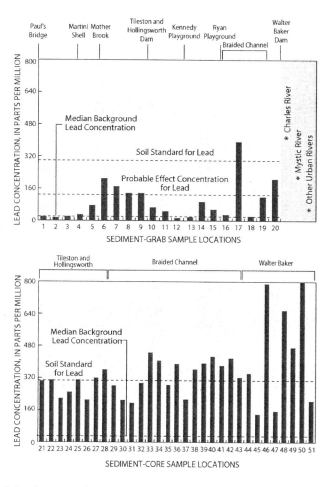

Figure 2. Lead concentrations measured in sediment samples collected from the Neponset River and median (*) lead concentrations measured in sediment -grab samples collected from the lower Charles and upper Mystic Rivers of Ma ssachusetts and other urban rivers from across the United States. The direct -contact, exposure -based soil standard and probable effect and background concentrations for lead are also shown.

Unlike trace elements, organic compounds do not generally occur naturally in uncontaminated sediment; the detection of an organic contaminant is considered the result of human activities. Some polyaromatic hydrocarbons (products of incomplete combustion) were detected in most of the samples. Some PCBs (manufactured organic compounds) were also detected in most of the samples (fig. 3). Organochlorine pesticides, however, were not detected in the grab samples, but there were many detections of these pesticides in the core samples. Most notably, Chlordane, and the pesticides DDT, DDD, and DDE were detected in many core samples. The detection of these compounds in many core samples and not in any sediment-grab samples is expected. Because the use and disposal of these organic compounds were more prevalent in the past than in recent years, the products from past uses are buried in the sediment. The general use of DDT was banned in the United States in 1972. In 1988, the USEPA banned all uses of Chlordane.

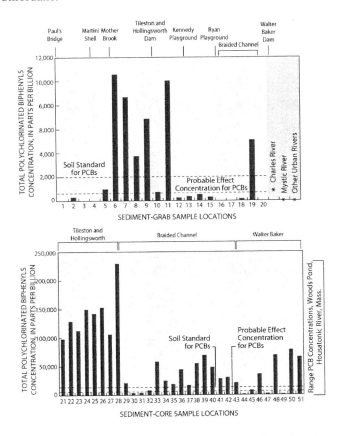

Figure 3. Total PCB concentrations measured in sediment samples collected from the Neponset River and median (*) total PCB concentrations measured in sediment-grab samples collected from the lower Charles and upper Mystic Rivers of Massachusetts and other urban rivers from across the United States. The range of PCB concentrations measured in a sediment-core samples collected from Woods Pond, the first impoundment downstream of a heavily PCB-contaminated area on the Housatonic River, Massachusetts, probable effect concentration, and the direct-contact, exposure-based soil standard for PCBs are also shown.

3.2 With the Exception of PAHs and PCBs, Contaminant Concentrations are Similar to Those of Other Urban Rivers

Grab-sample data were compared with data from other urban rivers throughout the United States (Rice, 1999); the data were also compared with data from the lower Charles (Breault et al., 2000) and upper Mystic Rivers (Breault, unpublished data), two other urban tributaries to Boston Harbor. Sediment-core PCB data were also compared to data from the Housatonic River, a river in western Massachusetts with known PCB contamination (United States Environmental Protection Agency, 2003).

Concentrations of trace elements, like lead, in grab samples from the Neponset River are generally equal to or less than concentrations in sediment collected from other urbanized, free-flowing rivers (Rice, 1999; fig. 2). In contrast, organic compound concentrations, particularly PAHs and PCBs, are much higher in grab samples from the Neponset River. For example, median PCB concentrations measured in grab samples from the Neponset River are more than 125 times greater than those median concentrations in sediment samples from other rivers across the United States (Rice, 1999).

Generally, concentrations of many contaminants in sediment are lower in the Neponset River than they are in either the Charles or the Mystic Rivers. In contrast, concentrations of PCBs in sediment in the Neponset are similar to those in the Charles and greater than those in the Mystic River. Some parts of the Neponset River are highly contaminated with PCBs, particularly within current impoundments and former impounded areas (fig. 3).

To assess the contamination severity, it is useful to compare PCB concentrations measured in sediment cores collected from current impoundments and former impounded areas of the Neponset River to concentrations in sediment samples collected from the Housatonic River (fig. 3). Although concentrations are higher in the Housatonic River than those in the Neponset River, concentrations in sediments of the first impoundment (Woods Pond) downstream of the heavily contaminated area in the Housatonic River are comparable to the concentrations in the sediments of the Tileston and Hollingsworth impoundment in the Neponset River (U.S. Environmental Protection Agency, 2003).

3.3   Sediment Contaminant Levels May Adversely Affect Human Health and Aquatic Life

River-restoration efforts often focus on bringing people and river together. When a river is restored, access to it improves, recreational opportunities increase, and more people fish, swim, and boat in the river. As people interact with the river, they will likely come in contact with sediment.

Information about human health risks associated with contaminated sediment would allow water-resource managers to plan their river-restoration efforts. Potential human health effects can be indirectly assessed by comparing sediment-contaminant concentrations with exposure-based guidelines. These guidelines, however, do not exist for aquatic sediment, but they do exist for contaminated upland soil (Massachusetts Department of Environmental Protection, 1996).

Soil standards exist for many of the constituents tested; but in the sediment samples, only a few of the constituents were detected at concentrations near or above the soil standard. More importantly, however, PCBs and some PAHs, when considered in relation to the direct-contact, exposure-based soil standards, may represent the greatest human health risk in the Neponset River (fig. 3).

By comparing contaminant concentrations to sediment-quality guidelines, known as probable effects concentrations, sediment quality is quantifiable in terms of potential adverse biological effects to benthic organisms, which live and feed in the river bottom and are an important food source for much of the other life in the river (Ingersoll et al., 2000). The average estimated toxicity potential to benthic organisms ranged from about 15 percent to 80 percent among the grab samples (fig. 4). For example, a single sediment sample with an estimated potential toxicity of 15 percent indicates that 15 out of 100 toxicity tests on that single sample likely will show toxicity. In this study, trace metals and PAHs accounted for some potential toxicity. More importantly, PCBs were most responsible for predicted toxicity (Ingersoll et al., 2000).

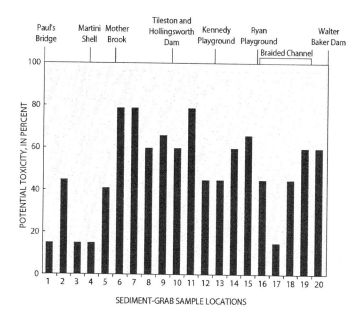

Figure 4. Estimated average potential toxicity for the top 2 –4 in. of bottom sediment collected from the Neponset River. For example, a sediment sample with an estimated potential toxicity of 15 percent means that 15 out of 100 toxicity tests are likely to show some level of toxicity for the contaminant concentrations measured in that sediment sample.

3.4 PCBs Continue to be Released into the Neponset River

PISCES samplers, deployed in the water column, absorb PCBs from the water across a semipermeable membrane into a sample of hexane, in which they dissolve and accumulate. Sample analysis shows a characteristic pattern of individual PCB compounds, known as congeners, which can be characteristic of a PCB source, in the same way that fingerprints are unique to an individual. The degree of difference between these patterns in samples collected from adjacent sampling locations can be analyzed with a statistic known as the root mean square difference (RMSD). Statistically, different RMSD values indicate a measurable difference in congener pattern between sampling locations and thus potentially different PCB sources (Breault et al., 2004).

The most likely source of dissolved PCBs to the Neponset River, as indicated from PISCES data, is bottom sediment downstream from Fairmont Avenue. This PCB-contaminated sediment was likely transported from a historic source and deposited in this part of the river. The point of maximum change in PCB congener pattern measured in bottom sediment coincides with that of the PISCES samples (fig. 5). These data are consistent with a historic source located just upstream from Fairmont Avenue. Alternatively, the change in congener pattern could be caused by anaerobic degradation of the historic PCB contamination in the depositional zone in that part of the river. In this case, the historic source would be located farther upstream (Breault et al., 2004).

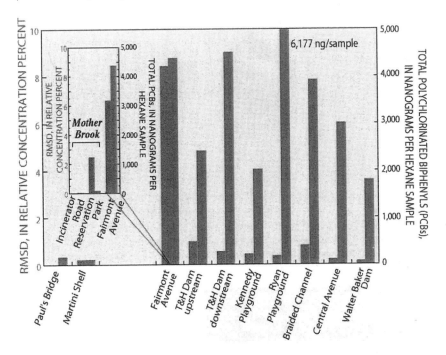

Figure 5. Mean PCB concentration measured in passive -water samplers (PISCES) and the degree of the change in PCB congener pattern (or fingerprint) along the Neponset River —as indicated by the statistic root mean square difference (RMSD). The most likely source of PCBs to the Neponset River is somewhere between the Martini Shell and Fairmont Avenue sampling locations or between Reservation Park and Fairmont Avenue.

3.5   PCBs in the Water May Have Toxic Effects

Because PCBs affect fish, wildlife, and human health, environmental management agencies, including the USEPA, set maximum acceptable values for PCBs in water. The current acceptable value for PCB concentrations in freshwater is 14.0 parts per trillion (ppt).

PISCES sample data indicate that PCB concentrations in Mother Brook (average estimated PCB concentration equal to 1.2 ppt) or in the Neponset River upstream of Fairmont Avenue (2.8 ppt) would not be expected to cause adverse biological effects. On the other hand, exposure to PCBs downstream of the Fairmont Avenue sampling location (66 ppt) may adversely affect aquatic organisms or wildlife (fig. 6).

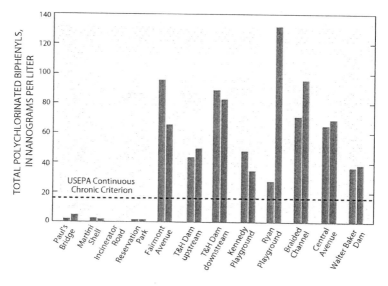

Figure 6. Concentrations of PCBs dissolved in the water estimated from duplicate PISCES samplers at each location. Water concentrations increase sharply downstream of the Fairmont Avenue PISCES sampling location and generally decrease downstream of that location. Estimated concentrations of PCBs dissolved in the water downstream (left to right) of the Fairmont Avenue PISCES sampling location exceed the U.S. Environmental Protections Agency's (USEPA) freshwater concentration limit (continuous chronic criterion) for PCBs dissolved in water.

Although PCB concentrations in aquatic organisms were not directly measured, they were estimated from the PISCES data (Breault et al., 2004). Average PCB concentrations in predatory-fish fillets (edible fish parts) were estimated to range from about 0.04 parts per million (ppm) to 0.3 ppm upstream of the Fairmont Avenue sampling location and to be about 0.9 ppm downstream of the sampling location. These concentrations are similar to those measured by the Massachusetts Department of Environmental Protection (MDEP) in 1994 (Massachusetts Department of Environmental Protection, 1994). On the basis of the MDEP findings, the Massachusetts Department of Public Health issued a fish-consumption advisory for brown bullhead collected from the Neponset River between the Hollingsworth and

Vose Dam (Walpole, MA) and the Tileston and Hollingsworth Dam (Hyde Park, MA).

The organic compound dioxin is among the most toxic of substances found in the environment. Of the 209 PCB congeners, 13 are considered dioxin-like; that is, they cause toxic effects similar to those caused by dioxin. Because of this similarity, the toxicity of these congeners can be assessed in relation to the toxicity of dioxin. This assessment is referred to as toxic equivalency (TEQ).

Estimated PCB water-column concentrations in the Neponset River, when expressed in terms of TEQ, were generally near or slightly greater than the USEPA human-health standard for dioxin (0.000005 ppt), except for the sample from the Incinerator Road location. The largest total TEQ was calculated for the samples collected just downstream of the Tileston and Hollingsworth Dam. At this sampling location, the TEQ was estimated to be about 10 times greater than the standard.

4.   CONCLUSIONS

The data presented here will help environmental managers evaluate the advantages and limitations of fish-passage alternatives and sediment-management options. Possible options include dredging and removal of contaminated sediment, channel restoration with stabilization of contaminated sediment, and breaching and removal of dams allowing for redistribution of contaminated sediment downstream. Knowledge of existing concentrations and distribution patterns of contaminants in the river will help guide the selection of the most cost-effective and environmentally beneficial river-restoration strategies. Restoration of the lower Neponset River will preserve this valuable resource and ultimately contribute to the protection of the Neponset River Estuary and Boston Harbor.

REFERENCES

Babbitt, B. 1998. A river runs against it —America's evolving view of dams in Open Spaces Quarterly. 4.

Breault, R.F., Cooke, M.G., and Merrill, Michael. 2004. Sediment quality and polychlorinated biphenyls in the lower Neponset River, Massachusetts, and implications for urban river restoration and dam removal: USGS SIR 2004-5109.

Breault, R.F., Reisig, K.R., Barlow, L.K., and Weiskel P.K. 2000. Distribution and potential for adverse biological effects of inorganic elements and organic compounds in bottom sediment, Lower Charles River, Massachusetts: USGS WRIR 00-4180.

Ingersoll, C.G., MacDonald, D.D., Wang, N., Crane, J.L., Field, L.J., Haverland, P.S., Kemble, N.E., Lindskoog, R.A., Seve rn, C., and Smorong D.E. 2000. Prediction of sediment toxicity using consensus -based freshwater sediment quality guidelines; Chicago, IL EPA 905/R-00/007. June 2000.

Massachusetts Department of Environmental Protection. 1994. The Neponset River Watershed 1994 resource assessment report, Boston, MA. Massachusetts Executive Office of Environmental Affairs.

Massachusetts Department of Environmental Protection. 1996. Massachusetts contingency plan, Boston, MA. Massachusetts Department of Environmental Protection.

Oliver, B.G. and Niimi, J.A. 1985. Bioconcentration factors of some halogenated organics for rainbow trout—limitations in their use for pr ediction of environmental residues. *Environ. Sci. Technol*. 9, 842-849.

Rice, K.C. 1999. Inorganic -element concentrations in streambed sediment across the conterminous United States. Environ. Sci. and Technol. 15, 2499 -2504.

U.S. Environmental Protection Agency, 2003, Boston, MA. (February 3, 2004); GE/Housatonic River Site in New England. http://www.epa.gov/region1/ge/thesite/restofriver-maps.html.

CHAPTER 24

A FRAMEWORK FOR RIVER CLEANUP DECISION MAKING
Foundation Principles for Managing the Neponset River Watershed

David F. Ludwig[1], Stephen P. Truchon[1], and Carl Tammi[2]
[1]Blasland, Bouck & Lee, Inc., 326 First Street, Suite 200, Annapolis, MD 21403 -2678; [2]ENSR International, 2 Technology Park Drive, Westford, MA 01886

Abstract: The Neponset River drains a landscape that has been disturbed by human beings for hundreds, possibly thousands, of years. The watershed is now intensely developed and modified to accommodate built environments. Deliberations regarding restoration and cleanup of the Neponset (and other rivers in urban landscapes) will benefit from a set of principles to guide activities and a decision context to provide a framework for competing considerations. This paper proposes a draft suite of eight principles that, together, offer social, scientific, and engineering foundations for decision making. Those principles are consolid ated in a decision model incorporating trade offs among the value of the habitat in place, the potential value of restoration activities, habitat destruction inherent in remediation technologies, risks associated with exposure to indust rial chemicals now in the ecosystem, and the potential risk -reduction benefits of chemical remediation. We conclude that optimal management decisions must account for and balance the considerations inherent in each of these decision categories.

Key words: river cleanup, stewardship, remediation, environmental management , habitat restoration

1. INTRODUCTION

The Neponset River watershed resides in a landscape with a long history of intense human disturbance (Cronon, 2003). Before European colonists arrived in the region, Native American residents cleared forest plots by burning for agriculture, and foraged in the riparian ecosystems (Krech,

1999). With European settlement, environmental alteration increased rapidly in diversity and degree. Sequential episodes of deforestation, regrowth, industrialization, urbanization, and suburbanization resulted in the watershed conformation we have today.

The fact of ubiquitous human disturbance does not, in itself, tell us anything about the environmental quality of the drainage basin. There is nothing inherently "good" or "bad" about an ecosystem state—how we perceive the environment in a particular place and time depends on what we wish the ecosystem to be "doing," and whether the extant system meets those *desiderata*. In effect, we establish inherent "goals" for the environment, and judge environmental quality relative to those goals. In general, humans "manage" the environment by default—that is, our influence is universal, but it is not directed. We usually do not make decisions about what we want the ecosystem to be or what services we want it to provide in any larger context than that pertaining to a specific parcel of land (defined by ownership and/or local zoning constraints). However, if we are to survive and prosper as a species, and if we are to inhabit a healthy and prosperous biosphere, we must redress the disconnect between our technological ability to modify the environment and our general lack of planning for what those modifications should entail. In other words, it is time for us to stop managing the biosphere by default and start managing it by design. The emerging science of restoration ecology and the related field of landscape/watershed management offer a foundation on which we can build goal-oriented decision making and directed management activities.

The objective of this paper is to provide some context for deliberations concerning the rehabilitation of the Neponset River watershed. Before we launch into technology evaluations and site assessments, as we begin the arduous process of documenting impairments and potential enhancements, as we consider how to quantify losses of natural resource services and the benefits of services restoration, it is useful to think about environmental restoration as more than a "tool chest" of methodologies. What is the basis of restoration as a human endeavor, and how does that basis affect the context in which we will make decisions about the Neponset River?

To answer these questions, we present a suite of general principles that could provide a basis for guiding the field of environmental restoration, and a model illustrating the application of those principles for decision making regarding river ecosystem restoration. We discuss the latter specifically in light of the application of restoration science to the Neponset River basin.

2.   PRINCIPLES FOR ENVIRONMENTAL RESTORATION: WHAT ARE THE SOCIAL, SCIENTIFIC, AND ENGINEERING FOUNDATIONS FOR ECOLOGICAL STEWARDSHIP?

Our power to alter the environment brings the responsibility of stewardship—if it is our technology that impacts, we must be the ones to constrain, control, and target the resulting effects. But what standards can we set to guide our control? Extreme objectives—return to "pristine" (that is, pre-human technology) conditions, and the converse (unrestrained development and resource extraction) are unlikely to be useful and are unrealistic in application. Between these extremes, we have generally defaulted to a patchwork of varying standards, criteria and objectives, resulting in intense ecosystem management (and preservation) in some limited areas and the creeping influence of non-management in most of the biosphere. In the long run, such a piecemeal approach to management is bound to fail, because of the human tendency to apply technologies as they come into existence.

As the science of restoration ecology matures, it may be useful for practitioners to consider developing, debating, and ultimately, agreeing on a suite of principles to guide restoration applications in the context of global environmental stewardship. Such a set of principles need not be a fixed, final, unchanging monolith. Rather, it will likely be an evolving, growing statement, stimulating debate and discussion, keeping pace with the growth of the field.

In that spirit, we present one possible approach in Table 1. We label these "draft" as a deliberate demonstration that their primary purpose is to initiate debate and discussion and not to claim any degree of fervent belief. With that said, we do believe that the eight draft principles presented here provide a reasonable foundation in natural science, social science, and engineering to guide environmental restoration as a general endeavor and to specifically support assessment and management of the Neponset River basin.

Table 1. Draft Principles for Environmental Restoration

1. Manage by design, not by default.
2. First, do no harm.
3. Manage adaptively (ecosystems are not robots).
4. Accommodate human prosperity.
5. Everybody shoulder the burden.
6. Cross-apply experiences, techniques, and approaches.
7. Sound management yields good stewardship .
8. Make the commons a triumph

The first principle calls for design-directed management, versus the "management by default" that has largely represented human interactions with the biosphere. Viewed simply, this means that we should ascertain what we want from the ecosystem and work to establish the corresponding environment. There is a school of thought, reflected in the extreme attitudes of some nongovernmental environmental groups, that humans have neither the knowledge nor ability to determine and control desirable ecosystem states. According to this philosophy, humans must work to minimize environmental impacts but not attempt to "control nature." We believe that this viewpoint is short-sighted and unrealistic. Human impact is so pervasive, our control of the landscape so complete, and our population levels so high that we cannot help but "control nature." The only question is whether we "control" it by design or by default. It seems obvious that the default approach is no option at all, given our collective history.

The second principle adopts a fundamental call to "do no harm" in the first instance. Contrary to what many people believe, any environmental management action has negative consequences that must be balanced against positive results. Our quandary is deepened by the fact that "no action" alternatives come with similar negative and positive "baggage"—we can't escape the need to decide, for "no decision" can have as many consequences as any more active approach. It is likely that ecologists, trained in whole-systems thinking and consideration of indirect effects, will understand that some clear thinking about the overall effects of a decision will enable us to avoid causing gratuitous injury to the biosphere.

The third principle takes us into the "nuts and bolts" of environmental management. Adaptive management, in which practitioners "learn by doing" as they proceed and alter their processes in response to that learning, is becoming a technique of choice for large-scale environmental projects and in particular for watershed management (Salafsky et al., 2004). The vast numbers and kinds of connections in ecosystems render complete predictive certainty impossible. Rather than establish rigid "pass/fail" criteria for management, we can accommodate the ecosystem through adaptive tools.

The fourth principle takes us into the controversial realm of human economy vs. ecological quality. In fact, because all human wealth flows from the ecosystem, economics (from the Greek *oikos*, the house, and *nomos*, to measure) and ecology (*oikos* plus *logos*, to reason or plan) are aspects of the same fundament. Given the ubiquity of human alteration of the biosphere, and the intensity of human development in regions like the northeastern United States, accommodating human prosperity in environmental decision making is *sine qua non*. The trick is to make certain that we account for all benefits and losses associated with an activity, not those related to a small group of people or to a very localized area. Our ability to do this is growing, but needs to be nurtured.

The fifth principle is related to the fourth. In terms of environmental impact, there is no "them or us." We are all responsible for the ecosystem. The polluter may pay in the short run, but people who buy the polluter's products, work in the polluter's facilities, and collect the polluter's taxes all pay in the long run. We all eat, breathe, move, and use tools and shelter. In the wealthy developed world we do more of these things than many in the developing world, but the nature of life makes us all "polluters" because we are, by virtue of our evolutionary history, consumers. Just as we must accommodate human prosperity in our management, we must also all take responsibility for our ecological impact.

The sixth principle returns us to the mechanics of the restoration process. While this may seem less fundamental than the other principles here, in fact we view this as a very high-order process. Human beings have, through diligent application of the tools of science and philosophy, learned an enormous amount about many things. However, our institutions tend to keep us each in a narrowly defined "box" on a day-to-day basis. Professors of history do not routinely lunch with those in the zoology department, investment bankers do not share knowledge with environmental consultants, and water quality regulators are not sought out by those enforcing labor laws. Yet, in the globally linked biosphere that we all share, ecological connections assure that in nature (as opposed to the human institutions of business and government), lessons learned in one field are likely to have at least some applicability in another. As a very young science, and an integrative one at that, the discipline of environmental restoration can and should take advantage of the full breadth of technical skills and experience from all fields of endeavor.

The seventh principle is, at the moment, more a hopeful statement than a direction or truth. The idea is that the cumulative effect of "good" decisions on a project-by-project basis will be effective, long-term, high-quality environmental stewardship. We believe the idea has merit, but in practice there has yet to be a sufficient accumulation of "good" decisions on focused issues to test it at a realistic scale.

The eighth and final principle relates, of course, to Garrett Hardin's concept of the "tragedy of the commons" (Hardin, 1977), whereby shared resource use dilutes responsibility and encourages environmental degradation. In fact, that tragedy can be converted. If we all share the responsibility, manage collectively, and accommodate the facts of human existence and its ecological dependence, the environmental commons will become a showcase for our ongoing stewardship, rather than an artifact of our short-sightedness.

Taken together, and refined, revised, and modified by debate and experience, a set of principles like this can provide a "back stop" for decision making, something to which we can refer to help us answer unique and important questions. Such a foundation would help turn environmental restoration into the universally important and broadly acknowledged global discipline that it seems destined to become.

3.   A DECISION CONTEXT FOR RIVER MANAGEMENT

Acting on the stewardship principles shown in Table 1 requires that we step into a world of specifics. For decisions regarding river management, we can aggregate these specifics into categories based on conditions and outcomes. Figure 1 shows such an aggregation scheme applicable to cleanup and restoration decisions for rivers in settled landscapes. The decision categories represent status and projected conditions that must be balanced in the context of particular actions that might be implemented. Taking each in turn, "extant habitat value" represents the quality of the ecosystem in place, as it exists before actions are taken. If the quality is high, it is likely that a diverse biota is present and potential impacts of chemical exposure and habitat degradation are minimal. It also indicates that destructive activities (such as dredging, capping, channelizing, redirection, etc.) would have a great impact. In a sense, the converse of this is "habitat restoration potential," which indicates how much scope there is in the present ecosystem for habitat improvement. For example, in rivers with highly urbanized watersheds, there are generally severe habitat constraints, apart from any chemical exposures. A related concept is "remedial habitat destruction," or the degree and extent of impact that a contemplated management activity will have. For example, dredging is essentially completely destructive of the habitat in place, while capping can accommodate habitat enhancements that can sometimes mitigate some of the destructive consequences. In light of this, "extant chemical risks" reflect the negative impacts of exposure of people and other organisms to the quantities of anthropogenic chemicals in the ecosystem. Conversely, "risk reduction benefits" represent the increase in environmental quality that can be expected to accrue from activities that reduce chemical exposures. The Venn diagram model of Figure 1 demonstrates that optimal decision making reflects the simultaneous issues and concerns represented by the intersection of the decision categories.

What does all this mean as a practical matter? Consider the Neponset River (Note: the following brief discussion is very general and illustrative. Data on conditions in the Neponset Riveer are sparse, and before decisions are taken and actions implemented, considerable data collection needs to be undertaken in support.) Extant habitat values are impaired throughout the drainage, by the nature of the watershed and its history. However, there appears to be quite a range in habitat quality, with some reaches supporting much more diverse biota. The habitat restoration potential in the most severely impaired habitats is very high, less so where habitat quality is presently high. The potential for remediation activities such as dredging and capping to impact habitats is greatest in the areas of highest habitat quality. We know little about extant chemical risks, but it must be remembered that the mere presence of measurable quantities of industrial chemicals does not mean there is necessarily "risk." From the little information available, it appears that concentrations of many chemicals are likely to vary, and are not generally at overtly toxic levels. In this context, risk reduction benefits (of possible chemical remediation) will be related to exposure conditions and chemical concentrations on a location-specific basis throughout the river.

In balancing these decision categories, it is clear that habitat restoration would have the greatest return-on-investment in the most impaired river reaches, and chemical risks are likely to be highest where high habitat quality overlaps with high chemical concentrations. Such conditions-the overlap of high habitat quality with high chemical exposures-render the balance between the destructive effects of remediation and extant chemical risk very delicate. It is a balance that cries out for information on which to base decision making. But, with due consideration for principles and context of the decision matrix, we can at least initiate the data-gathering exercise, identify potential funding sources, and start down the road to comprehensive management of the Neponset River. When successful, that program will likely serve as an example for the management of rivers in settled landscapes nationally, and perhaps world-wide.

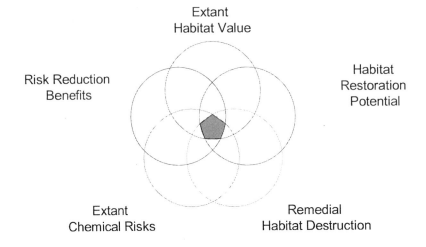

Figure 1. Decision Context for River Cleanup and Restoration

LITERATURE CITED

Cronon, W. 2003. Changes in the Land: Indians, Colonists, and the Ecology of New England. New York, Hill and Wang.

Hardin, G. 1977. Managing the Commons. New York, W.H. Freeman Co.

Krech, S. III. 1997. The Ecological Indian. New York, W.W. Norton & Company.

Salafsky, N., R. Margoulis and K. Redford. 2004. Adaptive management: a tool for conservation practitioners. Available at: http://fosonline.org/resources/Publications/Adap ManHTML/Adman_1.html.

CHAPTER 25

NEPONSET RIVER WORKSHOP
Overview of Remedial Options

Dale W. Evans, P.E.
CEO, Metrix, Inc17301 West Colfax Ave, Suite 170, Golden, CO 80401

Abstract: Environmental restoration of the Neponset River will represent numerous complexities due to chemical impacts and the existing urban and suburban infrastructure coupled with natural geomorphic processes. The purpose of this paper is to provide a perspective on these complexities to aid in development of a sound strategy for remediation and restoration activities.

Key words: dam removal; remediation; PCBs; geomorphology; sediments; stormwater; groundwater; watershed.

1. INTRODUCTION

Dam removal programs are inherently complex, as they represent a broad range of publicly sensitive issues. Since dam removal and the accompanying environmental restoration encompasses many factors, ranging from environmental impacts and socioeconomic issues to wildlife, recreation, and aesthetics, these programs may often suffer from misperceptions due to the unbalanced focus created by broadly varied perspectives. Effectively merging engineering, scientific, and public interest factors on these projects is a complex challenge that must be met in order to ensure that remediation and restoration processes proceed in a responsible and efficient manner. This requires careful orchestration of public programs to ensure that the expectations created can be technically fulfilled during the remediation and restoration program developed.

2.   RIVER RESTORATION OVERVIEW

Any river system is a complex living element of the current geomorphic platform. Besides carrying water, rivers become transit ways for fish, as well as for sediment during naturally occurring erosion. A river's function within developed urban and suburban watersheds inherently becomes additionally influenced by the presence of human populations and resulting industrial effluent discharges, stormwater, groundwater, and septic tank-related impacts, the presence of dams, or forced channelization meant to achieve geomorphic stability. Mitigation of these impacts to provide a more natural setting requires thoughtful planning to balance the actions proposed with the realities associated with human development. Such balance provides the basis for a sound strategy that will meet the broadest range of interests within the watershed.

The issues associated with the Neponset River are varied and require the careful application of technical responses to ensure that environmental conditions can be improved with the least degree of impact. Such impacts may be inevitable, as construction processes, even with controls, are likely to lead to some disruption within the existing conditions. In essence, the ideal solution will provide an optimal degree of environmental recovery with the least degree of impact.

3.   NEPONSET RIVER ISSUES

In the interest of providing a template for future definition of a watershed-based remedial strategy for the Neponset River, this paper outlines some of the important factors that should be considered, including:

- Technical
- Regulatory
- Financial

Each of these factors represents a host of oftentimes overlapping considerations that must be considered holistically to ensure that the project meets the ultimate objective of the program. It is important to consider that in as complex an environment as a watershed, it is likely that not all individual goals will be met. As an example, these goals must consider realities such as hydraulic conditions within a stream that may not support kayaking, fish passage, aesthetics, and flood control. To achieve the best degree of environmental balance, some compromises may thus be necessary.

At a minimum, a well-organized approach to the process must be used to help organize an effective strategy. Such an approach should consider each of the key boundary conditions that will influence or be influenced by the resulting program. A summary of a selected group of these boundary conditions is provided as Table 1.

Table 1. Boundary Conditions for Reservoir Restoration

Hydrology	
Annual hydrograph	Design storm and event
Flow rate variations	Weather factors
Design flow rate	Return period for significant events
Annual predictability of flow rates	Velocity profiles
Channel Characteristics	
Profile	Bed material
Channel cross-section	Critical shear stress
Overbank conditions	Impedances/obstructions
Sinuosity	Channel structures
Thalweg	Proximity to man-made features
Bank stabilization	
Extraordinary Events	
Man-made channel restrictions	Channel restrictions during flooding
Ice flows	Tributary effects
Ice scour	
System Stability	
Structure scour	Overbank erosion
Bed scour	Aggradation
Bank scour	Vegetative re-establishment
Ecosystem Considerations	
Impacted areas	Fish habitat
Construction	Benthic invertebrate populations
Cultural resources	Riparian areas
Endangered species	Sediment load
Wetlands	Wildlife
Fish passage	Vegetation
Political Factors	
Regulations	Recreation
Environmental impacts	Economic development
Post-restoration land use	Historic preservation
Stakeholders	
Federal agencies	Irrigation districts
State agencies	Downstream dams
County governments	Public-at-large
Planning districts	

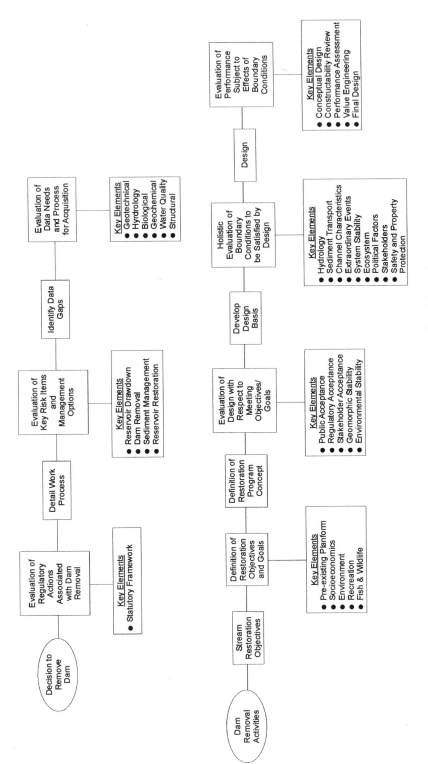

Figure 1. Restoration Planning and Design Schematic

By considering these boundary conditions, a well-scoped process for design of remediation and restoration processes can be employed to accommodate project-specific factors. Based on previously successful efforts, a schematic has been developed to help guide dam removal decision-making and the associated dam removal activities, as shown on Figure 1.

4.   ADDRESSING NEPONSET RIVER ISSUES

As noted previously, the remediation issues associated with the Neponset River inherently require consideration of many factors. However, specific issues related to the presence of PCBs in the sediment have been considered as outlined in the following paragraphs.

5.   MANAGEMENT AND CONTROL TECHNOLOGIES

Remediation of PCB impacts must obviously consider numerous factors such as the concentrations and distribution of contaminants, hydraulic and hydrologic factors, and habitat related issues. Effective consideration of these elements requires definition of a sound site model that considers each potential boundary condition.

Within the context of remediation, management of PCB-impacted sediments will rely on either leaving the materials in place, or removing the impacted materials for disposal at an appropriate location. While the number of management options is limited, the methods of implementation can be rather broad, depending upon the site-specific conditions. If the conditions warrant, PCBs may be isolated from future environmental exposure through the use of capping, provided geomorphic stability is considered within the impacted reach of river. Removal can also be used provided the materials handling methods chosen offer containment against impacted sediment migration, and further, that the limits of the impacts can be adequately defined prior to removal.

6.   MERITS OF REMEDIATION TECHNOLOGIES

Remediation technologies for sediment management are not likely to be significant with regard to overall ecological restoration or fish passage. The key element for consideration is more related to the methods available for dam removal and either isolation or removal of sediment coincident with the dam removal activities. As mentioned previously, a broad array of boundary conditions must be addressed to satisfy key environmental criteria.

7.   THE APPROPRIATE DEGREE OF CLEANUP

Very few projects will have the luxury of providing total restoration. Therefore, the degree of cleanup must be carefully weighed against site-specific conditions, the practical constraints associated with remediation, and the desired degree of environmental improvement that may be reasonably achieved. A key element associated with defining the degree of cleanup required will be the ability to validate the effects of remediation through monitoring. Such efforts are especially relevant within urban corridors, where environmental impacts may be from a variety of sources.

8.   SUCCESS STORIES

Many projects have been performed where impacted sediments have been removed or isolated. However, it is important to consider that the available body of data is limited to mostly projects involving clean sediments, or where impacted sediments were removed, the volume of materials managed were relatively small. Further, in the vast majority of work performed, only minimal monitoring has been prescribed to evaluate post-removal environmental factors. Accordingly, the processes used to guide future projects need to address such monitoring as a means to define the degree of success in meeting remedial goals.

CHAPTER 26

OPTIONS FOR THE NEPONSET

Danny D. Reible, Ph. D., PE
The University of Texas at Austin, Bettie Margaret Smith Chair of Environmental Health Engineering, Civil Engineering Department, 1 University Station C1700, Austin, TX 78712 - 0283

Abstract: Remedial approaches and options for the Neponset River in Southeaster Massachusetts are described. The advantages of combining restoration , in particular restoration of the river as an ecological resource, with remediation is identified. Dredging and capping are identified as possible remedial approaches. Dredging in the small areas be hind the dams where PCBs are concentrated was identified as potentially very effective. Capping over larger areas requiring remediation was also identified as effective and also as contributing directly to improvement in habitat values for th e river.

Key words: Contaminated sediments, remediation, restoration, dredging, capping

1.   INTRODUCTION

The Neponset River in southeastern Massachusetts has faced a long history of municipal and industrial development. As a result it has been faced with a variety of human challenges and efforts are underway to restore the river to its ecological and scenic potential. Chief among the challenges is PCB contamination along portions of the river, especially in the soft sediment behind two dams that separate the lower reaches from the upper river. In this paper, the options for remediating and restoring the Neponset will be investigated. Only general options will be considered given that the river and sediment characterizations necessary to make informed decisions are not yet available.

Sediment remediation is difficult and expensive and there are no default or obvious solutions that can be applied to Neponset River. Instead a formal decision-making framework involving all stakeholders should be appropriately applied to the river. A six-stage framework for decision-making by the Presidential/Congressional Commission on Risk Assessment and Risk Management was recommended by the NRC (2001). The six stages of the framework are:

1. Define the problem in context and establish risk management goals.
2. Analyze the risks associated with the problem.
3. Examine options for addressing the risks.
4. Make decisions about which options to implement.
5. Take actions to implement the decisions.
6. Evaluate the actions and results.

The NRC (2001) emphasized that central to each of these stages of the framework is stakeholder involvement. Nowhere is stakeholder involvement more important than in identifying goals. It is important to identify and agree on common goals early in a project. Failure to reach agreement as to the key concerns or the ultimate objectives of any effort may doom that effort to failure. It is important that all stakeholders, including potentially responsible parties and regulators as well as the affected public, reach a common vision for the site undergoing remediation and/or restoration.

The framework is designed to develop and implement solutions that minimize risk. It is important to recognize, however, that risks are more broadly defined than risk to human health and the environment. There are economic and social risks that may be difficult to express and quantify but may prove far more important to the stakeholders. It is also important to recognize that narrow goals focused strictly on reduction of risk to human health and the environment may not be considered successful if other issues are paramount. Often projects that satisfy multiple goals and demands are viewed as far more successful than those that are limited strictly to environmental remediation. Consequently, projects that simultaneously *remediate* the human health and environmental risks and *restore* human and ecological conditions are likely to be more universally accepted and considered successful. Restoration may include economic improvements, such as remediation projects that simultaneously provide economic benefits (e.g., expansion of a shipping port or other development project), or provide valued natural benefits (e.g. improvements in scenic character or ecological diversity of the site).

The advantages of simultaneous remediation and restoration are especially important in the Neponset. Currently, there are no identified potentially responsible upland parties to fund remediation. But, support for large expenditures of funds requires the identification and support of such parties for the realization of common goals. Success likely requires both remediation and restoration of the river characteristics valued by stakeholders. As mentioned previously, development of a common vision for the river is important to the success of *any* remedial and restoration effort but is critical to the success of one dependent upon public support.

Although I would hope to facilitate such efforts, I am not a stakeholder and can do little to identify the nature of that shared vision. It is believed, however, that some remediation of PCB-contaminated sediment will be required to achieve that shared vision. A limited number of remedial options, that is, options aimed specifically at reducing risk to human health and the environment, are available for application to contaminated sediments. The balance of the paper will be focused on identifying those options and their potential applicability to the Neponset River.

2.   REMEDIAL OPTIONS

There remain remarkably few options available to effectively control risks to human health and the environment from PCB-contaminated sediments. In principle, natural recovery with or without institutional controls, *in situ* treatment, *in situ* containment and removal, and *ex situ* treatment or containment are all available to manage risks from PCB-contaminated sediments. In reality, however, natural recovery for a persistent compound such as PCBs is largely due to containment by the natural deposition of clean sediments. The amount and quality of currentdeposition is not known and thus the rate of natural recovery cannot currently be determined. *In situ* treatment of PCBs is largely unavailable due to the refractory nature of PCBs and the inability to effectively deliver reagents or catalysts (e.g. nutrients or specific microbial populations) to sediments.

Thus dependable risk reduction is largely limited to removal by dredging or *in situ* containment by capping for the sediments of the Neponset River unless natural recovery by clean sediment deposition can be demonstrated to be sufficiently rapid and effective.

Removal of contamination via dredging has the potential to completely eliminate much of the contamination from the system. Dredging, however, exhibits a number of potential problems that may limit its effectiveness, namely:

- Resuspension – a fraction of the sediment dredged is resuspended and lost from the area around the dredge site leading to migration and exposure at points downstream. A typical value for resuspension from a well-controlled environmental dredging project may be of the order of 1% of the sediment dredged.

- Residual – a significant fraction of the sediments targeted for removal are often left at the sediment water interface. A useful tool for estimating this residual in the absence of site specific data is to assume that the surficial concentration post-dredging is the average of the layer dredged. Hence the ability to overdredge (cut deeper than necessary) into clean sediment is one of the most effective ways to reduce the residual concentration.
- Dredged material handling – any dredged material must be treated or disposed of elsewhere. In most cases, disposal in a landfill is the only economically viable option and transportation and disposal gives rise to public concern and some risk either during handling or after disposal. Dredged material also contains significant amounts of water that must be removed and returned to the waterway. Hydraulic dredging might produce 4-10 times as much water as sediment while mechanical dredging might produce 1-2 times as much water as sediment. The treatment of the produced water is costly and limited in its effectiveness. The return of the treated water to the waterway may exceed its ability to assimilate the residual contamination.

These problems do not necessarily preclude the use of dredging but they require assessment before implementation, and monitoring during implementation, to insure that the risk of remedy does not exceed the advantages of remediation and restoration. In the Neponset River the presence of a significant fraction of the PCB-contaminated sediment in small areas upstream of dams suggests that local dredging in these areas may provide significant benefits. Dredging is preferred when small areas of relatively high contaminant concentration contribute significantly to the overall risk.

An alternative remedial approach is *in situ* capping of the contaminated sediments. A layer of clean sand or other material could be placed above the contaminated sediment to:

- Armor the contaminated sediment and insure that it is not resuspended by high flow events
- Separate the contaminated sediment from the benthic organisms that typically populate only the upper 5-15 cm of sediment and which would not occupy the upper portion of the cap layer
- Reduce the contaminant migration from the sediment to the overlying water by posing as a resistance to mass transfer
- Serve as a tool to improve habitat by control of the material that makes up the upper portion of a cap or by controlling water depth and flow through design features

Because an engineered cap could be designed to improve habitat it may be particularly useful in restoration activities, in this case to improve the ecological value of the Neponset River. Capping is particularly cost-effective when large areas of river bottom must be remediated or when dredging is not feasible or poses too great a risk. Often capping may be necessary to control a residual left behind by dredging, so capping (or backfilling as it is called in that application) may be useful to develop habitat even if dredging is the primary remedial option selected. Conducting both dredging and capping, however, adds to the cost and exhibits both the advantages and disadvantages of these approaches.

Engineering a cap requires understanding the hydraulic forces that may act upon a cap to insure that it would remain in place under such conditions. Conditions that may encourage erosion of a cap include ice jams, high water flow events and navigation by highly powered craft in shallow water. If a cap can be designed to remain in place, however, it can be extremely effective for strongly solid-associated contaminants such as PCBs. Even a one or two inch layer of sand separating PCB-contaminated sediment from the layer influenced by benthic organisms is usually sufficient to provide large reductions in exposure and risk. A more detailed description of the design and implementation of a cap can be found in Palermo et al. (1998).

In some cases, however, conventional capping with an inert material like sand may not be sufficiently protective to meet remedial and restoration goals. In the Neponset significant groundwater upwelling into the river, for example, may reduce the effectiveness of a cap. In such instances, amendments that encourage degradation or sorption can be added to the cap material to slow or eliminate any contaminant migration through the cap. A variety of techniques have been investigated for this purpose but only a few have been demonstrated in the field. A demonstration of active capping approaches is being conducted in the Anacostia River of Washington DC. Of the approaches being employed in the demonstration, a permeability control agent, Aquablok, and a carbon sorbent, Coke, may be effective for the conditions of the PCB-contaminated sediments in the Neponset River. Aquablok is a clay surrounding a granular core that swells when placed underwater creating a tight, low permeability barrier. If groundwater upwelling into the river is a major source of exposure and risk, use of the Aquablok in key contaminated areas could eliminate or reduce this risk. Coke or other sources of organic carbon, including activated carbon or even organic-rich topsoil, can be used to effectively slow migration of contaminants through the capping material. Organic contaminants such as PCBs tend to sorb strongly to organic material in the sediments and any organic amendments in a cap will encourage sorption and retardation of the

PCBs. Activated carbon may be especially effective in that contaminants sorbed onto activated carbon tend to be bound very tightly and with less potential for desorption than from natural sources of organic matter. Additional information on the Anacostia active capping demonstration project can be found at www.hsrc-ssw.org.

3.   CONCLUSIONS

Although all the necessary information is not yet available to fully define a remedy for the Neponset, this paper attempted to identify the importance of setting common goals and outlined some of the options for meeting those goals. The most successful approach to the Neponset was identified as one that simultaneously met remedial goals, reduction of PCB risks, and restoration goals. Restoration goals should be identified by cooperation among all stakeholders but may very well involve restoration of the ecological values of the river. Dredging in small areas behind dams and capping in other more diffuse contaminated areas were identified as means of achieving remediation. In addition, capping provides many opportunities for restoration of habitat in that the needed characteristics for habitat improvement can be incorporated into the cap design.

REFERENCES

National Research Council (NRC), 2001.   A Risk Management Strategy for PCB - contaminated Sediments, National Academy Press, Washi ngton, D.C.

Palermo, M. R., S. Maynord, J. Miller, and D. Reible, 1998. Assessment and Remediation of Contaminated Sediments (ARCS) Program, Guidance for *In situ* Subaqueous Capping of Contaminated Sediments, EPA 905-B96-004.

Reible, D.D., D. Hayes, C. Lue-Hing, J. Patterson, N. Bhowmik, M. Johnson, and J.Teal , "Comparison of the Long-Term Risks of Removal and *In situ* Management of Contaminated Sediments in the Fox River,", Journal of Soil and Sediment Contamination, 12(3) 325-344 (2003)

CHAPTER 27

SORBENT-AMENDED "ACTIVE" SEDIMENT CAPS FOR IN-PLACE MANAGEMENT OF PCB-CONTAMINATED SEDIMENTS

Gregory V. Lowry[1], P.J. Murphy[1], A. Marquette[2], and D. Reible[2,3]

[1]*Carnegie Mellon University, Department of Civil and Environmental Engineering, 119 Porter Hall, Pittsburgh, PA 15213-389;* [2]*Lousiana State University, Department of Chemical Engineering;* [3]*University of Texas, Austin, Department of Civil and Environmental Engineering.*

Abstract: *In situ* capping manages contaminated sediment on-site without creating the removal, transportation, and disposal exposure pathways associated with dredging. PCB mass, a perceived surrogate for risk posed by PCB-contaminated sediment, remains on site during *in situ* capping projects creating concerns over the 1) duration of chemical i solation provided by sediment caps in advection dominated sediment systems and 2) the potential for PCB mass reduction by biotic and abiotic means during periods of chemical isolation. This study characterizes and compares commercially available sorbents to amend traditional sand caps by measuring properties relevant to advective-dispersive transport through porous media and uses these properties as inputs to a numerical model that predicts the performance of thin sorbent layers as sediment cap media in ad vection and diffusion dominated systems. Thin layer (1.25cm) sorbent amendments include thermally altered carbonaceous materials (coke and activated carbon) and soil. Sorbents provide chemical isolation of PCB-contaminated sediment in the same order (activated carbon – coke and soil - sand) as measured (highest to lowest) sorption Freundlich constants and sorption capacity. All sorbents outperform sand, coke and soil behave similarly. Thin (1.25cm) activated carbon layers chemically isolate PCB-contaminated sediment in advection -dominated systems for hundreds to thousands of years allowing time for natural attenuation.

Key words: *In situ* remediation; polychlorinated biphenyls ; ecological restoration

1.   INTRODUCTION

The proper management of PCB-contaminated sediment is a wide spread, complex, and costly issue. A National Research Council (NRC) committee on the remediation of PCB-contaminated sediment proposed a risk management framework, which attempts to minimize short- and long-term risk to human and ecological endpoints (NRC, 2001). Reible et al. have proposed the Surface Area Weighted Average Concentration (SWAC) as a useful surrogate risk metric (Reible, et al., 2003). SWAC is the average concentration in the biologically active portion of the sediment; this surficial zone in fresh and salt water systems is estimated to have a mean depth of 10 ±5cm (Thibodeaux, 1996). SWAC considers the biologically available fraction of PCBs when assessing exposure pathways and resulting risk. This surrogate risk metric assumes that PCBs buried below the biologically active portion of sediment are less likely to enter the aquatic food chain.

In situ capping (ISC) is a commonly employed risk-management technology to minimize SWAC. This technology eliminates contact between the benthic community and contaminated sediment with a physical barrier of clean material, usually sand. The objective of ISC is to minimize SWAC by isolating contaminants, stabilizing the contaminated sediment, and reducing contaminant flux into the biologically active portion of the sediment. ISC manages the risk of contaminated sediment without many of the disadvantages associated with dredging, including the generation of large volumes of dredged wastewater requiring treatment, contaminated sediment resuspension and transport, residual contamination, and the creation of new exposure pathways during *ex situ* treatment, transport, and disposal.

Thin sand caps can isolate contaminants for years and decrease the contaminant flux rate into the bioactive zone (Thoma, et al., 1993). The upper layer of sand caps can be tailored to promote the recolonization of native benthic organisms. Adding sorbents to sand caps will prolong contaminant isolation by retarding contaminant movement. Potential sorbents include organic rich soils and sediments, thermally altered carbonaceous material (TACMs), and engineered sorbents such as activated carbon. The tendency for hydrophobic organic compounds (HOCs) to strongly adsorb to TACMs and activated carbon and make them less bioavailable, is well documented (Grathwol, 1990; Acardi-Dey and Gschwend, 2002; Talley, et al., 2002). Sorption provided by organic rich soils or by TACMs will retard contaminant transport and delay contaminant breakthrough, allowing more time for inherently slow natural attenuation processes (e.g. deposition of clean sediment and biodegradation of contaminants buried in sediment). Both sorption strength and capacity are important parameters affecting a sorbent's ability to retard contaminant transport.

Thin layers (cm) of sorbent may provide sufficient capacity and retardation to meet management goals, but traditional placement methods, e.g. particle broadcasting, can not accurately deploy such thin layers. Fines, and low specific gravity materials (e.g. coke and activated carbon) that are resistant to settling, present an addition placement concern. Geotextiles are porous, synthetic fabrics that may allow for the accurate placement of thin material layers. The use of geotextiles to deploy thin layers of sorbent is promising, but field experience in subaqueous placement is limited.

The study objectives are to i) measure the physical properties relevant to advective-dispersive transport through four types of material (e.g. equilibrium partition coefficient, effective porosity, sorption capacity), ii) compare the PCB flux rate through a capping layer of each material under flow conditions expected in a sediment environment, and iii) develop and test a method to place thin layers of capping media. Laboratory batch and column studies were used to measure the physical properties of promising sorbents including soil, coke, and activated carbon. A numerical model is developed and used to predict the contaminant retardation provided by each sorbent. A thin (1.25 cm) sorbent-filled geotextile was developed, placed in the Anacostia river, and monitored for its effectiveness.

2. MATERIALS AND METHODS

1,2 Dichlorobenzene (99% Aldrich), Hexane (Fisher) and Tetrachloroethylene (99% Acros Organics) were used as received. A surface area reference material (24.1 +/- 0.6 m^2/g) for black carbon was used as received for quality control (Micromeritics – Norcross, GA). Four sorbents were used in this analysis: coke (U.S. Steel Clairton Works, Clairton Pa), F-200 activated carbon (Calgon Carbon, Pittsburgh Pa), organic-rich soil (3.2% OM) (Sestili Nursery, Pittsburgh, Pa), and silica sand (AGSCO Corporation - Hasbrouck Heights, NJ).

2.1 Material Characterization

The specific surface area of soil, coke, and a black carbon surface area standard were measured by N_2 BET analysis using a Nova 2100 (Quantachrome). Total organic content of soil and Anacostia river sediment (typical sediment) was determined by ASTM D2974-87 using a commercial analytical laboratory (UEC Labs, Monroeville, Pa).

2.2   Equilibrium Sorption Isotherms (K_f, $1/n$)

Isotherms were measured for coke and soil. Sorbents were pulverized, sieved, and dried for 12 hours at 105 °C to remove H_2O ($d_p < 0.5mm$). A measured mass of sorbent was added to 130 ml of 1,2 DCB solution (2-40 mg/L water) and 30 ml of headspace. After 10 days, 200 µL headspace samples were manually injected onto a Hewlett Packard 6890 Gas Chromatograph (GC) with HP-5 column and µECD to determine the concentration of 1,2-DCB in the headspace. The dimensionless Henry's Constant for 1,2 DCB in DI water at 25°C was measured using a previously described method (McAuliffe, 1971) and used to determine the aqueous phase DCB concentration. The sorbed mass was then determined by difference. Sorption isotherms where fit to a solubility-normalized Freundlich Isotherm (Carmo et al. 2000).

2.3   Physical and Hydrodynamic Sorbent Properties

Sorbent filled columns were used to assess the sorption capacity (1, 2 DCB) and effective porosity (ε) (sodium chloride tracer) of each material. Influent (~20 ppm) and effluent 1,2 DCB samples (1 mL) were extracted into hexane (1 mL) and analyzed by GC/µECD. An interstitial fluid velocity of 0.14 $^+/.$ 0.03 cm/min (sorbent porosity dependent) was used. The sorption capacity of each sorbent was determined by numerical integration of the area between the influent and effluent 1,2 DCB measurements until complete breakthrough. Effective porosity of each sorbent was determined by measuring the time for 50% sodium chloride tracer breakthrough.

Column breakthrough data were fit (visual inspection) to a one-dimensional advective-dispersive transport analytical solution to quantify hydrodynamic dispersion (α). (van Genuchten and Alves, 1982.) Sorbent bulk density (ρ_b) was measured gravimetrically, by weighing dried material in a 500 ml container. A molecular diffusion coefficient was estimated by the molar volume of the compound and the viscosity of water at 25°C using the method of Hayduk and Laudie.

2.4   Environmental System Properties

A bioturbation depth (B_d) of 15 cm (Thibodeaux 1996), a biodiffusion coefficient (D_b) of 10 cm^2/yr (Palermo et al. 1996), and a benthic boundary layer mass transfer coefficient (BBL_{mtc}) of 1 cm/hr (Palermo et al. 1996) were assumed. A uniformly contaminated sediment depth of 50 cm was used and allowed to deplete over time as contaminants transport through the cap or degrade. Consolidation of sediment and capping materials is not considered.

2.5   Model Development

Chemical fate and transport is analyzed for a contaminant initially residing in a sediment layer which is capped by a thin layer active cap (h_{cap1}) and an overlying sand layer of thickness h_{cap2} (Figure 1). The upper portion (10cm) of the sand layer (h_{bio}) is considered the bioturbation zone. The benthic activity reworks the surficial sediments at an effective biodiffusivity of $D_{bio.}$

Diffusion driven by a concentration gradient in the porewater, and groundwater seepage, contribute to the flux within the sediment and cap layers. Slowed migration due to contaminant sorption onto solids is described by a non-linear Freundlich equilibrium relationship between the solids and porewater. Contaminant degradation is represented as an xth-order homogenous reaction in the source term. Algebraic calculation of PDE coefficients are solved by Matlab v6.5 then transferred to Femlab 3.0a, finite element solver, to solve the governing PDE's. Built-in iteration functions return flux and concentration point data for specified temporal and spatial coordinates.

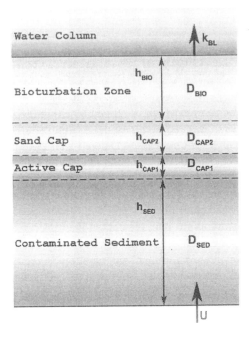

Figure 1. Proposed Design for Thin Active Layer in Sand Cap.

3.   RESULTS

3.1   Sorbent Properties

The measured or estimated properties for each of the sorbents are provided in Table 1. The calculated retardation factor for each sorbent assumes 2,4,5-trichloro-PCB as the sorbate.

3.2   PCB Transport through a thin (1.25 cm) layer of "active" cap.

The estimated flux of 2,4,5-trichloro-PCB through a 1.25-cm layer of each of the sorbents evaluated is shown in Figure 2. These simulations assume a sediment PCB concentration of 1 mg/kg and a seepage velocity of 1 cm/d, and thus represents an advection dominated system (Peclet number = 2).

Table 1. Measured and Estimated Sorbent Properties used in Model Simulations

Property	Units	Sediment[1]	AC	Coke	Soil	Sand
Log K*$_{Fr}$	Mg/kg	3.67	6.06[2]	3.95	4.13	0.1
Log K$_{Fr}$ (2,4,5 PCB)	(mg/kg)((L/mg)$^{1/n}$)	3.92	6.18[2]	4.66	4.9	1.35
N	(--)	1	0.28[2]	0.84	0.94	1
Sorption Capacity	(mg 1,2 DCB/g sorbent)	NM	260.7	0.39	0.38	0.0016
ε	(--)	NM	0.53	0.48	0.47	0.29
ρ_b	(g/cm^3)	1.95	0.66	0.72	0.99	1.65
α	(cm)	2	0.4	0.5	0.7	0.15
F_{oc}	(--)	0.024	NM	NM	0.016	0
SA	(m^2/g)	/	919	7 $^+$/. 5	6.6	<1
Cap height	(cm)	50	1.25	1.25	1.25	15

NM = Not Measured

[1] "Typical" Sediment assumed with K*$_{Fr}$ estimated based from Anacostia River Sediment Sample   F$_{oc}$ measurement using (Karickoff et al. 1979)

[2] Published F-100 Calgon Carbon Activated Carbon Properties (Kleineidam et al. 2002)

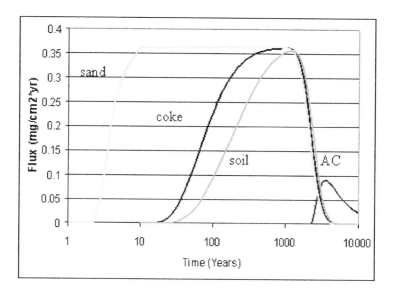

Figure 2. Flux of 2,4,5 PCB vs. time at top of cap (hcap = 1.25 cm). Seepage velocity = 1cm/day; initial contaminated sediment porewater concentration = 1mg/kg.

4.   DISCUSSION

The Freundlich partition coefficient (K_{Fr}), a measure of sorption strength, and sorption capacity were significantly higher (2-3 orders of magnitude) for activated carbon than that measured for soil and coke (Table 1). Soil and coke demonstrated similar K_{Fr} values and sorption capacity (Table 1), suggesting that these materials will have similar performance as a capping media. The similarity of coke and soil is a surprising result given previous work (Kleineidam et al. 2002; Jonker and Koelmans 2002) showing other TACMs (including soot, fly ash, and charcoal) to have orders of magnitude higher Freundlich constants than soils. This may be because the specific surface area of the coke used in this study is lower than TACM's used in previous studies, and is under investigation. Soil and coke exhibited higher K_{Fr} and sorption capacity values than the traditional capping material, sand. Sand provided negligible sorption strength and capacity. The Freundlich exponent (n) is a measure of sorption nonlinearity, indicating lower sorption strength at higher sorbent and aqueous PCB concentrations. Activated carbon (.28) exhibited more sorption non-linearity compared to coke (.84) and soil (.94). Other sorbent properties measured, porosity (ε), bulk density (ρ_b), and hydrodynamic dispersivity (α), are similar for all sorbents analyzed and are not expected to alter the ordering of material performance based on sorption properties (K_{Fr} and sorption capacity). Based on sorption properties, AC carbon is a significantly better sorbent than coke or soil, but all three sorbents have the potential to significantly improve the retardation capacity of sand, the traditional capping material.

The ordering of chemical isolation in Figure 2 (longest to shortest) is the same as the ordering of Freundlich constants (Table 1) for sorption isotherms and sorption capacity (highest to lowest). Figure 2 provides an estimate of expected chemical isolation provided by thin layers (1.25cm) of sand (years), coke and soil (decades to centuries), and activated carbon (thousands of years) in a sediment system with an upward groundwater seepage velocity of 1cm/day. In the absence of seepage, these chemical isolation times will be significantly higher.

The isolation time afforded by *in situ* capping, i.e. the decades to centuries before PCBs in the underlying sediments begin to migrate into the bioturbation zone, provides time for natural attenuation processes to occur. This includes deposition of clean sediment and biodegradation of PCBs in the underlying sediment. Both processes will further decrease the flux of PCBs into the benthic community and overlying water column. The deposition of clean sediment as a function of the isolation time provided by activated carbon and coke are shown in Figure 3.

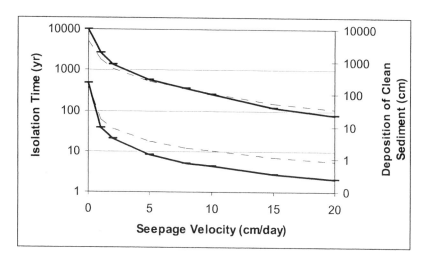

Figure 3. Isolation Time and Deposition of Clean Sediment at isolation time vs. Seepage
Velocity for AC and Coke.

By the time PCBs begin to breakthrough the thin activated carbon layer
(breakthrough is arbitrarily defined as 10% of the initial contaminated
sediment porewater PCB concentration), sediment deposition, assuming a
deposition rate of ~4mm/year, buries PCBs under ~1m of clean sediment
(for a seepage velocity of 1 cm/d). This further reduces the SWAC, and the
risk, associated with PCB-contaminated sediment.

The isolation time provided by sorbent-amended sediment caps can also
provide time for natural attenuation of PCBs in the underlying sediment
through biological dechlorination. The extent of dechlorination will depend
on the rate at which microbial dechlorination occurs in the underlying
sediment. The percent of PCBs degraded as a function of the PCB
degradation half- life time is shown in Figure 4.

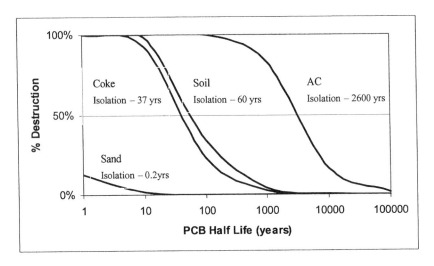

Figure 4. PCB mass removal as a function of 1/2= -lifetime of PCB congener for the isolation time provided. Isolation time represent the time for 10% b reakthrough of 1.25 layer of material with a seepage velocity of 1cm/day.

A well-accepted chemical half-life for destruction of PCB congeners in anaerobic sediment is not available in the literature. Such uncertainty makes predicting the expected extent of potential PCB mass destruction of capped sediment difficult.   Half-life times of decades to centuries have been reported in the literature so some natural attenuation is likely. If methods to stimulate PCB degradation *in situ* in underlying sediments become available, an engineered system designed to contain PCBs while they are degraded in the underlying sediment is certainly possible, and would be an attractive remedial option.

4.1   Field placement of thin sorbent-amended "active" sediment caps

Thin sorbent-amended sediment caps can effectively manage PCB-contaminated sediments, but methods to place thin layers of sorbent must be developed.   One approach is the use sorbent-filled geotextiles.   Sorbent-filled geotextiles, reactive core mats (RCM), were recently developed with CETCO (Arlington Heights, IL) and placed in the Anacostia River. The coke-filled RCM placed in the Anacostia River (Washington, D.C.) in April 2004 is shown in Figure 5 and Figure 6. The RCM sunk readily, was placed without incident (unrolled using crane and diver), and fines were not visually detected. A sand layer (12") was placed above the RCM to secure it and to provide a habitat for benthic organisms to colonize without compromising the integrity of the cap. Organics and metals found in solid coke and leachate were below the appropriate federal regulations (NOAA's Effect Range Low and EPA's Criterion Maximum Concentration, respectively). Sediment cores from this capping demonstration site will be used to assess the performance of sorbent (contaminant isolation) and RCM (placement, sustained permeability, and structural integrity).

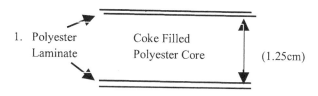

1. Polyester Laminate   Coke Filled Polyester Core   (1.25cm)

Figure 5. Schematic of RCM

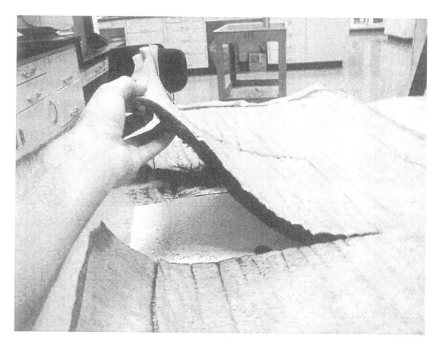

Figure 6. Picture of RCM

5.   CONCLUSIONS

The sorbents evaluated, soil, coke, and AC, can isolate contaminants from the biologically active zone and overlying water column for decades to centuries, allowing time for the deposition of clean sediment and the biodegradation of buried contaminants. In high seepage areas, e.g. >2 cm/d, these layers of activated carbon can still provide decades to centuries of protection before contaminants migrate through the cap. Thin layers of alternative sorbents can be accurately deployed in RCMs. The material overlying the RCM can be chosen to ensure cap stability, and to provide opportunities for recolonization of the benthic zone. These thin layers may provide sufficient capacity and retardation to allow for significant attenuation of contaminants in the underlying sediments. This technology can deliver both in-place containment of PCBs, and *in situ* reduction of PCB mass, provided that PCB biodegradation is occurring at a sufficiently rapid rate, or that a means to stimulate contaminant degradation in the underlying sediments is available.

ACKNOWLEDGEMENTS

This work was funded by the Hazardous Substance Research Centers through a research grant to Dr. Lowry (R139634) and by Alcoa, Inc. (Pittsburgh, PA). The authors also thank Jim Olsta and Chuck Hornaday of CETCO for constructing the RCMs.

REFERENCES

National Research Council. 2001. "A risk-management strategy for PCB-contaminated sediments." Washington, D.C.

Reible, D.D., Hayes, D., Lue-Hing, C., Patterson, J., Bhowmik, N., Johnson, M., and Teal J. 2003.   Comparison of the long-term risks of removal and *In situ* management of contaminated sediments in the Fox River. *J. Soil and Sed. Contam.,* 12(3) 325-344.

Thibodeaux, L. J. 1996. Environmental chemodynamics, 2 nd Ed., Wiley-Interscience, New York.

Thoma, G. J., Reible, D. D., Valsaraj, K. T., and Thibodeaux, L. J. 1993. Effi ciency of capping contaminated sediments *in situ* 2. Mathematics of Diffusion/Adsorption in the capping layer. *Envion. Sci. Technol .,* 27, 2412-2419.

Grathwohl, P. 1990. Influence of organic matter from soil s and sediments from various origins on the sorption of some chlorinated aliphatic hydrocarbons: implications on Koc correlations. *Environ. Sci. Technol.,* 24, 1687-1693.

Accardi-Dey, A., and Gschwend, P.M. 2002. Assessing the combined rol es of natural organic matter and black carbon as sorbents in sediments . *Environ. Sci. Technol.,* 36, 21-29.

Talley, J. W., Ghosh, U, Tucker, S.G., Furey, J.S., Luthy, R.G. 2002. Particle -scale understanding of the bioavailability of PAHs in sediment. *Environ. Sci. Technol.,* 36, 477-483.

McAuliffe, C.D., 1971. GC determination of solutes by multiple phase equilibration. *Chem. Tech.* 1, 46-51.

Carmo, A. M.; Hundal, L. S.; Thompson, M. L. 2000. Sorption of Hyd rophobic Organic Compounds by Soil Materials: Application of Unit Equivalent Freundlich Coefficients. *Environ. Sci. Technol.* 34(20); 4363-4369.

van Genuchten M. Th. And W. J. Alves, 1982. Analytical solutions of the one -dimensional convective-dispersive solute transport equation. USDA ARS Technical Bulletin Number 1661. U.S. Salinity Laboratory, 4500 Glenwood Drive, Riverside, CA 92501.

Palermo, M., Maynord, S., Miller, J., and Reible, D. 1996. Guidance for *In situ* Subaqueous Capping of Contaminated Sediments. EPA 905-B96-004, Great Lakes   National Program Office, Chicago, IL.

Karickhoff, S.W., Brown D.S. , and Scott T.A. 1979. Sorption of Hydrophobic Pollutants on *Natural Sediments. Wat. Res..,* 13, 241-248.

Kleineidam, S., Schuth, C., and Grathwohl, P. 2002.   Solubility -normalized combined adsorption-partitioning sorption isotherms for organic pollutants. *Environ. Sci. Technol. ,* 36, 4689-4697.

Jonker, M.T.O., and Koelmans, A.A. 2002 . Sorption of Polycyclic Aromatic Hydrocarbons and Polychlorinated Biphenyls to soot and soot -like materials in the aqueous environment: mechanistic considerations. *Environ. Sci. Technol.,* 36, 3725-3734.

CHAPTER 28

RIVER RESTORATION: A VIEW FROM WISCONSIN

Mark Velleux[1], P.E. and Edward Lynch[2], P.E.

[1]*Department of Civil Engineering, Colorado State University, Engineering Research Center, Fort Collins, CO 80523-1372;* [2]*Wisconsin Department of Natual Resources (RR/3), 101 S. Webster Street, Masison, WI 53701*

Abstract: River cleanup and restoration efforts pose significant legal and technical challenges for state and other regulatory agencies. Efforts are often conducted under different authorities, particularly CERCLA (Superfund) and NRDA rules. Wisconsin is home to a number of cleanup and restoration sites. The experience of project managers who worked for the Wis consin Department of Natural Resources on the Fox River and other cleanup and restoration site projects is presented. A number of common issues affect restoration projects. At many sites, human use has significantly affected the landscape and the presence of dams will have altered the function and habitat of river systems. Dam removal is a tool of choice for restoration but the appropriateness of removal can be influenced by: the presence of contaminated sediments , public acceptance, publi c safety and floodplain management. Consequently, restoration is typically more difficult than just dam removal . The presence of contaminated sediments is a big hurdle to project implementation. To move forward with cleanup and restorat ion where contaminated sediments exist, it may be necessary to work under CERCLA, NRDA, and other processes. Regardless of the processes used, project managers should use a diverse range of methods that collectively document the need and best approach for action at the site. Most importantly, regulatory agencies should combine CERCLA and NRDA processes for synergy and leverage but be aware of pitfalls where these processes can conflict. The combined results of CERCLA and NRDA efforts can create leverage for settlement by demonstrating that slower, less complete cleanup actions can escalate damages and associated costs. This can in turn provide an economic incentive for more extensive cleanup in order to reduce overall liabilities in the form of resources dam ages and promote development of remedial actions that facilitate overall restoration efforts.

Key words: river restoration; cleanup; contaminated sediment ; CERCLA; NRDA.

1.   INTRODUCTION

Neponset River ecological restoration efforts are at the beginning of what may be a long, challenging process to address contaminated sediments. Such restoration efforts often require that state regulatory agencies be active, if not leading, participants in processes that inevitably involve federal agencies, one or more potentially responsible party (PRP), local government, and the public. The experience of cleanup project managers at state agencies such as the Wisconsin Department of Natural Resources (WDNR) may be of value to other state agencies as they embark on restoration efforts.

Wisconsin is home to a number of cleanup and restoration sites. Issues associated with implementation of cleanup and restorations at sites have been, or are being, addressed using a range of different processes including the Comprehensive Environmental Restoration, Compensation, and Liability Act (CERCLA), known as Superfund, Natural Resources Damage Assessments (NRDA), as well as state spills laws. The Fox River is perhaps the most well-known site because of its size and the presence of contaminated water, sediment, and fish, as well as the technical and legal challenges associated with response and restoration efforts.

Beyond the Fox River, other project areas across the state may also be appropriate analogs for Neponset River efforts. Possible analog sites include: 1) Cedar Creek and the Milwaukee River; 2) Sheboygan River and Harbor; 3) Pine Creek, Hayton Millpond, and the South Branch Manitowoc River; and 4) the Baraboo River. Projects at these sites address a wide range of issues including: contaminated sediment; dams, dam removal, and fish passage; urbanization and channelization; stream bank and bed instability; floodplain loss. Collectively, consideration of the lessons learned from these possible analogs sites may be useful for guiding the development and implementation of restoration efforts for the Neponset River.

With the goal of sharing some of the lessons we have learned, this paper provides a staff-level perspective of a few of the pathways and pitfalls associated with contaminated sediment cleanup and restoration efforts. Before proceeding, it should be noted that the views and opinions presented below are those of the authors. These views were developed based on our experience as front-line staff responsible for developing and managing key parts of the technical effort for Fox River remediation as well as other project sites. The opinions presented do not represent the official policy of the Wisconsin Department of Natural Resources, the U.S. Environmental Protection Agency (USEPA), or U.S. Fish and Wildlife Service (USFWS).

2.   THE CLEANUP PROCESS: MANAGE IT OR BE MANAGED BY IT

There are a number of common issues that affect restoration projects. At many sites, 150 years or more of human use has significantly affected the landscape. With respect to river systems, the presence of dams will have significantly altered the function and habitat of the system. As a result, dam removal is often a tool of choice for restoration, particularly for fish passage. However, the appropriateness of dam removal can be strongly influenced by the presence of contaminated sediment, public acceptance, public safety and floodplain management concerns. The role of the public in the process cannot be underestimated. Riparian landowners are often particularly concerned about dams and their removal and often will ask how a removal will affect their ponds and shoreline. Consequently, habitat or ecological restoration is typically far more difficult than just dam removal. In a sense, dam removal is a means to an end and not the end itself, especially with contaminated sediments. This is also important to note because, in a sense, the on-the-ground restoration action that may go on at a site can differ from the legal definition of restoration as presented under NRDA rules.

The presence of contaminated sediments is almost always the biggest hurdle to project implementation. There are several reasons for this. One is that cleanup and restoration processes may fall under different jurisdictions such as CERCLA, NRDA, and other spills laws (both state and federal). As a result, different actions may be needed under multiple processes. Another is that the requirements for different processes, are sometimes similar but are rarely, if ever, the same. Despite the differences in processes, a sound database and extensively documented decision-making are needed. This is particularly true when PRPs for site contamination are identified as part of CERCLA or NRDA efforts. Once PRPs are involved the legal landscape and character of a project will be profoundly altered.

PRP involvement can bring, or rather unleash, the full force of the legal workings of the CERCLA and NRDA processes. When this happens, there are only two outcomes for a state agency: manage the process or be managed by it. Ultimately complex legal processes, such as CERCLA and NRDA can be their own worlds. In particular, it can be a high stakes game if liabilities are large and will almost inevitably be highly contentious.

State agencies need to be very aware that the CERCLA and NRDA process can be misused to delay and prevent cleanup and restoration actions. The ability of a well-organized PRP effort to delay significant action has in some instances been measured in decades. The reason for this is simple economics and boils down to the time value of money. The costs associated with contaminated sediment cleanup and restoration efforts are often so large that, in our experience, the view of most PRPs, and especially those with liabilities at multiple sites, is that it is more cost-effective to oppose and delay cleanup. In fairness, it is also worth pointing out that, whether intentionally or unintentionally, state agencies can also use the system to impede action.

Nonetheless, in the face of resistance, state agencies must be prepared to lead or be led. At least in our experience for CERCLA sites where PRPs are present, the first tactic that will be used is that the PRPs will bypass the state and approach USEPA with an offer to lead the remedial investigation and feasibility processes. Again, in our experience this has always meant delay and substandard work designed solely to minimize PRP liabilities rather than address sites contamination concerns. For NRDA efforts, a different tactic is likely and PRPs will approach the state to assert its authority as a trustee. As part of a package deal to induce the state to assert its trustee status, PRPs may offer to perform economic and assessment studies. However, such studies will generally be in the same vein as those performed under CERCLA and designed with the intent of minimizing liabilities regardless of the damage that may exist at the site.

To be clear, it is worth stating that state agencies should, in fact, work with PRPs wherever possible. At the same time, work done with PRPs should not be used as a means to drive a wedge between state and federal regulators. So, when working with PRPs, regulators need to remain aware that not everything may go smoothly. Be prepared for all variety of legal maneuvers and other bumps in the road. Always keep in mind that both CERCLA and NRDA are legal processes. These processes have defined steps but are not "cookbook" approaches to site cleanup and restoration. The best approach may be to attempt to minimize the potential for litigation by being as well prepared as possible to make sure that all products produced under CERCLA or NRDA meet or exceed all legal requirements. Failure to do so will likely only encourage delay.

3.   CERCLA VS. NRDA VS. RESTORATION

Cleanup and restoration processes are different. It is important to understand these differences, as well as where the processes can overlap, in order to move a project forward. CERCLA targets remediation and management of site risks. NRDA targets compensation for value of past, present, and future resource damages, and often at the system level rather than for a localized site. In contrast, restoration (again defined in an operational, on-the-ground sense) targets the return of function of a site to some pre-disturbance condition.

These processes overlap but will not necessarily fit together unless the efforts are designed that way. Coordination of state agencies with USEPA, USFWS, and other regulators and trustees is critical to integrate efforts. And, just as important as cooperation between regulators, the public cannot be ignored. Public input is not only required under CERCLA and NRDA rules, it can be critical to the ultimate success or failure of cleanup and restoration efforts. A cleanup effort is unlikely to proceed in a timely manner if the community affected does not accept it.

The proper degree of coordination between regulators is difficult to precisely quantify. However, some pitfalls we have seen at least demonstrate what to avoid. One pitfall is that CERCLA is not restoration. For example, CERCLA cannot be used to order restoration actions unless those actions materially address site risks managed under CERCLA. An example from Wisconsin is the case of the Hayton Millpond dam. Dam removal would certainly aid overall site restoration. However, since the dam was physically sound, owned by a party outside of the primary PCB-contaminated sediment cleanup effort for the site, and contaminated sediments could be addressed without removing the dam, CERCLA could not be used to order dam removal. Another pitfall is that NRDA does not necessarily provide site restoration at local scales. Even if a link from substance release to resource damage is established, local restoration is not guaranteed as funds can be used for off-site activities. This is both good and bad in sense: good because a holistic approach to balance remediation and restoration at the system level can be developed and bad because site-specific restoration may not be possible.

No matter the case, state regulators must be flexible in order to find ways to combine processes while avoiding pitfalls. To do so, it is important to consider the processes and the tools available for them. While it should go without saying, regulators (both state and federal) must always bear in mind that sound field data are the key to complete technical work to support cleanup or restoration. Without sound data, regulators invite legal challenges that can derail cleanup and restoration efforts. With respect to data use, make every attempt to go beyond the minimum requirements of the law. In particular, consider the widest possible spectrum of tools with a diverse range of underlying assumptions for sound decisions. Beyond field studies, possible tools include detailed examinations of contaminant trends in water sediment and fish, site-specific chemical transport and bioaccumulation models, and evaluations of sediment bed stability and contaminant depth horizons. As an example, see the White Papers WDNR used to document the CERCLA decision-making process it used for the Fox River. In our opinion, we strongly recommend that the angle of regulatory agencies be to combine resources and activities wherever possible for synergy. At the same time, it has been our experience that the PRP angle will likely be to keep processes separate to minimize their liabilities. Despite differences in approach or mechanism to do so, the key point is that regulators need to be sure to build tools to promote development of comprehensive settlements in order to move site cleanup and restoration towards implementation.

4. PUTTING SYNERGY INTO YOUR PROCESS

Under the right conditions, it is possible to put synergy into projects involving CERCLA and NRDA efforts. To do so, project managers will need to consider stakeholder goals. Obviously, different groups will often have different goals. Sometimes those goals will be mutually exclusive. Assuming that a state agency is leading the efforts, this puts the burden on that agency to find commonality in the processes used. Obviously, the ultimate commonality between PRPs, regulators, and the CERCLA and NRDA processes is the need to reach mutually acceptable and beneficial agreements to provide liability release and funds for meaningful site work. The best way to do this is to design efforts under CERCLA and NRDA to provide sound data for both processes. The data sets the table for settlement discussions. This means that state and federal agencies should coordinate efforts. More importantly, if possible, CERCLA and NRDA processes should be combined.

More specifically, some level of synergy can be achieved when the studies that underlie the CERCLA and NRDA processes are designed to look at all alternatives and build a common base of information. When efforts are combined under a state lead, it will be up to the state to make sure that due consideration is given to all possible alternatives. Presumptive remedies that restrict consideration of no action or monitored natural recovery alternatives are inherently prone to increased litigation risks. Regardless of which agency has the lead, good working relationships among staff are critical. Similarly, regulators most foster good working relationships with PRPs. At the same time agencies must ease public fears that less adversarial relationships indicate that the negotiating process is somehow improper. Nonetheless, while noting that the ultimate commonality is the link between liability release and desire for full-scale project implementation, full synergy will be difficult to achieve unless regulators and PRPs alike consider liability release and remedy permanence for different cleanup options, as well as the feasibility of implementing specific restoration plans and their likelihood of success. Biased assessments in this regard can not only shortchange site cleanup and restoration but can lead to significant delays in project implementation.

It is again worth noting that coordination between regulators can make the difference. This is more than just sound public policy. Further, this is more than just the balance between remediation and restoration for a site. By bringing forward the results of CERCLA and NRDA processes at the same time regulators can create leverage needed for better cleanup efforts. The combined results of CERCLA and NRDA efforts can demonstrate that slower, less complete cleanup actions escalate damages and associated costs. This in turn can provide an economic incentive for more extensive cleanup in order to reduce overall liabilities in terms of resources damages. This can then further promote development of remedial actions that facilitate and contribute towards overall restoration efforts.

5.   SUMMARY AND CONCLUSIONS

In summary, Wisconsin is home to several possible analog sites that may be relevant to ongoing and upcoming actions for the Neponset River ecological restoration. Because of its PCB-contaminated sediments, the Fox River may be particularly relevant. Based on that Fox River experience, when dealing with contaminated sediments, it may be necessary to work under CERCLA, NRDA, as well as other processes. Our most important lesson there, and one that may be applicable the Neponset River and other sites is that, regardless of the processes used to drive cleanup and restoration efforts, project managers should use a diverse range of characterization and assessment methods that collectively document the need and best approach for action at the site. Most importantly, regulatory agencies should combine CERCLA and NRDA processes for synergy and leverage but be aware of pitfalls where these processes can conflict.

The authors express their thanks to conference and workshop organizers for providing this opportunity to share our views.

CHAPTER 29

CHARACTERIZATION OF CONTAMINATED SEDIMENTS FOR REMEDIATION PROJECTS IN HAMILTON HARBOUR

Risk Assessment and Remedial Approaches Towards Restoration and Management of Contaminated Rivers

Alex J. Zeman and Timothy S. Patterson
Environment Canada, Aquatic Ecosystem Management Research Branch, National Water Research Institute, 867 Lakeshore Road, PO Box 5050, Burlington, Ontario, L7R 4A6 Canada

Abstract: Hamilton Harbour is located on the western end of Lake Ontario and has an area of approximately 31 square kilometers. The harbour has been designated by the International Joint Commission (IJC) as one of the 42 Areas of Concern (AOCs) within the Great Lakes. Most fine-grained sediments in the harbour exceed sediment quality guidelines at the severe effect level due to contamination by both metals and organic compounds such as PAHs and PCBs. Current investigations are concentrated on two areas of the harbour called Randle Reef and Windermere Arm. The Randle Reef "hot spot" contains the most highly contaminated sediment for PAH concentrations in the harbour. Extensive information on sediment physical and chemical properties was collected by coring and offshore boreholes. Bioassays were carried out to determine sediment toxicity. A range of remediation alternatives has been considered, including removal and *ex situ* treatment. The current preferred alternative is an engineered containment facility (ECF), which will contain *in situ* contaminated sediments within the footprint of the structure. In addition, dredged sediment from other contaminated sites in the harbour will be placed in the ECF. Windermere Arm is a 50-ha narrow channel situated in the southeast portion of the harbour. Contamination in Windermere Arm is not as severe as that found in Randle Reef. Recent sediment surveys in the area, however, yielded higher PCB values in surficial sediments than previously reported. Sediments in Windermere Arm are also subject to considerable physical disturbance due to extensive ship traffic. For this reason, historical sediment contamination occurring in deeper sediments also has to be considered as a potential risk to the aquatic environment.

Key words: Sediment characteristics and quality , sediment mapping, toxic metals, PAHs, PCBs

1.   INTRODUCTION

Hamilton Harbour is located on the western end of Lake Ontario and occupies an area of approximately 31 square kilometers or 3,100 hectares (Fig. 1). The harbour has been identified as one of the 42 Areas of Concern (AOCs) by the International Joint Commission (IJC) due to pollutant concentrations exceeding both the (Ontario) provincial water quality objectives (PWQOs) and the provincial sediment quality guidelines (PSQGs) with respect to nutrients and toxic pollutants such as PCBs, PAHs and metals. In several areas of the habour, contaminated sediments exceed the PSQGs at the "severe effect level" for metals and persistent organic compounds.

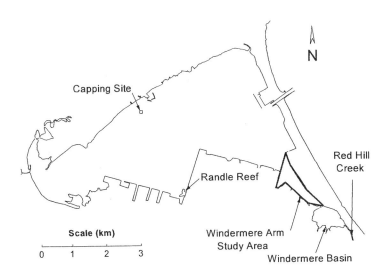

Figure 1. Study areas in Hamilton Harbour

Industrial and municipal loading of suspended solids and associated contaminants have decreased appreciably in recent years due to improved source control; however, the non-point sources of pollution from urban and agricultural land uses, airborne contaminants deposited through the water column, sediment disturbance due to waves, currents and ship traffic, and possible historical sources of pollution may all contribute to the present severity of sediment contamination.

The eastern shore of the harbour is comprised of a narrow sand bar used primarily as a transportation corridor. The northern shore has a mix of land use, being mainly residential and recreational. The southern shore of the harbour borders mainly on industrial land. This pre-industrial marshland was filled in to make room for industrial properties between 1862 and 1926. Steel mills began operating in 1910 (Rodgers et al., 1992).

The surficial sediments of the entire harbour were mapped in the 1980s and 1990s (Poulton, 1987; Rodgers et al., 1992; Rukavina and Versteeg, 1996). The results suggest that black fine-grained sediments cover the entire deep-water (water depth of 8 – 25 m) area of the harbour. The highest concentrations of toxic metals occur in the deepest area of the harbour, as pollutants are mixed and gradually deposited in the deep central basin (Fig.2). The distribution of PAHs in the sediments (Fig.3) indicated that the steel mills are the main source and that several STPs discharging into the harbour are the minor source. Very high elevations of PAHs were found in the Randle Reef area near the south shore of the harbour and elevated levels of PCBs in the Windermere Arm area.

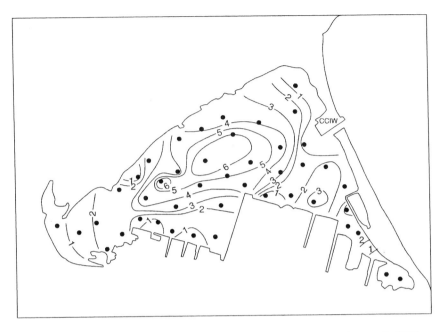

Figure 2. Zinc concentration (mg/g) in Hamilton Harbour sediments (Rodgers et al., 1992)

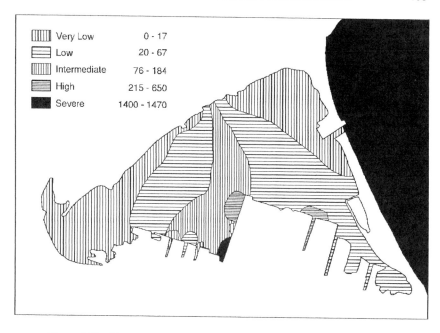

Figure 3. Total PAHs in surface sediments of Hamilton Harbour, grab samples, concentrations in µg/g (Murphy at al., 1994)

This paper describes both the multidisciplinary methodology used for the assessment of sediment contamination and the sediment remediation strategy, both of which are being developed within the framework of the Remedial Action Plan (RAP) for the harbour.

2. PHYSICAL AND CHEMICAL MAPPING OF CONTAMINATED SEDIMENTS

2.1 Physical Mapping and Testing

The texture of sediments is an important factor for mapping sediment beds as contaminants are usually found in much higher concentrations in finer sediments. A map of surficial sediment texture may be a good indication of where contaminated sediments are more likely to be found, and may show areas of sediment where core samples can be more easily retrieved.

Over the years, various technologies have been used for mapping marine and freshwater beds using acoustical equipment, primarily for navigational and fisheries purposes. An example of this is the acoustic marine-bed classification system known as RoxAnn™, which has been successfully used within the Great Lakes on various projects in determining surficial sediment type (Rukavina and Caddel, 1997). RoxAnn works on the principle that various textures of surficial sediment reflect acoustic signals differently. The type of echo received by RoxAnn is interpolated and instantly classified within the software system used in conjunction with the unit.

Cone penetration testing may be conducted on *in situ* sediments to obtain a quick and economical general understanding of shear strength vertical profiles. A device known as a free-fall penetrometer, called STING™, has been used for determining approximate shear strength values at Hamilton Harbour sites where contaminated sediments exist (Rukavina, 2002). The penetrometer is dropped from a boat and allowed to fall and sink into the sediment bed. An electronic measuring device attached to the instrument measures the rate of reduced speed upon impact of the tip with the sediment. This gives an indication of the shear strength of the sediment, which can aid in the interpretation of sediment grain size. For example, if the penetrometer penetrates through 50 cm of sediment with an abrupt stop at 50 cm, it may be assumed that the upper 50 cm sediment layer is comprised of soft, fine-grained sediments underlain by a different type of sediment with much higher shear strength (e.g., an overconsolidated glaciolacustrine sediment). A general knowledge of the site sediment stratification is necessary, however, to avoid misinterpretation. Both sand and hard clay layers will greatly slow down or stop the penetrometer.

Where *in situ* testing is deemed to be too costly or impractical, laboratory testing on samples is required to determine geotechnical parameters. One of the most basic classification tests performed on sediments is grain size. At NWRI, particle-size distribution is determined using the sieve and sedigraph methods (Dalton, 1997). Shear strength can range from < 1 kPa for very soft sediments, to > 80 kPa for hard clays. Fall-cone test measurements (Hansbo, 1957) have been used to measure undrained shear strength of very soft Hamilton Harbour sediments.

2.2   Chemical Mapping

Chemical mapping of contaminated sites is best done in three dimensions, which requires the sampling of sediment below the sediment/water interface. This is done by taking core samples, whereby a tube usually made of clear acrylic is either dropped or pushed through the sediment. Core samples may then be cut and subsampled at any depth measured from the top of the core. Where duplicate core samples of the upper sediment from the same location are desired, a large box corer with a retractable bottom is often used to scoop up almost 1 m^3 of sediment. Two or more short core tubes are then pushed through the sediment by hand and pulled out.

The mapping of chemical compounds within contaminated sediments has proven to be a challenge at many sites. Deposited compounds from spills or dumped material tend to be very site-specific and initially cover a small area, often spreading out into a thin layer of only several millimetres thick. Over several years, natural sedimentation will bury such a layer. Subsampling cores at intervals that are too far apart may result in contaminated layers being missed, especially if the sediment colour and texture are very similar to the neighboring sediment within the core. Contaminated sediments also tend to be very fine, and are thus easily disturbed by currents, propeller wash, and anchor dragging from passing ships. In industrial areas, contaminant spills not only come from various sources, but usually differ in chemical makeup, frequency of occurrences and volume. As a result, the composition of sediments in industrial areas is often very heterogeneous and, in extreme cases, can significantly vary among sediment samples collected only several metres away. Accurate three-dimensional chemical mapping is thus very difficult in such cases, and the site sediment can only be properly mapped with a high density of sampling locations. The constant disturbance, migration, and settlement of sediments results in the rendering of chemical maps obsolete over several years, depending on the rate of change.

Samples taken from an initial set of sampling sites usually reveal areas of sediment that require more detailed mapping, and thus a secondary set of samples. Once areas of contaminated sediments are sufficiently mapped to reveal calculated volumes and measured concentrations, remedial options may be considered. Volumes are determined by setting a contaminant concentration limit (often determined by financial resources available), and measuring the vertical range in each core where concentrations meet or exceed the limit, in addition to the horizontal distances between the cores. Computer software programs may then be used to calculate the volumes and to plot the contour maps and profiles of contaminated sediment deposits.

3.   SEDIMENT TOXICITY

For areas of Canadian jurisdiction in the Great Lakes, the Ontario Ministry of the Environment (MOE) sets the PSQCs for sediment toxicity. Determining what can be classified as toxic sediment in legal terms is no simple task. Toxicity must be determined in relation to the effects contaminants have on organisms for any given area. The obvious disadvantage to this methodology is the fact that different species of organisms react differently to contaminants. As a result, the MOE does not rely on single-species data to develop PSQGs. Based on the long-term effects that contaminants in sediment have on sediment dwelling organisms, the MOE classifies sediment into three categories (Persaud et al., 1994):

No Effect Level – Fish and sediment dwelling organisms are not affected.
Lowest Effect Level – Majority of sediment dwelling organisms are not affected.
Severe Effect Level – Likely to affect the health of sediment dwelling organisms.

In the early 1990s, sediment toxicity in Hamilton Harbour was investigated using the zooplankter *Daphnia magna*, mayfly *Hexagenia limbata* and a sediment contact bioassay with *Photobacterium phosphoreum* (Brouwer et al., 1990; Rodgers et al., 1992; Murphy et al., 1994). These studies delineated toxic hot spots in the harbour associated with high concentrations of PAHs in the vicinity of two large steel companies Stelco and Dofasco (Fig. 3). Maps of sediment toxicity to *Daphnia* and *Photobacterium* showed the sites nearest the steel mills and municipal storm water discharges to be the most toxic.

Investigation of biological significance of contaminants in sediments from Hamilton Harbour (Krantzberg and Boyd, 1992; Krantzberg, 1994) used as test organisms mayfly nymphs (*Hexagenia limbata*), three- to four-month-old juvenile fathead minnows (*Pimephales promelas*) and egg-sac-stage rainbow trout (*Salmo gairdneri*). Growth, mortality, and bioaccumulation of contaminants were the measured end points. Sediment samples were collected from five locations in the harbour on two separate occasions. Most sediment samples did elicit sublethal and/or lethal responses in bioassays species. Results of analyses of tissue residues in bioassay organisms implicated trace metals as contributing to sediment toxicity. Metal bioavailability (measured by weak acid extraction, metal bioaccumulation by fathead minnows, and sediment toxicity) was greater in sediment collected in the fall than in sediment collected in the spring. In addition, for some stations there was evidence that PAHs were responsible for the detected deletrious effects. High sediment oxygen demand contributed to the restricted benthic community *in situ* and some of the toxicity observed *in vitro*. The distribution of benthic invertebrates in the harbour appears to be limited by both sediment-bound contaminants and high-sediment oxygen demand.

An approach developed by researchers at the National Water Research Institute (NWRI) classifies sediment toxicity based on the effects observed on four benthic invertebrate species (*Chironomus riparius, Hexagenia spp.,*

Hyallela azteca and *Tubifex tubifex*). The assessment process utilizes benthic invertebrates present in the sediment, as these animals are the most exposed and potentially most sensitive to contaminants occurring in harbour sediments.

Assessment of sediment quality is based on multivariate techniques using sediment physical and chemical data, as well as the functional responses of laboratory organisms of benthic invertebrates in toxicity tests. Community impairment and sediment toxicity are assessed using the BEAST (Benthic Assessment of Sediment) methodology (Reynoldson et al., 2000). The departures of tests sites from unimpaired reference sites are then indicators of community alteration or sediment toxicity. Using this method, sites are classified in four bands as: a. unstressed/non-toxic, b. possibly stressed/possibly toxic, c. stressed/toxic, and d. very stressed/severely toxic. Data from the test sites are also compared to the above-mentioned sediment quality criteria developed by the MOE.

Results of two recent sediment biological surveys in Hamilton Harbour found that the sediment "severe effect level" is exceeded at numerous test sites.

The first survey (Milani and Grapentine, 2004) assessed the sediment quality for the entire harbour to define the general status of sediment contamination for metals as well as for PAHs and PCBs. Out of 44 sites sampled, the sediment "severe effect level" was exceeded for Mn (29 sites), Zn (22 sites), Fe (20 sites), Cu (17 sites), Pb (11 sites), Cr (9 sites), Hg (8 sites), As (4 sites) and Ni (2 sites). Total PAHs in sediments range from 0.5 to 498.7 µg/g (median 26.6), with highest values occurring at sites in the Randle Reef area. Total PCBs range from 0.07 to 6.17 µg/g (median 0.46), with the highest values in the Windermere Arm area. Strong evidence of benthic community impairment was determined at 35 sites. Eighteen sites showed strong evidence of sediment toxicity (Band 3 and Band 4).

The second survey (Milani and Grapentine, 2003) assessed the sediment quality for 80 sites in the Randle Reef area and for 20 sites in the Windermere Arm area; i.e. for the areas that are known to be highly contaminated from previous surveys. Only two toxicity tests (*Hexagenia spp., Hyallela azteca*) out of four were used to reduce cost. These two bioassays showed the strongest relationship to toxicity and were therefore considered sufficient to provide required information. Eight Randle Reef sites and 13 Windermere Arm sites had two or more trace metals elevated above the SEL. Total PAH and PCB concentrations in the Randle Reef area were in the ranges of from 1 to 9048 µg/g and from below detection (20 sites) to 1.4 µg/g, respectively. The ranges obtained for the Windermere Arm area were from 5.1 to 715 µg/g for PAHs and from 0.1 to 2.3 µg/g for PCBs. Highest PAH concentrations were found in the Randle Reef area and highest PCBs were found in the Windermere Arm area. The survey determined strong evidence of toxicity for 31 sites in the Randle Reef area and for 12 sites in the Windermere Arm area.

4.   SEDIMENT REMEDIATION STRATEGY IN HAMILTON HARBOUR

A 200-ha large area in the main basin of the harbour contains PAH concentrations > 50 µg/g and high metal concentrations (e.g. Zn > 3000 µg/g, Fig.2). Due to the large volume of contaminated sediment, the Remedial Action Plan (RAP) for the harbour does not recommend immediate remediation of this area, as it may be prohibitively costly and technically impractical. Sediment removal by dredging on a large scale would also likely have a severe environmental impact.

Two *in situ* remedial methods, *in situ* treatment and *in situ* capping, have been tried in Hamilton Harbour on a pilot scale in the deep-water basin. *In situ* sediment treatment was also tested in the Randle Reef area and the Windermere Arm area.

4.1   *In situ* Bioremediation of Organic Contaminants

The *in situ* treatment of contaminated sites (Murphy et al., 1994, Murphy et al., 1995) has been tried at three different locations in the harbour, using oxidant (calcium nitrate) and nutrients (a proprietory organic amendment). In total, about 2 hectares of medium to severely contaminated sediments were treated. These pilot projects took place in 1992 and 1993.

The treatment site in the main basin measured 200 m x 16 m. After one treatment about 80% of the hydrogen sulphide was oxidized. Oxidation resulted in precipitation of about 98% of the porewater Fe in the surface 15 cm of the sediment but the concentration of most trace metals was unchanged. The high metal concentration did not appear to limit bioremediation.

The second treatment site was located in the Randle Reef area and measured 150 m x 24 m. Nitrate injections in several stages resulted in a rapid (within two weeks) increase in redox potential and the oxidation of reduced sulphur to sulphate. Biodegradation of large molecular PAHs resulted in the production of naphthalene.

The third treatment site was located in the Windermere Arm area (the southwestern end of the Dofasco Boat Slip shown in Fig. 9) and measured 150 m x 32 m. The site was treated twice in 1992 with calcium nitrate. In

1993, it was treated once with calcium nitrate and once with the mixture of calcium nitrate and organic amendment. In the 1992 treatments, biodegradation of organic contaminants varied from 79 % for low molecular weight compounds (BTXs), to 25 % for oils (TPHs), to 15 % for 15 of the 16 priority pollutant PAHs. In the 1993 treatments, 94 % of naphthalene and 57 % of the TPH biodegraded.

It was concluded that this remedial method may require time, particularly at badly contaminated sites, and multiple treatments would be required. For some sites it should be possible to bioremediate sediments for about 20 % of the cost of dredging and storage in a confined disposal facility.

4.2   *In situ* Capping

A one hectare (100 m x 100 m) pilot size capping project was completed near the north shore of the harbour in 1995 (Zeman and Patterson, 1997). The level of contamination with metals and organic contaminants was representative of the deeper portion of the main basin. Water depths at the site range from about 12 to 17 m (International Great Lakes Datum 1985). Bottom sediments consist of about 30 cm of very soft black silty clay (contaminated from industrial sources) underlain by very soft greyish brown silt and clay (natural harbour sediments). A sand (medium to coarse sand, average grain size 0.5 mm dia.) cap was successfully placed over very soft contaminated sediments with minimum disturbance. The average cap thickness was 34 cm. Because of the sand cap, post-capping cores had to be obtained by vibracoring rather than by gravity coring. The cores showed a very sharp sediment/cap interface and a significant difference in trace metal concentration between the sediment and the sand cap (Fig. 4). The PCB and PAH profiles also showed sharp breaks between low chemical concentrations in the cap and substantially higher concentrations in the underlying sediment.

Metal Concentrations

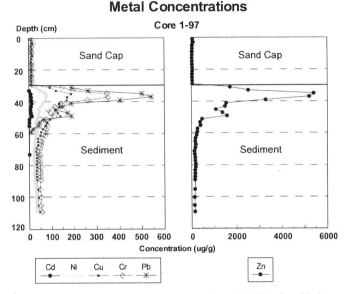

Figure 4. Metal concentration in a core from the capping site in Hamilton Harbour (Zeman and Patterson, 1997)

Multibeam echo-sounding surveys were carried out at the site before, during and after cap placement (Fig. 5). The surveys consisted of sounding intervals that were approximately 5 m apart, together producing a grid pattern of depths. Changes in bathymetry were most evident within the capping area. The placement of the sand-sized cap on the fine-grained bottom was clearly delineated by the acoustic marine-bad classification system known as RoxAnn[TM] (Rukavina and Caddell, 1997; Zeman and Patterson, 2000).

BATHYMETRIC CHANGES AFTER CAP PLACEMENT
October 1994 to October 1995

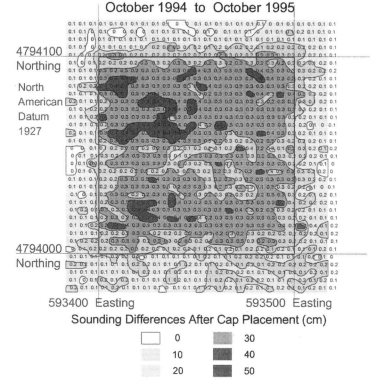

Sounding Differences After Cap Placement (cm)

	0		30
	10		40
	20		50

Figure 5. Bathymetric changes after capping (Zeman and Patterson, 1977)

5.   RANDLE REEF SITE

The Randle Reef "hot spot" is located near the south shore of Hamilton Harbour (Fig. 1). It is considered to contain the most highly contaminated sediment for PAH concentrations in the harbour. Although the hot spot is generally referred to as "Randle Reef", it should be noted that the actual geomorphic feature is located to the north/west of the hot spot. Historical industrial discharge created the contaminated sediment site which contains a high concentration of coal tar. Sediment containing coal tar can be a source for long-term contamination to the water column due to the slow release of organic solutes (Environment Canada, 1999). A large number of sediment cores and borehole samples was collected at this site starting as early as mid-1970 (Zeman and Patterson, 2003 a).

In the late 1990s, a geographic area containing the most highly contaminated sediment was mapped out, and is known as the "hot spot" or "fish tail" (Fig. 1). Mapping of the hot spot was completed with reference to a polygon map of sediment PAH concentrations, which was produced using linear interpolation between adjacent points. The hot spot was delineated on the map based on sediment PAH concentrations of 800 μg/g (minus naphthalene) and higher. Due to the volatility of naphthalene, a decision was made to exclude this compound from PAH concentration values.

The stratigraphy at the Randle Reef site is very mixed and complex. Sediment cores retrieved just metres apart from each other have often been found to be significantly different in composition, texture and layering. This suggests a man-made origin of the deposit and combination of sediment deposition, mixing and disturbance that will not likely ever be fully understood. The widely differing geotechnical and chemical data obtained from the first few sets of cores resulted in the continuation of sediment sampling in more closely spaced core and grab sample sites. In 1999, nine offshore boreholes ranging in length from 2.9 to 4.9 m were put down at the site in 1999 to obtain information on stratigraphy and contamination at greater depths. The varying sets of sediment samples collected over the years has led to Randle Reef being the most sampled area of the harbour to date.

Subsamples for PAH analyses were taken at varying intervals from the cores and boreholes and concentration values were then mapped using a contouring software program (Fig. 6). Since the sediment at the site was found to be extremely heterogeneous, chemical maps of the site showed widely varying concentrations within a relatively small area. The maps significantly changed appearance depending on how the extent of contamination was expressed (e.g. maximum PAH concentrations found vs. maximum depth of highly contaminated sediment). The maximum concentration of PAHs found at the site was 17 000 μg/g (minus naphthalene) near the middle of the hot spot area.

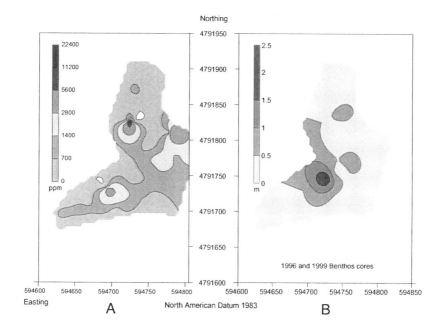

Figure 6. Maximum PAH concentrations regardless of depth (A), and maximum depth of
PAH concentrations >700 ppm (B), Randle Reef study site

6.   RANDLE REEF REMEDIATION PROJECT

In 1995, the Randle Reef site was identified in the Hamilton Harbour
RAP as a high priority area for remediation. Various alternatives have been
considered, including sediment removal by dredging and *ex situ* sediment
treatment. It was proposed to remove approximately 20 000 m³ of the most
toxic sediment with PAH concentrations greater than 700 µg/g (minus
naphthalene). Several *ex situ* sediment treatment options have been assessed
for applicability (Environment Canada, 1999).

A final remedial project for the Randle Reef site has yet to be
undertaken, although stakeholders (committee members) have tentatively
agreed to a confinement and capping plan. The proposed remediation
involves the construction of a dry-cap, diked Engineered Containment
Facility (ECF) consisting of a 8-ha peninsula attached to Pier 16 in addition
to a 1.5-ha triangular extension of Pier 15 (Fig.7). The ECF will cover about
130,000 m³ of PAH-contaminated sediments *in situ* and will contain
additional 500,000 m³ of contaminated sediments that will be dredged from

the general Randle Reef area not covered by the proposed ECF and possibly will include sediments from other Hamilton Harbour hot spots. Two-thirds of the ECF area will be used for commercial port facilities and operations, one-third of the ECF is designated as naturalized open space. The port facility will be suitable for ships of Seaway draft (Great Lakes St. Lawrence Seaway System). The construction and dredging required for project completion is expected to take 5– 10 years.

Final Conceptual Design accepted by PAG Dec. 8, 2003
9.5 ha to contain estimated 500,000 m3 of dredged sediments

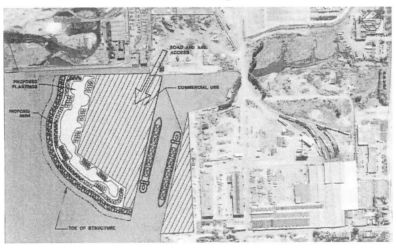

Figure 7. Conceptual design, Engineered Containment Facility, Randle Reef site (Great Lakes Sustainability Fund, 2003)

7.   WINDERMERE ARM SITE

Windermere Arm is a narrow channel situated in the southeast portion of the harbour, covering an area of approximately 80 hectares (Fig. 1). The western shoreline of the Arm is industrial. The northern section of the Arm is a well used shipping corridor. Much of the eastern shoreline borders on a confined disposal facility, which is located between the Arm and a major highway. The Red Hill Creek flows into a basin which empties into the Arm from the south.

Although sediment contamination in the Arm is not as severe as that found in Randle Reef, concentrations of PAHs and PCBs are well above the MOEEs "lowest effect level" guidelines, whereas concentrations of at least seven trace metals exist above the "severe effect" level. The sediment in Windermere Arm is not as heterogeneous as sediment found at Randle Reef, but chemical contamination is still extensive, and chemical compounds often reach their peak values independent of each other at various depths in the sediment.

In a recent survey (Zeman and Patterson, 2003 b), surficial sediments in the Arm were found to have lower metal concentrations than deeper sediments (Fig.8), probably due to the reduction of industrial point discharges. Apart from the two main boat slips, concentrations of PAHs were relatively low and in agreement with previously reported values. In contrast, new results yielded higher values of PCB concentrations than previously reported, and 21 sub-samples of surficial sediment analysed yielded values grater than 1 µg/g (Fig. 9). Results suggest that the majority of recent PCB concentrations are entering the Arm from nearshore industrial spills or dumping. Pronounced sediment disturbance noted in several cores suggest that contaminated historical sediments within this area are at risk of being disturbed and re-exposed to the water column.

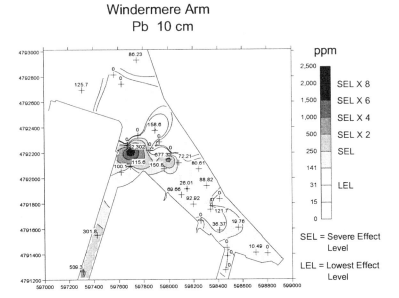

Figure 8. Lead concentrations at a sediment depth of 10 cm (Zeman and Patterson, 2003b)

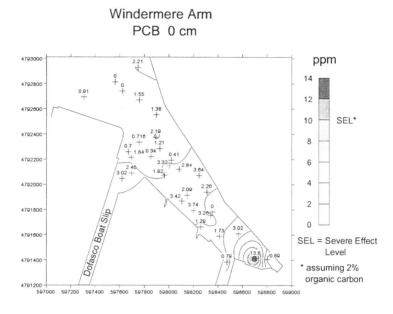

Figure 9. Total PCB concentrations in surficial sediment (Zeman and Patterson, 2003b)

The issue of PCB contamination in Windermere Arm is under further current investigation. Eleven sediment surveys conducted from the 1980s until the present indicate typical concentrations in Windermere Arm ranging between 1 and 10 µg/g, and isolated hot spots with total PCB concentrations exceeding 10 µg/g. Existing sets of PCB data in samples of water, sediment and biota are being examined. The data will be used in a mass balance model integrated with a food chain model in order to understand the sources, transfer and fate of PCBs. This modeling exercise will attempt to answer whether Windermere Arm is an important source of PCBs for the rest of the harbour, and, if it is, what would be the effect of sediment remedial measures and resulting reduced loadings to the rest of the harbour. This information is essential for the application of sound management decisions, which are being developed as a component of the RAP for Hamilton Harbour.

8.   CONCLUSIONS

The cleanup of AOCs and other sites with contaminated sediments require adequate mapping and characterization of contaminants within the sediment. This is to ensure that the volume of contaminated sediments and its geotechnical and geochemical properties are known. From this information, the appropriate remedial technologies may be selected for any cleanup plan, subject to budget limitations. Authoritative sediment toxicity guidelines are necessary to determine whether contaminant levels in sediments at any site warrant a cleanup plan.

From the three sites examined in Hamilton Harbour, it is clear that acoustical mapping along with geotechnical and geochemical testing have provided significant information for site delineation and sediment composition. Accurate placement of the sand cap for the 1995 capping project was confirmed through both acoustical mapping and cores collected at the site. The tentative cleanup strategy for Randle Reef was based largely on sediment data collected from the site. Ongoing sampling and testing of Windermere Arm sediments is expected to identify areas within the Arm to be selected for a likely future cleanup project.

ACKNOWLEDGEMENTS

Funding for most sediment assessment and remediation projects described in this paper was provided by the Great Lakes Sustainability Fund of Environment Canada.

REFERENCES

Brouwer, H., Murphy, T. P. and McArdle, L., 1990, A sediment -contact bioassay with *Photobacterium phosphoreum, Envir. Toxicol. Chem.* 9: 1353-1358.

Dalton, J. 1997, Methodologies of testing procedure in the Sedimentology Laboratory. In house laboratory report, NWRI, Burlington, Ontario.

Environment Canada, Environmental Protection Branch – Ontario Region, 1999, *Randle Reef Sediment Remediation Project, Environmental Screening Report* , Draft Report, Downsview, Ontario.

Great Lakes Sustainability Fund, 2003; http://sustainabilityfund.gc.ca/scoping_e.html

Hansbo, S., 1957, A new approach to the determination of the shear strength of clay by the fall-cone test, *Proc. Royal Swedish Geotechnical Institute* , No. 14, 46 p.

Krantzberg G., 1994, Spatial and temporal variability in metal bioavaiability and toxicity of sediment from Hamilton Harbour, Lake Ontario, *Envir. Toxicol. Chem.* 13(10): 1685-1698.

Krantzberg G., and Boyd, D., 1992, The biological significance of contaminants in sediment from Hamilton Harbour, Lake Ontario,. *Envir. Toxicol. Chem.* 11: 1527-1540.

Milani, D and Grapentine, L. C., 2003, Identification of acutely toxic sites in Hamilton Harbour. 2nd Draft, NWRI Contribution (in preparation), Burlington, Ontario.

Milani, D. and Grapentine, L. C., 2004, The app lication of BEAST sediment quality guidelines to sediment in Hamilton Harbour, Final Draft, NWRI Contribution (in preparation), Burlington, Ontario.

Murphy T. P., McArdle, L., Brouwer, H., Moller, A, and Krantzberg , G. 1994. Spatial variation in sediment toxicity and potential methods of *in situ* treatment, *Verh. Internat. Verein Limnol.*, 25: 2036-2042.

Murphy, T. P., Moller, A., and Brouwer, H., 1995, *In situ* treatment of Hamilton Ha rbour sediment, *Journal of Aquatic Ecosystem Health* , 4(3): 195-203.

Persuad, D., Jaagaumagi, R. and Hayton, A., 1994, Guidelines for the Protection and Management of Aquatic Sediment Quality in Ontario, ISBN 0 -7729-9248-7, Ontario Ministry of the Environment, Water Resources Branch, Toronto, Ontario.

Poulton, D. J., 1987, Trace contaminant status of Hamilton Harbour , *J. Great Lakes Res.*, 13: 193-201.

Reynoldson, T. B., Day, K. E., and Pascoe T., 2000, The development of the BEAST: a predictive approach for assessing sediment quality in the North American Great Lakes . In: *Assessing the Biological Quality of Fresh Waters. RIVPACS and Othe r Techniques*, J. F. Wright, D. W. Sutcliffe, and M. T. Furse, eds., Freshwater Biological Association, U.K., pp. 165-180.

Rodgers, K., Vogt, J., Cairns, V., Boyd, D., Simser, L., Lang, H., Murphy, T. P., Painter, S., 1992, *Remedial Action Plan for Hamilton Harbour*, Stage I Report, Environmental Conditions and Problem Definition, 243 p.

Rukavina, N. A., 2002, Sediments and bathymetry of the Windermere Arm, Hamilton Harbour: A progress report, *Research and Mon itoring Report, 2001 Season, Remedial Action Plan for Hamilton Harbour* , Burlington, Ontario, pp. 125 -133.

Rukavina, N. A., and Versteeg, J. K.,1996, Surficial sediments of Hamilton Harbour: Physical properties and basin morphology. *Water Qual. Res. J. Canada* , 31(3): 529-551.

Rukavina, N. A. and Caddel, S., 1997, Applications of an acoustic seabed classification system to freshwater environmental research and remediation in Canada, In: *Proc. 4th Conf. Remote Sensing for Marine and Coastal Environments* , Orlando, FL, pp.I-317-I-329.

Zeman, A. J. and Patterson, T. S.,1997, Results of *in situ* capping demonstration project in Hamilton Harbour, *Proc. Inter. Symp. Engineering Geology and the Environment* , Athens, Greece, A. A. Balkema, Rotterdam, pp. 2289 -2295.

Zeman, A. J. and Patterson, T. S., 2000, *Results of Physical, Chemical and Biological Monitoring at the In situ Capping Site in Hamilton Harb our*, Contr. No. 00-088, National Water Research Institute, Burlington, Ontario.

Zeman, A. J. and Patterson, T. S., 2003 a, *Sediment Sampling at Randle Reef, Hamilton Harbour*, Contr. No. 03-172, National Wa ter Research Institute, Burlington, Ontario.

Zeman, A. J. and Patterson, T. S., 2003 b, *Sediment Survey of Winderemere Arm, Hamilton Harbour*, Contr. No. 03-171, National Water Research Institute, Burlington, Ontario.

PART IX: SITE ASSESSMENT

CHAPTER 30

EVALUATION OF SOLVENT PLUME DISCHARGE TO A WETLAND STREAM USING AN INNOVATIVE PASSIVE DIFFUSION SAMPLING METHODOLOGY

Lucas A. Hellerich, Ph.D., P.E.[1], John L. Albrecht, L.E.P.[2], and Richard C. Schwenger, P.Eng.[3]

[1]*Project Engineer. Metcalf & Eddy, Inc., 860 North Main Street Extension, Wallingford, CT 06492, (203) 741-2821, lucas.hellerich@m-e.com;* [2]*Project Manager. Metcalf & Eddy, Inc., 860 North Main Street Extension, Wallingford, CT 06492, (203) 741-2826, john.albrecht@m-e.com;* [3]*Regional Reclamation Manager, Noranda Inc., P.O. Box 200, Bathurst, N.B Canada E2A 3Z2*

Abstract: This paper describes the first phase of a multi-phase project currently underway to delineate a chlorinated solvent (primarily trichloroethene [TCE]) plume in groundwater downgradient of a former industrial site. The TCE was introduced to the subsurface environment at the site, resulting in a groundwater plume extending from the on-site source to a wetland-stream complex, the primary discharge point for groundwater migrating from the site. To date, the plume has been delineated from the source to the upgradient edge of the wetland via quarterly groundwater sampling through a network of monitoring wells. At the upgradient edge of the wetland, TCE concentrations in groundwater remain elevated, while products of reductive dechlorination are present. Additionally, surface-water sampling indicates that the solvent plume is discharging to a small stream that drains the wetland.

The first phase of the plume delineation project consisted of a recently completed solvent plume discharge evaluation (SPDE), with the objective of evaluating the locations of solvent plume discharge to the wetland stream. The SPDE was cost-effectively conducted by deploying passive diffusion samplers (PDSs) in the wetland stream sediments at 45 locations along a 1,900-foot length of the stream. Once equilibrium between the PDSs and the sediment pore water was attained, the concentrations of chlorinated ethenes in the PDS samplers were measured. The analytical results were mapped as a function of stream length and indicate a definitive point of plume discharge to the stream and a possible plume fringe. The results were also correlated with sediment type, water-quality parameters, and piezometric measurements obtained during this work. The ratios of degradation products to TCE were greater in the PDS samplers than ratios observed at upgradient monitoring locations, indicating an increasing natural attenuation potential as the plume migrates through the wetland. The results of the SPDE were used to focus the next phase of plume delineation, currently being conducted. Preliminary results of the plume evaluation compare well to the SPDE results.

Key words: chlorinated ethenes, groundwater, passive diffusion samplers, wetland, sediments, groundwater discharge

1.   INTRODUCTION

1.1   Objective

The overall objective of the project was to collect data that could be used to determine area(s) of discharge in a wetland stream and focus future solvent-plume evaluation activities at a site where solvent contamination migrates in groundwater from beneath a former manufacturing building location, and into an unnamed wetland/stream complex downgradient of the former industrial facility. Although solvent contamination was previously detected in surface-water samples collected from the unnamed wetland/stream complex, the location(s) of solvent plume discharge to the small stream was not known. However, the discharge is likely isolated to the stream due to the hydrogeologic setting in which the valley resides. Discharge locations in the stream were evaluated by analyzing for the presence of solvent contamination in the stream sediment pore water using passive diffusion samplers (PDSs), and various stream water-quality and hydrologic parameters, with the assumption that the streambed acts as a point of discharge for the groundwater solvent plume. By determining the solvent plume discharge location(s) in the stream, future placement of additional groundwater monitoring wells might be more efficiently placed to evaluate the extent of the plume.

1.2   Site Setting and Background

The study area consists of: a ±12 acre parcel (site) with a concrete slab remaining from a manufacturing facility that was razed in 1996, a paved driveway and parking areas, grassed and wooded areas, and a ±6 acre pond; and a ± 40 acre wooded wetland/stream complex (wetland) located downgradient of the site. Copper tubing heat exchangers were manufactured on the site from 1947 to 1986, with manufacturing operations ceasing in 1986. There are networks of groundwater monitoring and vapor extraction wells on the site. A site plan showing these features is provided as Figure 1. Groundwater and surface-water samples are collected quarterly from the monitoring well network and the nearby wetland stream. The wetland stream is located east and downgradient of the site within a valley.

During manufacturing activities, chlorinated ethene solvents (primarily trichloroethene [TCE]) were released into the subsurface through floor drains on the site and migrated into groundwater. Information gathered during site investigations from 1986 to 1993 documented the presence of chlorinated solvent contamination in groundwater both on-site, and in the upgradient edge of the downgradient wetland. In addition, chlorinated ethenes have been detected in the surface-water wetland stream.

Figure 1. Site Plan

2.   PRELIMINARY CONCEPTUAL SITE MODEL (CSM)

2.1   Contaminants of Concern (COCs)

TCE and to a much lesser extent, tetrachlorethene (PCE) and the chlorinated ethene degradation products, are the primary COCs at the Site. Historic TCE concentrations in groundwater range from trace levels to over 260 mg/L of TCE. The three dichloroethylene (DCE) isomers, *cis*-1,2-DCE, *trans*-1,2-DCE, and 1,1-DCE, and vinyl chloride, chlorinated ethene degradation products, have also been detected in groundwater samples obtained at the site, but are present at lower concentrations than PCE and TCE.

2.2   Contaminant Distribution

Information gathered during previous investigations and quarterly groundwater and surface-water monitoring has documented the presence of TCE and PCE in both overburden and bedrock groundwater. The concentration of TCE in samples of groundwater obtained from overburden wells in the vicinity of the site has varied from non-detect to 260 mg/L. Overburden groundwater samples with the highest concentrations of TCE have been historically obtained from wells installed within and downgradient (northeast) of the suspected on-site source area (see Figure 1). The concentrations of TCE in samples of groundwater obtained from bedrock wells located in the vicinity of the site have varied from non-detect to 82 mg/L. The concentrations of TCE in samples of groundwater obtained from overburden wells located downgradient of the site in the wetland have varied from non-detect to 21 mg/L. Of these wells, groundwater samples with the highest concentrations of TCE have been historically obtained from wells GW-2 and GW-3 located northeast of the former building at the upgradient edge of the wetland. Based on surface-water testing to date, TCE, PCE, and their associated degradation products are present in surface-water samples obtained from the stream. The maximum concentration of TCE was 270 µg/L, measured in surface-water samples from location SW-6, located approximately 500 feet east of the Site. PCE has been historically detected in surface water at low levels at sampling locations SW-1 and SW-6. The maximum concentration of PCE was 3.8 µg/L, measured in a surface-water sample from location SW-6.

Based on the results of surface water and groundwater samples collected from wells in the upgradient edge of the wetland, solvent-contaminated groundwater is discharging to the stream and wetland complex. Due to the localized nature of the detectable concentrations of solvents in surface water and the nature of the fractured bedrock in the area, contaminated groundwater is believed to be discharging from bedrock and overburden to the stream and wetland complex. However, the location(s) of these discharge areas have not been delineated. The objective of this plume discharge evaluation is to further identify the plume discharge areas to the stream.

2.3   Geologic Setting

The study area exhibits characteristics of glaciation such as hummocky terrain, erratics (large boulders) on the land surface, and extremely heterogeneous surficial deposits.

2.3.1   Overburden

The unconsolidated materials encountered during previous investigations on the western portions of the site are interpreted as dense till. Although localized laminae of fine sand and silt were encountered in the till under the building, no other bedding was apparent.

Unconsolidated materials in the wetland east of the site are interpreted as stratified drift. The stratified drift and till deposits grade into one another somewhere near the eastern property boundary. Material described as "flowing sand" was encountered during previous drilling activities in the wetland. This material is believed to be stratified drift deposits and is assumed to be relatively permeable with a high contaminant transport potential. The stratified drift deposits overlie glacial till and weathered Gneiss bedrock with an approximate thickness varying from 20 to 30 feet. Competent Gneiss bedrock underlie the weathered bedrock surface.

2.3.2   Overburden Hydrogeology

Groundwater is encountered at the site in the till layer, typically between 5 to 15 feet below ground surface; however, seasonal fluctuations as great as 8 feet have been measured in individual wells. Groundwater flow through the site is to the east-northeast, toward the wetland. The till is thought to be functioning as an aquitard that partly confines the weathered bedrock and bedrock aquifer and separates it from an upper aquifer in the stratified drift. However, east of the site the till does not appear to be present, allowing water from the weathered bedrock and bedrock aquifers to discharge upwards to the wetland complex. The depth to groundwater in the wetland is approximately 1 foot below the ground surface.

2.3.3   Bedrock

A layer of weathered fractured bedrock was identified during drilling at several locations. The weathered bedrock is composed of sand and weathered rock fragments and appears to be a continuous layer that ranges in thickness from one foot in the western portion of the site to five feet in the eastern portion of the site. This layer is thin and thought to be hydraulically connected to the underlying bedrock and, where present, to the overlying stratified drift.

2.3.4   Bedrock Hydrogeology

Seasonal fluctuations in piezometric head measured in the bedrock are less pronounced than those reported in the overburden aquifer. Downward and upward vertical gradients have been observed between the overburden and bedrock/weathered bedrock aquifer during different seasons. This may suggest that discharge/recharge of groundwater between the bedrock and overburden aquifers are seasonally variable.

3.   APPLICABILITY OF PASSIVE DIFFUSION SAMPLERS

Two common types of PDSs, water-water and water-vapor, have been historically used to collect groundwater samples from saturated environments, based on the principle of achieving concentration equilibrium across a membrane through diffusion of contaminants from a contaminated water source to an uncontaminated water source. Water-water PDSs utilize a membrane separating an initial deionized (DI) water and groundwater. Water-vapor PDSs utilize a diffusion membrane separating DI water and groundwater, with the addition of an air space (vapor phase) between the DI water and the groundwater, resulting in partitioning between the vapor and DI water phases. Although water-water PDSs have been more commonly utilized to obtain VOC concentrations from groundwater monitoring wells (USGS, 2001a & 2001b, 2002a), water-vapor PDSs have also been successfully used to obtain VOC concentrations within sediment pore water spaces (Savoie et al., 1999; USGS, 2002b; and Vroblesky and Robertson, 1997). This SPDE study utilizes water-water PDSs to collect representative samples of stream sediment pore water. The use of water-water PDSs compared to water-vapor PDSs allows for a more direct determination of the concentration of VOCs in the aqueous phase by eliminating the need to account for partitioning of VOCs to the vapor phase.

The stream downgradient of the site is a known receptor of solvent contaminated groundwater based on previous groundwater and surface-water testing. However, the discharge areas of the plume were not well understood as contaminant transport occurs in both fractured bedrock and overburden, and contaminants are detected in the stream water over a significant length.

The water-water PDS method was selected to identify plume discharge areas along the stream length so that future efforts could be cost-effectively focused on the plume pathways and discharge areas. This method was thought to be effective due to the elevated concentrations (~ 9 - 10 mg/L) of total chlorinated VOCs in groundwater monitoring wells GW-2 and GW-3 and detections of solvents in surface water. With concentrations in this range it was anticipated that discharge areas (indicated by detectable [and possibly elevated] concentrations of VOCs) to the stream could be easily distinguished from non-discharge areas (indicated by non-detectable concentrations of VOCs). In addition, pore-water contamination levels are likely more representative of discharge areas due to the transient nature of contamination levels in surface water.

Minimum detection limits for water-water PDSs are based on the equilibrium concentration within the sediment pore-water space. Concentrations of various chlorinated ethenes of less than 5 µg/L have been detected at a number of study sites using water-water PDSs to sample groundwater monitoring wells (USGS, 2001b & 2002a).

Prior to performing the full deployment of PDSs, a test deployment of five PDSs was conducted to refine field techniques and evaluate if the method would be effective at the site. These five PDSs were placed at various locations within the stream sediment to evaluate if detectable concentrations of chlorinated ethenes would result in the PDSs over a 20 day timeframe. VOCs were detected in one of the five PLT PDSs, at concentrations for PCE and TCE of 0.9 µg/L and 19.9 µg/L, respectively. Based on this test case, this method appeared effective, and full deployment of the sampling program was initiated. In addition, a pre-deployment equilibrium study was performed to test the time needed for equilibrium to be met between the PDS and surrounding water by immersing PDSs filled with laboratory DI water in a closed container of GW-3 groundwater. PDS construction methods are detailed in section 4.3. After accounting for dilution, the data showed that equilibrium between the GW-3 groundwater and the PDSs was approximately reached within 7 days for the site COCs. The PDSs for full deployment were allowed a minimum of two times this test case duration to ensure equilibrium with surrounding sediments.

The full deployment of PDSs was recovered after 15 days of deployment (Section 4.3). Based on the literature reviewed (USGS Fact Sheet 088-01; USGS, 2001a), equilibrium across a PDS membrane with the surrounding pore water typically occurs within approximately two weeks after deployment.

Quality assurance/quality control (QA/QC) samples for the passive diffusion sampling consisted of a pre-deployment trip blank (prepared at the laboratory), an equipment blank (prepared in the field), a water blank (laboratory water tested at the laboratory), a trip blank, duplicates, and a materials test of the VOA vials used to construct the PDSs, in accordance with USGS recommended QA/QC practices for PDS sampling (USGS, 2001a). No VOCs were detected in the pre-deployment trip blank, laboratory water blank, and in the materials analysis performed by Eagle Pilcher, the manufacturer of the VOA vials.

4. SOLVENT PLUME DISCHARGE EVALUATION METHODS

Field work included sampling of sediment porewater using the PDSs, sampling of groundwater monitoring wells as part of the regularly scheduled September quarterly groundwater monitoring event, installation of piezometers in the stream, and sampling of the wetland/stream waters and sediments.

4.1 Groundwater Sampling

The quarterly groundwater sampling was conducted in conjunction with sediment and surface-water sampling being conducted for this SPDE. On September 17, 2003, the following monitoring wells were sampled: MW-1, MW-2, TP-4, MW-12, MW-105, MW-106, MW-107, GW-1, GW-2, GW-3, GW-4, and GW-5. Sampling locations are shown on Figure 1.

Groundwater samples were collected in accordance with USEPA low-flow sampling methods as described in *Low Stress (*low-flow*) Purging and Sampling Procedures for the Collection of Groundwater Samples from Monitoring Wells* (USEPA, July 30 1996). Prior to sampling, depth-to-water measurements were taken using an electronic water-level measuring tape. Water-level measurements were utilized to evaluate groundwater flow direction and gradient as part of this study (Section 5.1). Wells were purged and representative samples collected at a low-flow rate using a variable speed peristaltic pump. Groundwater quality instrumentation was calibrated according to the manufacturer's guidelines prior to use. The following groundwater quality parameters were measured by M&E in the field: pH, specific conductance (mS/cm), temperature (°C), oxidation-reduction potential (mV), turbidity (NTU), and dissolved oxygen (mg/L). Wells were purged at a low-flow rate until these parameters stabilized. Groundwater samples were then collected directly into clean, new, laboratory-provided containers and preserved with hydrochloric acid (HCl). A Connecticut Department of Public Health (CTDPH) certified analytical laboratory, analyzed the samples for volatile organic compounds by EPA Method 8260b. Analytical results are presented in Section 6.1.

4.2   Piezometers

Piezometers, consisting of 1-inch diameter polyvinylchloride (PVC) pipe with a ±0.5 feet screened interval, were installed at a depth of ±1 feet within the stream sediments to evaluate the vertical hydraulic gradient at 12 locations (see Figure 1). The level of the surface water and the groundwater inside piezometers were measured using an electronic water-level tape. The results of the two measurements were compared and the vertical hydraulic gradient at each point was estimated. The piezometers and ten additional control points were surveyed for location and elevation and were used as control/measuring points to locate PDSs and sediment and surface-water samples. Piezometric measurements are discussed in Section 5.4.

4.3   Passive Diffusion Samplers

PDSs were used to collect pore water samples from within shallow sediments (top ± 10 cm of sediment) in the stream at 45 locations (see Figure 1), distributed approximately 40 feet (PDS-1 through PDS-30) to 50 feet (PDS-31 through PDS-44) apart along the stream length starting downstream (location CP-1) and moving upstream. PDS-45 was placed in a tributary stream that was located between SW-5 and the power lines which cross the wetland stream. The PDSs were constructed from standard laboratory-certified 40 mL glass volatile organic analyte (VOA) vials filled with distilled water provided by the laboratory. A plastic membrane was stretched across the mouth of each vial such that no air remained in the vial. The plastic membrane was then secured by screwing on a septum-free vial cap over the membrane and vial mouth. The membrane consisted of a piece of 2-inch by 2-inch square, 0.04-inch thick polyethylene layflat tubing (AIN Plastics, Mount Vernon, NY). The vials were constructed in triplicate in the event that one of the vials became damaged and to provide sufficient sample volume to top off the vial to be analyzed. The triplicate set of vials was held together with a zip-tie fastener. A survey pin flag was secured to each triplicate set of vials to facilitate their recovery. The sampler was inserted into the sediments by creating a small opening in the sediment using a clean trowel or trenching shovel, depending on the nature of sediments encountered. The sediment characteristics were logged during PDS deployment. Based on observations made of the stream during previous surface-water sampling events, sediments appear to generally consist of organic silt and sand.

The PDSs were constructed on August 18, 2003 in a controlled indoor environment, deployed (placed in the stream sediment) on August 19, 2003, and recovered on September 3, 2003. The PDSs were collected in two rounds with an outside air temperature of approximately 55 to 60 °F under overcast conditions.

At the end of each round, the following steps were performed: the plastic membrane was carefully removed from the vial; the vial was topped off with sample water from another vial in the triplicate set; the sample was preserved by adding hydrochloric acid (HCl 1% to 2% v/v); and the vial was capped with a standard VOA cap and teflon septum. The samples were iced at the end of the recovery and preservation of the second collection round of PDS samples. The samples were analyzed by the CTDPH certified analytical laboratory for VOCs by USEPA Method 8260b. Quality assurance/quality control (QA/QC) samples for the passive diffusion sampling consisted of pre-deployment trip blanks (prepared at the laboratory), equipment blanks (prepared in the field), water blanks (laboratory water tested at the laboratory), trip blanks, and duplicates. Analytical results are presented in Section 6.2.

Stream water-quality measurements (reduction-oxidation potential, pH, dissolved oxygen, specific conductivity, and temperature) were collected at each PDS sampling location prior to collection of the PDSs. These measurements were used to supplement the VOC data and hydraulic gradient measurements to help evaluate potential groundwater discharge points along the stream length. Stream water-quality measurements are discussed in Section 5.5.

4.4   Surface-water Samples

Surface-water sampling was performed on October 13, 2003 at 16 locations (see Figure 1) along the stream length. The locations consisted of the 6 SW-# locations sampled as part of the quarterly monitoring event, and additional 10 PDS locations selected based on the results of the passive diffusion sampling. Grab samples were collected in pre-preserved (HCl) laboratory-certified VOA glass vials by carefully placing VOA vials in the stream water and slowly filling the containers. The filling was completed by gently submerging the vials into the stream and topping off the vial by adding stream water to the VOA vials using the vial screw caps. The vials were then securely capped. These samples were analyzed for VOCs by USEPA Method 8260b. Analytical results are presented in Section 6.3.

5. HYDROGEOLOGY AND STREAM DISCHARGE RESULTS

5.1 Groundwater Elevation Data

Estimated groundwater elevation contours, incorporating data from the groundwater-monitoring wells and the piezometers, were developed. Based on the groundwater measurements, groundwater flows in an east-northeastly direction away from the site into the wetland area, with an approximate hydraulic gradient of 0.013 ft/ft. As groundwater flows closer to the wetland stream, groundwater flow is redirected in a southerly direction down the slope of the valley, with an approximate hydraulic gradient of 0.004 ft/ft.

5.2 Streamflows

The streamflows are estimated during each quarterly sampling event using the wetted cross-section and mean water velocity of the stream. The streamflows are generally on the order of 0.5 to 4 cfs when overland flow (rainfall events) is not significant. During or soon after rainfall events, the streamflow can increase significantly.

5.3 Stream Sediment Characteristics

The stream sediments along the length of the stream from PDS-1 upstream to PDS-34 generally consisted of fine sands and silt to coarse sands. The sediments north of PDS-34 consisted primarily of organic peat, silt and organic muck.

5.4 Piezometer Measurements

The water levels in the stream piezometers were measured on July 22, September 3, and November 6, 2003. Although these data alone do not measure the magnitude of the vertical hydraulic gradient, they do measure its direction (positive, negative, or flat) and indicate whether the stream is "gaining" (being fed by groundwater) or "losing" (discharging to groundwater) at that particular piezometer location at that moment in time. Vertical hydraulic gradients varied over the three dates that they were measured; however, discharge to the stream at several locations and timeframes was indicated based on this information.

5.5   Stream Water Quality Measurements

5.5.1   pH, ORP, and DO

No consistent trends could be determined in the pH or ORP data. DO values are variable; however, DO values are generally lower upstream of PDS-29 compared to downstream measurement locations. Depletion of DO may be the result of a higher organic matter content in the upper reach of the stream, enhancing biological activity and subsequent reducing conditions (lower DO).

5.5.2   Temperature

Temperature measurements as a function of stream length are graphically presented in Figure 2. Temperature data were collected over a period of several hours beginning with PDS-1 and heading north up the stream in numerical PDS sequence. As the graphs show, the overall trend of water temperature declines moving upstream, even as the ambient temperature would have been rising, indicating that stream temperature during data collection was not strictly dependent on the ambient temperature. Moving in an upstream direction, temperature decreased within the area between PDS-19 and PDS-31 indicating that cooler groundwater is discharging in this area. Temperature then increased briefly again before trending downward.

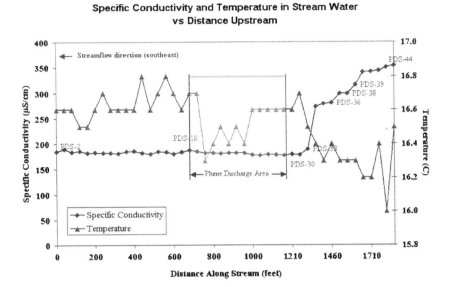

Figure 2. Temperature and Specific Conductivity in the Stream

5.5.3   Specific Conductivity

Specific conductivity measurements as a function of stream length are graphically presented in Figure 2. Moving upstream from PDS-1, the value of specific conductivity remained relatively steady until PDS-33. At PDS-33, this value increased rapidly and remained elevated for the remainder of the stream channel. This increase in stream-water specific conductivity value is possibly due to the type of sediments identified by M&E in the channel. Organic materials comprise much of the stream sediment upstream of PDS-34 and are likely less transmissive than the sandier sediments located downstream. These less-transmissive sediments may result in less groundwater recharge to the stream compared to downstream locations. Additionally, there is at least one secondary stream downstream of PDS-34 feeding the wetland stream, potentially increasing dilution within the stream. Therefore, at locations downstream of PDS-34, there may be more dilution of the stream water with groundwater of lower specific conductivity, resulting in lower specific conductivity values in the stream. Another possibility is that the organic-rich sediments directly affect the specific conductivity values in the stream water by being associated with higher quantities (mass/surface area) of ionic species than the sandier sediments.

6. ANALYTICAL RESULTS

6.1 Groundwater Monitoring Well Results

A summary of the concentrations of COCs along the plume length is presented as Figure 3. Based on the concentrations of the main COC, TCE, in wells GW-1 through GW-5 demonstrate the distribution of TCE in shallow groundwater at the upgradient edge of the wetland. The highest concentrations of TCE were detected in GW-2 (8,900 ug/L) and GW-3 (9,860 ug/L), while the lateral limits of the solvent plume were indicated by the significantly lower TCE concentrations in GW-1 (1.5 ug/L) and GW-5 (10.8 ug/L). Several other VOCs, PCE, *cis*-1,2-DCE, trichlorofluoro-methane, 1,4-dichlorobenzene, and 1,1,1-trichloroethane, were also detected, at much lower concentrations. The VOCs detected and their respective concentration ranges were generally consistent with historic contaminant detections. The QA/QC results for the duplicate, trip blank, and equipment blank samples indicated good data quality.

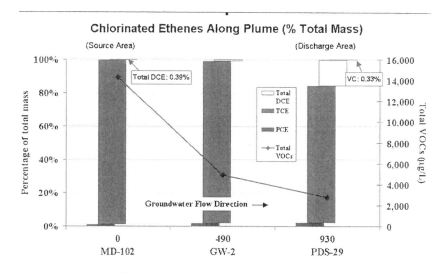

Figure 3. Chlorinated Ethenes Along Plume

6.2 Passive Diffusion Sampler Results

A summary of the analytical results for the PDSs is presented as Figure 4. The chlorinated ethenes PCE, TCE, *cis*-1,2-DCE, 1,1-DCE, *trans*-1,2-

DCE, and vinyl chloride were detected in 15 of the 45 PDS samples. Detectable concentrations of TCE ranged from 5.9 ug/L (PDS-28) to 2,280 ug/L (PDS-29), while detectable concentrations of *cis*-1,2-DCE ranged from 0.9 ug/L (PDS-27) to 741 ug/L (PDS-30). Detectable concentrations of PCE, 1,1-DCE, *trans*-1,2-DCE, and VC were generally lower than the concentrations of TCE and *cis*-1,2-DCE. QA/QC samples for the PDS sampling, consisting of a pre-deployment trip blank (prepared at the laboratory), an equipment blank (prepared in the field), a water blank (laboratory water tested at the laboratory), a trip blank, duplicates, and a materials test of the VOA vials used to construct the PDSs, in accordance with USGS recommended QA/QC practices for PDS sampling (USGS, 2001a) as discussed in Section 3.0, indicated good data quality.

Chlorinated Ethenes Along Stream (% Total Mass) - PDSs

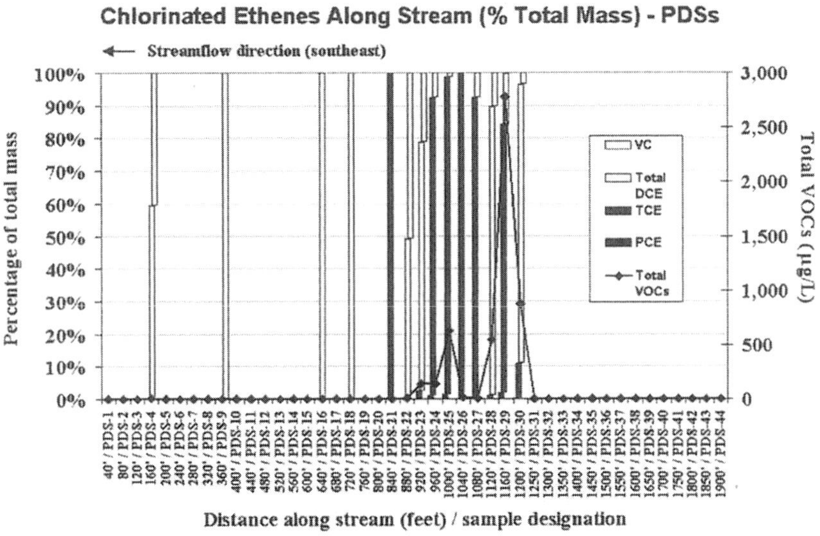

Figure 4. Chlorinated Ethenes Along Stream

6.3   Surface-water Results

A summary of the surface-water analytical results is presented as Figure 5. Vinyl chloride, PCE, *cis*-1,2-DCE, and TCE were detected in samples obtained from 1, 3, 13, and 13 of 16 surface-water sampling locations, respectively. TCE ranged in concentration from 0.8 ug/L to 53.7 ug/L. The analytical results from the QA/QC duplicate and trip blank samples indicated good data quality.

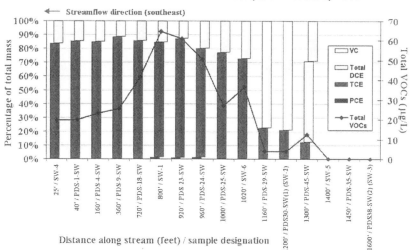

Figure 5. Chlorinated Ethenes in Stream Water

7.   DISCUSSION AND CONCLUSIONS

7.1   Evaluation of the Preliminary CSM

The preliminary CSM was updated using the site-specific data obtained from the SPDE. The on-site source of solvent contamination remains consistent with those described in Section 1.0. Regional groundwater appears to flow in an easterly direction from the source area towards the wetland and shifts to a southeasterly flow direction within the wetland to the consistent with the slope of the valley wall just east of the site.

The previous COC list included primarily the chlorinated ethenes TCE and PCE (Section 2.1). Based on the results obtained from the PDS samplers, this list may be expanded to include the three DCE isomers, cis-1,2,-DCE, trans-1,2-DCE, and 1,1-DCE, and vinyl chloride, which are present, and all potential products of the reductive dechlorination of TCE (Vogel et al., 1987). The physical and chemical characteristics of the additional COCs are similar to PCE and TCE. Potential migration pathways and receptors are unchanged.

Figures 3, 4, and 5 presents the distribution of TCE detected in the groundwater, porewater, and surface-water samples, respectively, analyzed as part of the SPDE. Groundwater and surface-water sample results are generally consistent with historically measured concentrations. The stream sediment areas affected by the solvent plume, and likely acting as points of solvent plume discharge to the wetland stream, are indicated on Figure 1. The predominant discharge area is from PDS-21 to PDS-30, covering an approximately 500 foot length of the stream. This area is located east-southeast of the former facility, placing it slightly cross-gradient (in a southerly direction) to the on-site direction of shallow groundwater flow, which is to the east-northeast from the facility. This discharge area is consistent with the decrease in temperature measurements in this area (Section 5.5). There are several individual PDS locations further downstream where COCs were detected in the samples obtained from the PDSs (PDS-4, PDS-9, PDS-16, and PDS-18), possibly serving as additional minor solvent plume discharge points through overburden or fractured rock flow. The downstream PDSs may also be the result of partitioning of chlorinated ethenes from the surface water, originating from upstream locations, back to the sediments.

Several geological/hydrogeological features are potential explanations for the location of the main solvent plume discharge area. As groundwater flows closer to the wetland stream, groundwater flow appears to be redirected in a southerly direction. This flow direction behavior is consistent with the topography of the wetland and with the groundwater levels measured at the time of this investigation. The wetland lies in a valley that slopes to the south. The plume was not found to discharge to the northeast since, as stated above, groundwater levels were lower than the stream level at the time of the investigation. Even if groundwater levels were higher than the stream during other seasons, it is likely that the organic-rich peat sediments north of PDS-34 have a much lower hydraulic transmissivity than the sediments to the south of PDS-34, possibly acting to some degree as a physical barrier to upward groundwater flow. Further, the organic-rich sediments likely will have relatively high sorption affinity for any VOCs that might migrate through the sediments, thereby stripping some of the VOCs out of the porewater.

Downward hydraulic gradients were measured in PZ-9, PZ-10, and PZ-11, located upstream of PDS-34, during the July and November, 2003 measurements. In contrast, the piezometric data also identify several areas of the stream as gaining at least part of the time between July and November, including areas east/southeast of the site at piezometer locations (PZ-5 and PZ-6) within the area where the highest solvent concentrations were identified. If part of the plume is fairly deep in the aquifer when it reaches the stream, it will flow down-valley beneath the stream as it rises through the aquifer to discharge. This downstream movement of groundwater may also contribute to the apparent offset of the plume discharge area to the southeast. Further, fracture patterns identified in regional bedrock during a previous investigation may contribute to preferential plume migration along fracture traces, causing the plume to travel slightly cross gradient to shallow groundwater flow, discharging to the stream in this area.

Chlorinated ethenes as a function of distance along both the plume and the stream are graphically presented as Figures 3 and 4, respectively. Chlorinated ethenes as % total mass along the plume length (Figure 3) indicate that transformation of PCE and TCE to DCE and vinyl chloride is occurring with increasing distance from the on-site solvent source. DCE is becoming a larger fraction of the total VOC mass. Chlorinated ethenes as % total mass along the stream length (Figure 4) indicate that transformation of PCE and TCE to DCE and vinyl chloride is occurring with increasing distance upstream and downstream from the main solvent-plume discharge location. Vinyl chloride and DCE are becoming a larger fraction of the total VOC mass. This increase in breakdown products is indicative of the dechlorination and natural attenuation of the primary COCs. Additionally, the plume fringe might be indicated by the increasing fraction of breakdown products at the upstream and downstream locations of the discharge area.

Chlorinated ethenes as % total mass in stream water are presented as Figure 5. Although some chlorinated ethenes, predominantly the breakdown products *cis*-1,2-DCE and vinyl chloride, are detected in the stream-water samples obtained upstream of the main discharge area, the highest concentrations of chlorinated ethenes are detected in the stream-water samples obtained from within or just downstream of the main discharge area. The highest fraction of the total VOC mass as breakdown products in the stream water samples is upstream of the main discharge area (PDS-45-SW) or at the upstream edge of the main discharge area, indicating that reductive dechlorination is occurring in this area.

7.2   Data Uncertainties

While the portion of the groundwater plume residing on-site and near the site is, to a certain extent, defined and appears to be characterized with relatively stable COC concentrations over time, the 3-dimensional extent of the solvent plume in the wetland is not known. This information will be critical in determining appropriate remedial actions in the future.

Based on contaminant concentrations observed in groundwater in the western edge of the wetland, the plume appears to be centered in the vicinity of GW-2 and GW-3 migrating from the southeast of the site. However the primary discharge area in the stream was identified to be just downstream of the power lines, south-southeast of the site. It is not known if the absence of

COCs in sediment pore water north and east of the site is the result of the plume succumbing to the southerly flow component of the valley prior to discharging to the stream, or if the plume is beneath the stream and is not detectable due to the low permeability of the stream sediments in this area, and/or the organic sediments capacity to dechlorinate the impacted groundwater prior to discharging to the stream, and/or a combination of these factors. Therefore, the presence, extent, and migration pathways of the potential COCs north and east of GW-2 and GW-3 warrant further investigation.

The presence of low levels of PCE and TCE daughter products in PDSs downstream of the main discharge area (e.g. PDSs-4, 9, 16, and 18) indicate transport in a southerly direction. However, the transport pathways are not known, but may be through overburden groundwater, weathered bedrock groundwater or bedrock fractures, a combination of these pathways, or may be the result of partitioning from the surface water to the sediments. It is important to understand the transport pathways so that remedial alternatives can be delivered to the affected area. Therefore, migration pathways, as well as the extent of COCs south-southeast of the main discharge area warrant further investigation. Important factors in the evaluation of the migration pathways in this area and throughout the plume are vertical gradients.

7.3   Next Phase of Groundwater Plume Evaluation

Further analysis of the nature and extent of contamination is required to address the data uncertainties. Installation and sampling of multi-level monitoring wells north-northwest, east-northeast, and southeast of the site are necessary to provide groundwater VOC concentration data to further evaluate the nature and extent of the solvent plume, while allowing for further definition of the potentiometric surface and hydraulic gradients in the vicinity of the plume.

In July 2004, the installation of additional monitoring wells, consisting of hand-driven shallow overburden, and drilled shallow and mid-depth overburden, deep overburden/weathered bedrock, and shallow bedrock wells at over 20 locations, was commenced. In most locations, wells are to be placed in clusters such that multilevel data and data from each of these geologic units can be collected. The actual number of wells installed at a specific location depends on the geologic features identified at the location.

Wells are being installed in stages to optimize well placement. Fourteen shallow hand-driven wells, screened from approximately 5 to 10 feet below ground surface, were installed prior to installing drilled wells. The objective of the initial stage of the plume evaluation is to better define the lateral limits of the shallow overburden plume as well as to collect additional data in the central portions of the plume. A preliminary round of groundwater sampling, consisting of the sampling of the hand-driven wells, was conducted following installation of the hand-driven wells.

The results of the preliminary groundwater sampling verify the CSM. Notably, chlorinated ethenes were detected in wells located within the suspected stream discharge area at concentrations over 1,000 µg/L, while no COCs were detected in shallow wells located along the stream both upstream and downstream of the suspected discharge area (See Figure 1). The distribution of chlorinated ethenes in the shallow overburden groundwater correlates well with the main discharge location in the stream. The plume appears to migrate from GW-2 and GW-3, turning sharply southeast, and discharging just south of the power lines in the main discharge location identified by the PDSs. Furthermore, the results of the sampling suggest that the predominant transport pathway for chlorinated ethenes to migrate into downgradient porewater (PDSs 4 and 9) is partitioning from the surface water to the sediments and not migration through the overburden groundwater.

7.4   Conclusions from the Solvent Plume Discharge Evaluation

The innovative deployment of the PDSs showed that a discharge area to a small wetland stream could be effectively delineated using the methodology described in this paper. The results from the PDSs also indicated that reductive dechlorination is occurring as the plume enters the stream. A plume fringe may be apparent at the upstream and downstream edges of the plume discharge area. Further, the estimated cost of collecting the 45 porewater data points using the PDSs is approximately 15% of the labor and materials cost of installing shallow groundwater-monitoring wells and conducting low-flow groundwater sampling to obtain this porewater data. Although the PDS data is limited to the sediment porewater, the PDS data were then used to focus the subsequent installation of monitoring wells.

REFERENCES

Savoie, J.G., Lyford, F.P., and Clifford, Scott, 1999, Potential for advection of volatile organic compounds in groundwater to the Cochato River, Baird and McGuire Superfund Site, Holbrook, Mass., March and April 1998: U.S. Geological Survey Water-Resources Investigations Report 98-4257.

USGS Fact Sheet 088-01 – Use of Passive Diffusion Samplers for Monitoring Volatile Organic Compounds in Groundwater.

United States Geological Survey. 2001a. User's Guide for Polyethylene -Based Passive Diffusion Bag Samplers to Obtain Volatile Organic Compound Concentrations in Wells. Part 1: Deployment, Recovery, Data Interpretation, and Quality Control and Assurance. Water-Resources Investigation Report 01 -4060.

United States Geological Survey. 2001b. User's Guide for Polyethylene -Based Passive Diffusion Bag Samplers to Obtain Volatile Organic Compound Concentrations in Wells. Part 2: Field Tests. Water-Resources Investigation Report 01 -4061.

United States Geological Survey. 2002a. Preliminary Analysis of Using Tree -Tissue Analysis and Passive Diffusion Samplers to Evaluate Trichloroethene Contamination of Groundwater at Site SS-34N, McChord Air Force Base, Washington, 2001. Water - Resources Investigation Report 02 -4274.

United States Geological Survey. 2002b. Guidance on the Use of Passive -Vapor-Diffusion Samplers to Detect Volatile Organic Compounds in Groundwater -Discharge Areas, and Example Applications in New England. Water -Resources Investigation Report 02 -4186.

United States Environmental Protection Agency. July 30 1996. Low Stress (low -flow) Purging and Sampling Procedures for the Collection of Groundwater Samples from Monitoring Wells.

Vogel, T.M., Criddle, C.S., and P.L. McCarty. 1987. Transformation of halogenated aliphatic compounds. *Environmental Science and Technology, 21* : 722-736.

Vroblesky, D.A., and W.T. Hyde. 1997. Diffusion samplers as an inexpensive approach to monitoring VOCs in groundwater: Groundwater Monitoring and Remediation, Summer 1997, p. 177-184.

CHAPTER 31

A PRELIMINARY ENVIRONMENTAL SITE INVESTIGATION FOR A BRIDGE OVER THE MISSISSIPPI RIVER AT MOLINE, ILLINOIS

C. Brian Trask

Illinois State Geological Survey, Environmental Site Assessments Section, 615 East Peabody Drive, Champaign, IL 61820

Abstract: A preliminary environmental site assessment along the alignment of I-74 and its bridge over the Mississippi River was completed by the Illinois State Geological Survey for the Illinois Department of Transportation in 2002. The purpose of the survey was to determine the presence of any environmental concerns, both natural and man -made, that the Illinois DOT might encounter during activities to build a new bridge to carry I -74 over the Mississippi River between Moline, IL, and Bettendorf and Davenport, IA. A preliminar y investigation of the project area from 27th Street in Moline to 67th Street in Davenport, using government databases, Sanborn Fire Insurance Maps, city directories, and a drive -through of the project area, identified a total of 127 sites that were believed to constitute a possible risk to the project. Following reduction of the project to just the Illinois side of the river, further investigation was conducted of 37 sites in Moline . The location of the project in part of downtown Moline and long-time development along the Moline riverfront by industrial and commercial operations offered a variety of parcels for investigation, ranging from corner gasoline stations to railroads and foundry sites. The dominant sites comprised current or former underground storage tank and leaking underground storage tank sites, automotive repair sites, and foundries and other sites where metals were handled. Railroads, junk yards, cl eaners, and spills made up the remainder of the sites investigated. During a limited subsurface investigation, heavy metals and volatile organic compounds indicative of petroleum were detected at several sites. Examples of some of these sites will be presented.

Key words: Environmental assessment; corridor assessment; transportation.

1.   INTRODUCTION

This project, completed in 2002 by the Illinois State Geological Survey (ISGS) for the Illinois Department of Transportation (IDOT), consisted of a Preliminary Environmental Site Assessment (PESA) investigation of possible environmental problems at sites along the right-of-way (ROW) of a corridor for reconstruction of Interstate 74 (I-74) and construction of new bridges over the Mississippi River from Moline, IL, to Bettendorf and Davenport, IA.   One of the bridges currently inplace had been constructed prior to 1938 to carry U.S. Route 6 over the river, and the second was built in the 1960s during completion of I-74.   The bridges are old and narrow and the sites of numerous accidents.   The focus of this project was to identify all parcels along the Illinois side of the alignment for the highway (Fig.1) and approaches to the bridges that might contain environmental hazards of concern to the Illinois Department of Transportation.   The objective of the study was to determine if any of the parcels proposed for acquisition or on which soil excavation was intended were sufficiently contaminated to require additional investigation by a commercial environmental consultant under contract to IDOT and to identify potential natural hazards that might have an impact on the proposed construction project.

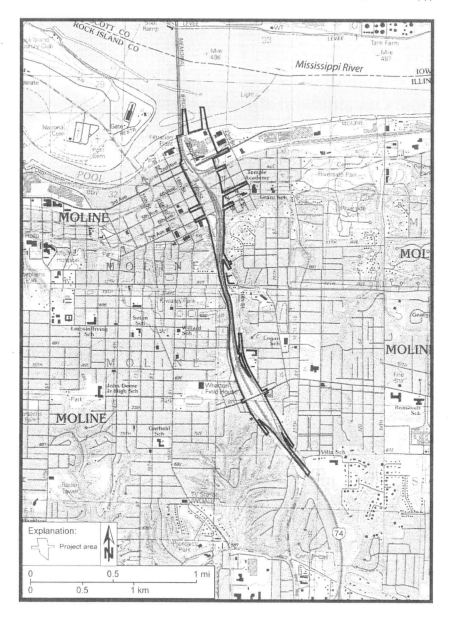

Figure 1. Project location map. Heavy dark line marks limits of project area.

2. PROCEDURE FOR COMPLETING PRELIMINARY ENVIRONMENTAL SITE ASSESSMENTS

The ISGS procedure for conducting environmental assessments for highway-construction projects consists of historical research into past uses of parcels along the ROW of the highway, followed by site visits and on-site interviews, and culminates in limited subsurface testing (Erdmann et al., 1996)—a modification of the American Society for Testing and Materials (ASTM) Phase II investigation (ASTM, 2003) of multiple parcels in a corridor. Several different sources of historical information were available for this assessment. The important sources were fire insurance maps, aerial photographs, city directories, and state and federal government databases.

2.1 Fire Insurance Maps

Fire insurance maps available for the project area were published by the Sanborn Map Company in 1886, 1892, 1898, 1906, 1912, 1950, 1957, and 1967 (Chadwyck-Healey, Inc., 1886, 1892, 1898, 1906, 1912, 1950, and 1957; University Publications of America, 1967). These fire insurance maps provided information concerning past uses of parcels, and delineated the former locations of underground storage tanks (USTs), foundries, railroad stations, automotive-repair facilities, and other features that might indicate possible environmental hazards along the project alignment.

2.2 Aerial Photographs

Aerial photographs, taken for soil surveys by the U.S. Department of Agriculture, were available for 1939, 1951, 1958, 1970, 1988, 1993, and 1998 from the University of Illinois Map and Geography Library. Information gathered from these photographs was combined with information collected from Sanborn Fire Insurance Maps to expand the knowledge of the history of development of industrial sites and aid in the identification of possible past locations of USTs, dispenser islands, electrical transformers, and other items of potential concern to the project engineers.

2.3 City Directories

City directories contain lists of parcels by street address, and can be examined for individual streets or portions of streets, as needed, in the project area. Parcels of potential concern to the project, and that might not be contained on Sanborn maps because of the large time gaps between consecutive maps, could usually be found in city directories, which are available on a year-by-year basis. City directories were examined for the years 1920, 1940, 1951, 1970, 1980, 1990, and 1998-99.

2.4   Government Databases

Several different types of government databases were used to complete this assessment. Of all the databases employed, the most valuable were the databases of underground storage tanks (UST) and leaking underground storage tanks (LUST), the Illinois Environmental Protection Agency (IEPA) Site Remediation Program database, and the U.S. Environmental Protection Agency's (USEPA) Comprehensive Environmental Response, Compensation, and Liability Information System (CERCLIS) database. The inventory database of the IEPA, Bureau of Land, and a Freedom of Information Act (FOIA) inquiry to the Illinois Emergency Management Agency may also reveal spill sites not contained in other databases.

UST database.   A database of underground storage tanks (UST database) is maintained by the Illinois Office of the State Fire Marshal (OSFM). This database contains the names and addresses of all businesses that have USTs and have registered them with the OSFM. The database provided valuable information concerning known USTs in the project area.

An electronic copy of the database was obtained from the OSFM and converted to a Microsoft Access database, which was searched using an Access search routine written specifically for the PESA process. This database gives the location of registered tanks by street address and company name. A FOIA request was then made to the OSFM to obtain copies of any documents contained in the agency's files concerning individual tank installations. These documents listed the type of tank(s) installed, the date of installation, date of removal (if a tank had been removed), and any problems that may have been encountered during removal, such as contaminated soils in the tank pit or holes in the tank. Some files contained maps showing the location of tanks installed (or formerly installed) on site.

LUST database.   The database of leaking underground storage tanks (LUST database) contains the name, address, and identification number of all sites with known leaking tanks reported to the IEPA. The database can be searched on-line at the website of the IEPA (http://epadata.epa.state.il.us/land/ust/index.asp). With company name, address, and identification number, a FOIA request was then made to the IEPA to obtain particular information that might be in IEPA files concerning what was known about the leak, any cleanup that might have occurred, and the current status of the event.

Site remediation program database. This database can be searched on-line at http://epadata.epa.state.il.us/land/srp/. The database contains sites that are undergoing voluntary remediation by the owner with IEPA oversight. Once a site had been identified, a FOIA request was made to the IEPA to obtain particular information that might be in IEPA files concerning what was known about the site, any cleanup that might have occurred, and the current status of the event.

CERCLIS database. The USEPA Superfund web site, located at http://www.epa.gov/superfund/, can be searched online to determine any CERCLIS sites that might be located along the ROW of a project. Once a list of such sites had been obtained, FOIA requests were made to the IEPA to obtain copies of IEPA files concerning the site.

Other databases. Other government databases used during the course of this investigation included the IEPA Bureau of Land Inventory, the IEPA Office of Chemical Safety Incident database, the U.S. Coast Guard National Response Center Database (http://www.nrc.uscg.mil/nrchp.html), the Resource Conservation and Recovery Act (RCRA) database published by the USEPA and the Illinois Waste Management Research Center, and the USEPA Toxics Release Inventory (http://www.rtknet.org/rtkdata.html).

2.5   On-Site Interviews

Following analysis of the historical data and government databases, interviews were conducted with current parcel owners and employees, adjacent businesses and residents, and the local fire department and public works agency. A resident historian in the fire department provided access to his collection of historical photographs of UST sites and spills along the alignment of the project. During the interviews, information was gathered concerning locations of USTs and other items of concern as well as recent histories of the parcels.

2.6   Field Investigation

A field investigation was then designed to provide limited testing of soils or groundwater on each parcel. Boreholes were emplaced using a Geoprobe Model 5400, a hand auger, or a manually pushed soil probe. Soil gas from 0.6, 0.9, 1.2, 1.8, and 2.4 m (2, 3, 4, 6, and 8 ft) in each borehole was tested for volatile organic compounds (VOCs) using a photoionization detector (PID) operating in methane elimination mode, and the headspace of soil and water samples was tested for VOCs using a portable gas chromatograph (GC). PID results were reported on a presence or absence basis, and specific VOCs were reported as identified by GC analyses. Soil samples were also collected, as necessary, to test the soil for the presence of metals, polycyclic aromatic hydrocarbons (PAHs), and polychlorinated biphenyls (PCBs). The relative risk to the project of metals and PAHs was judged based on the Tiered Approach to Corrective Action (TACO) Tier 1 objectives for residential properties, published by the Illinois Pollution Control Board (1997). The standards for residential properties were used, because they provide the worst-case scenario to IDOT for judging contaminants on a parcel.

3.   GEOMORPHOLOGY

The project area (Fig. 1) is located in the Mississippi River Valley and the uplands to the south that separate the Mississippi River from the Rock River. The downtown portion of the city of Moline is located on the valley floor, while residential and associated commercial areas to the south are situated on the Moline Upland. The relief across the project area is approximately 40 m (130 ft). Surface water in the project area flows mainly toward the north, though it is controlled by storm sewers throughout most of the project extent.

4.   GEOLOGY

Geology of the project area is based on Anderson (1980) and Hickerson and Anderson (1994), with terminology of geologic materials taken from Willman et al. (1975) as modified by Hansel and Johnson (1996).

4.1   Unlithified Sediments

Thickness of surficial deposits reaches as much as 30 m (100 ft) in the uplands south of downtown Moline. In the valley of the Mississippi River, total thickness of surficial deposits is 8 m (25 ft) or less. Thick deposits of wind-blown silt forming the Peoria Silt occur on uplands throughout Rock Island County. The Peoria Silt consists primarily of silt with lenses of fine sand, which originated from exposed sediment following melting of the last glaciers and prior to development of the current heavy cover of vegetation. In upland areas, the Peoria Silt reaches thicknesses of 4.6-15 m (15-50 ft). The Peoria Silt commonly sits on bedrock in these upland areas, while on slopes between uplands and the Mississippi River floodplain, glacial tills forming the Glasford Formation occur between the Peoria Silt and bedrock. Within the floodplain of the Mississippi River, the uppermost surficial deposit is the Cahokia Formation, which overlies the Henry Formation. The Cahokia Formation consists of silt, sand, and gravel deposited by currently active streams, particularly the Mississippi River. The underlying Henry Formation consists of glacial-outwash sand and gravel deposits. In the project area, these two units together are 8 m (25 ft) or less in thickness.

Sand in the project area provides conduits for transport of contaminants. Till, on the other hand, commonly causes contaminants to remain relatively stationary. Thus in an area where sand dominates, it is common to not detect materials such as petroleum compounds. In areas underlain by till, petroleum compounds commonly remain for years after removal of leaking USTs or deterioration of the storage container.

4.2   Bedrock

Bedrock in the project area was not penetrated in any of the boreholes probed for this investigation. The overlying blanket of unlithified surficial sediments commonly isolates the bedrock from surficial spills and leaks from buried storage containers. Therefore, bedrock was not investigated for this project.

5.   NATURAL HAZARDS

The project area is not subject to devastating earthquakes, and no major earth movements have been identified in the Moline area. However, its location along one of the major rivers in the world does make the project area susceptible to flooding.

5.1   Flooding

According to Flood Insurance Rate maps published by the Federal Emergency Management Agency (1980), the project route crosses the 100-year floodplain of the Mississippi River extending north from between 3rd and 4th Avenue at an elevation of approximately 175 m (575 ft). Flooding, standing water, and saturated soils may be encountered in this area, particularly during periods of high or extended rainfall or spring snowmelt.

6.   MAN-MADE HAZARDS

During an early examination of the project area, on both the Illinois and Iowa sides of the river, from south of 27th Street in Moline to north of 67th

Street in Davenport, 127 sites were identified for investigation. After a revision of the project restricted it to only the Illinois side of the river, this number was reduced to 63 parcels. Of these sites, 37 received a full, detailed investigation, including field sampling and testing, while the remaining 26 were examined and found to offer no risk to the project. More than half of the sites tested contained or had formerly contained underground storage tanks, three of which were leaking underground storage tanks (LUSTs). The second most common sites were those that handled metals–foundries, machine shops, and other such businesses. The remaining sites were approximately equally divided among railroads, spills, and transformers, with minor numbers of junkyards and roofing companies.

In the following discussion, examples of several of these site categories are presented. Due to the number of sites examined during the course of this investigation, not all can be described. The intent of the following discussion is to provide an overview of the complicated character of this investigation and to present information concerning typical or particularly interesting parcels along the proposed ROW of this project. The reader should have an understanding of how this complicated corridor was analyzed on a parcel-by-parcel basis to provide IDOT with the information needed to complete the construction process with minimal risk to life and health of site workers and residents in the project area during the construction phase.

6.1   Underground Storage Tank Sites

A total of 24 UST or former UST sites were investigated. These ranged from former automotive service areas that had or might have had USTs to active gasoline stations.

Figure 2. I-74 in downtown Moline.

Vacant lots. One former intersection was covered by the I-74 superstructure (Fig. 2) and surrounded by vacant lots and parking lots when this PESA was completed. Formerly, the intersection had been the site of residential neighborhoods with neighborhood businesses such as grocery stores, restaurants, and gasoline stations (Fig. 3). The northwest quadrant of this intersection contained a gasoline station that had been built between 1938 and 1950, according to Sanborn Fire Insurance Maps, aerial photographs, and city directories. Sanborn maps show three USTs south of the former building, in the northwest quadrant of the intersection. Aerial photographs show that a gasoline station occupied this quadrant through at least 1970. The building had been demolished by 1973 for construction of I-74.

The northeast quadrant of the intersection was occupied by a gasoline station from as early as 1950 through at least 1964, according to Sanborn maps, city directories, and aerial photographs. The parcel was vacant in 1973, the building having been demolished for construction of I-74. According to Sanborn maps, three USTs were located north of the former station building.

The southwest quadrant of the intersection contained a gasoline station from as early as 1950 and through at least 1967. Sanborn maps show three USTs north of the station building. The building had been demolished by 1973 for construction of I-74.

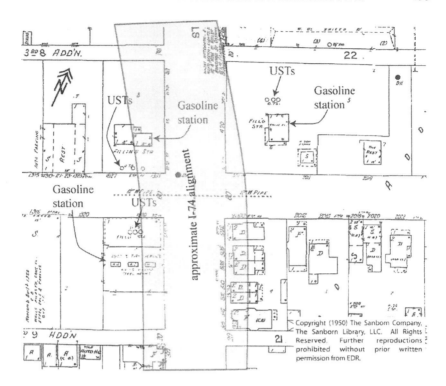

Figure 3. 1950 Sanborn Fire Insurance Map showing former locations of gasoline stations and approximate current location of I -74.

When I-74 was constructed through Moline, the three gasoline stations and surrounding residential structures were demolished and little remains of the former neighborhood. The areas around and beneath the highway have been converted to greenways, and the remains of the street can be seen beneath the highway (Fig. 2).

Because this intersection had formerly contained USTs, a magnetometer survey was completed to determine if any USTs might remain beneath the bridges in IDOT ROW. During that survey, magnetic anomalies were detected in the southwest quadrant of the intersection. These anomalies may suggest the presence of old abandoned USTs, or they may reflect steel in footings for the elevated highway that passes over the site.

During testing of the three parcels beneath the highway that contained the support structures for the current highway and that would be excavated to install supports for the proposed new highway, six boreholes were probed around this intersection adjacent to the former gasoline stations there. VOCs indicative of petroleum compounds were detected in the southwest quadrant of the former intersection.

Figure 4. Residential garage. Sanborn maps show an UST where the left -hand garage doors now exist.

Residential site. One former UST site in particular was interesting, because it has been a residential structure since its construction prior to 1912. This parcel was a private residence consisting of a single residential building with a garage at the rear along the north side of the alley south of the residence (Fig. 4). The garage was constructed of concrete block. Two shed additions had been added to the formerly nearly square garage. Sanborn maps from 1912 show an UST containing gasoline located on the west side of the garage (Fig. 5). The shed additions had been added to the building by 1950, and the western addition covered the former location of the UST.

This site was not found in any reviewed regulatory databases and there were no known records of product release at this site.

Potential hazards include VOCs from the UST shown on Sanborn maps. No VOCs significantly above background levels were detected in soil gas or the headspace of soil samples taken from two boreholes probed adjacent to this facility.

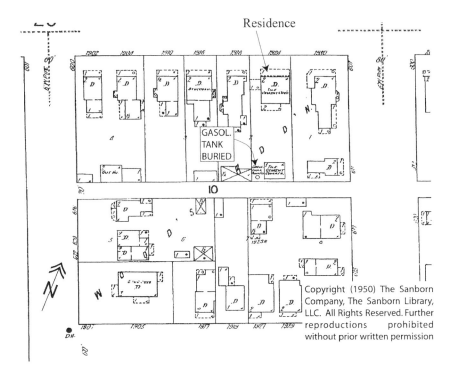

Figure 5. 1912 Sanborn Fire Insurance Map showing buried gasoline tank on west side of garage.

Office building. A building (Fig. 6) in one intersection contained a travel agency, a chiropractor, a political office, and a plumber. A rectangular concrete-block structure protruded from the rear of the building (Fig. 7). Examination of the parcel pavement revealed scars indicating the former location of two dispenser islands along the west side of the building (Fig. 7). According to Sanborn Fire Insurance Maps, city directories, and aerial photographs, a gasoline station had occupied this parcel from as early as 1940 until at least 1967. This was confirmed by the owner of the parcel, who stated that the gasoline station formerly on site ceased operations in the late 1960s. Since that time, he had constructed a new building around three sides of the old station building, mainly by addition of extensions on the north and south ends (Fig. 8).

Stained soil emitting a petroleum odor was collected from one of three boreholes probed adjacent to this parcel. VOCs indicative of petroleum compounds were detected in the headspace of soil samples collected from that borehole.

Figure 6. Travel agency building from northeast quadrant of intersection, looking southeast.

Figure 7. Travel agency building. Photograph on left shows former gasoline station building protruding from rear of travel agency. Photograph on right shows scar of former dispenser island in asphalt parking lot.

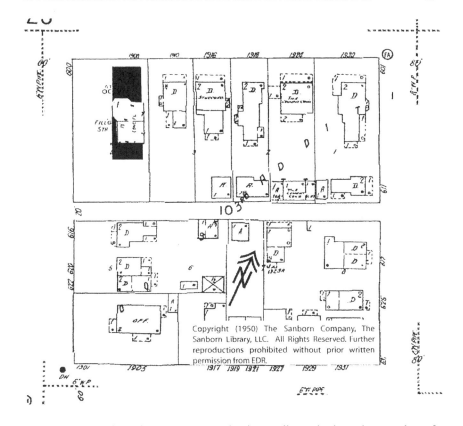

Figure 8. 1950 Sanborn Fire Insurance Map showing gasoline station in southeast quadrant of intersection. Black polygons show locations of additions to station building on north and south ends.

Gothic Revival cathedral and park. A large building located in the southeast quadrant of one intersection was an ornate Gothic Revival style building. Along the east side of the building was an asphalt parking lot. Separated from the parking lot by a cross street was a greenway adjacent to the elevated I-74. The greenway was covered with grass and contained several trees. These parcels had the appearance of offering no risk to this project.

The cathedral parking lot was a residential site in 1912. However, according to city directories, Sanborn Fire Insurance Maps, and aerial photographs, the parking lot formerly contained an automotive dealership (Fig. 10), which was present from the 1940s through at least 1973. The former commercial building on the parcel was demolished between 1973 and 1980.

The greenway to the east was residential in 1950. However, it contained a gasoline station in the 1960s and early 1970s (Fig. 10). The station building had been demolished by 1973, probably for construction of I-74, which passes along the east edge of the greenway.

Figure 9. Gothic Revival cathedral and parking lot are in right photograph. Left photograph contains parkland with several trees.

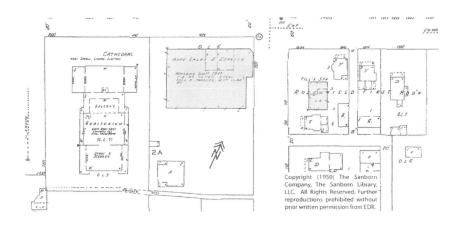

Figure 10. Sanborn Fire Insurance Map from 1957 showing Gothic Revival cathedral and automobile dealership in southwest quadrant of intersection and former gasoline station in southeast quadrant.

Due to the former presence of USTs on these two parcels, magnetometer surveys were conducted. A total of seven magnetic anomalies were detected—two adjacent to the cathedral parking lot and five in the greenway. These anomalies may indicate the presence of abandoned USTs that may have been used to store gasoline.

Three boreholes were probed adjacent to the cathedral parking lot, and two were probed on the greenway, which is already IDOT ROW. VOCs indicative of petroleum compounds were detected in soil gas and the headspace of soil samples at both sites.

6.2   Metals-handling sites

Metals-handling sites included active and former foundries, an elevator manufacturer, machine shops, and railroad facilities. Six metals-handling sites were investigated for this project. One site, now abandoned and demolished, was located just south of the Mississippi River.

This parcel was a vacant lot behind a visitor's center, which was formerly a passenger station for a railroad. The area north of the visitor's center contained concrete slabs indicating the former locations of several large buildings.

Figure 11. Sanborn Fire Insurance Map from 1957 showing the former arrangement of buildings at the foundry complex in the northeast quadrant of the intersection. Location of foundries is indicated.

According to Sanborn maps, a foundry was formerly located on this site. Sanborn maps show four foundries- three steel and one aluminum- a milling room, sand and casting storage, and other miscellaneous facilities at the site (Fig. 11). Transformers were located at the aluminum foundry. The current visitor's center was part of the foundry complex, having been used for office space. According to information received from the visitor's center, this facility began operations as a gray iron foundry in 1917 and ceased operations in 1992. Aerial photographs show the foundry buildings through 1988, and they had apparently been removed by 1993. When field work for this project was conducted, only the slabs of the foundries remained on site.

This site is listed in the OSFM UST database, the IEPA LUST database, and the IEPA BOL inventory. It is also included in the IEPA Site Remediation Program. Investigations of the large parcel for all of these events have detected petroleum, PCBs, and metals, particularly lead, chromium, and arsenic.

Four boreholes were probed on the parcel to test for VOCs, PAHs, PCBs, and metals. VOCs indicative of petroleum compounds and stained soil emitting a solvent odor were detected in the area of the three former iron foundries and above a benzene plume that has been mapped at the eastern side of the parcel. Nickel was detected at the eastern iron foundry site at levels above the total metals soil component migration to Class I groundwater value of the IEPA TACO Tier 1 residential standards. PCBs were also detected in the former transformer area.

6.3   Railroads

The main railroad line of the former Rock Island Lines passed through the project area just south of the Mississippi River. Though no large railroad facilities such as roundhouses, repair tracks, freight stations, and maintenance yards were located in the project area, all street crossings were protected by crossing guards. Control boxes for such crossing facilities commonly contain transformers and other electrical equipment to maintain the proper electrical current to operate the crossing guards and to continue operations during power outages. Most contain backup power provided by lead-acid batteries. Thus, testing was conducted at each of these railroad sites for lead and PCBs.

Figure 12. Railroad control boxes. Left box is acti ve, while right box has been taken out of service.

One railroad control box (Fig. 12) was located near two of the main ramps from and to I-74. The parcel contained an active control box and the remains of an older control box that had been disconnected. PCBs were detected in a soil sample collected from between the two control boxes.

6.4   Spill sites

Sites where spills had been reported or that might have been susceptible to spills were identified in the project area. These included a spill of petroleum on the Mississippi River, a former junkyard where petroleum might have leaked from scrapped vehicles onto the ground, a residence where lead paint had been deposited on the ground during sandblasting of an old house, and an I-74 accident site.

Mississippi River. Commercial traffic on the Mississippi River carries numerous types of cargo, ranging from grain to petroleum products and farm chemicals. Also, the area along the river has been contaminated with metals from more than 100 years of discharge of particulates from numerous foundries in Moline and distribution of these particulates by prevailing southwesterly winds. The project alignment over the river was tested by boreholes on both the south bank of the river and the island beneath the I-74 bridges (Fig. 13). Neither VOCs nor metals above TACO Tier 1 residential standards were detected in boreholes located on the Moline bank of the river or on the island in the river. However, polycyclic aromatic hydrocarbons (PAHs) were detected in a soil sample from the island.

Figure 13. Island in Mississippi River. Photograph taken looking south. Tall building at center rear of photograph is elevator test tower for elevator manufacturer. PAHs were detected in a soil sample from this island .

Former junkyard. The area of this former junkyard was occupied by parking lots for an elevator company with a greenway adjacent to the parking lots (Fig. 14). The lots were located beneath the approaches to the current I-74 bridges and between the elevator company and a Moline city water plant. The parcels were covered with concrete, asphalt, and grass.

According to Sanborn Fire Insurance Maps (Fig. 15) and city directories, a junkyard was located at this site from the 1940s through the 1980s. Aerial photographs show that the site had been cleaned and converted to its present use by 1988.

Four shallow boreholes were probed adjacent to these parcels to test for VOCs that might have been spilled on the ground surface. No VOCs were detected in soil gas or the headspace of soil samples collected from the boreholes.

Figure 14. Former junkyard. Currently occupied by parking lots and green space, junkyard formerly occupied entire area beneath a nd to left (west) of I-74 superstructure.

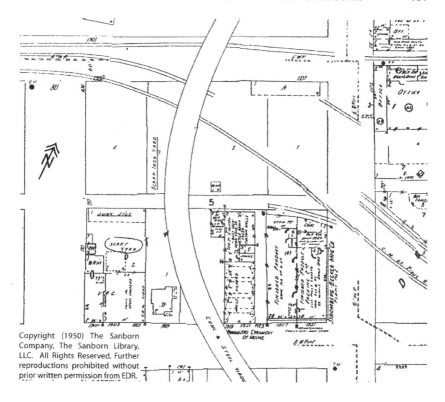

Figure 15. Sanborn Fire Insurance Map from 1957, showing location of junkyard (scrap iron yard, junk storage, and scrap yard).

Residence. A residence in the project area (Fig. 16) had been sandblasted several years prior to this investigation to remove old paint. Paint fragments had been scattered around the building on the property and on an adjacent property, and the paint was determined to contain lead. Lead was prevalent enough that a neighbor with a new infant was forced to move out temporarily. The paint chips were removed from the property by the IEPA, and the site was determined to be clean.

A soil sample was collected by the ISGS from along the street that passed adjacent to the house for lead analysis. Though lead was detected in the sample, it was at levels below the TACO Tier 1 residential standards.

Figure 16. Residence where lead-base paint was sandblasted. Soil sample for lead analysis
was collected from adjacent to highway.

I-74 accident site. Diesel fuel had been spilled at the side of the
highway following rupture of a fuel tank by debris. The accident site was
particularly interesting, because a map in IEPA files showed the shape of the
spill, and stressed vegetation in the spill area permitted identification of the
spill site. The accident had occurred on I-74 eastbound near the south end of
the project area. The site contained a drainage ditch along the west side of I-
74. The highway in this area had been excavated during construction to a
depth of approximately 6 m (20 ft) below the natural ground surface to allow
for construction of bridges at street level for streets passing over the
highway.

Two shallow boreholes were probed to test for VOCs. The boreholes
were positioned in the area of the stressed vegetation. No VOCs were
detected in the headspace of soil samples collected from the boreholes.

7.   SUMMARY

During examination of parcels in this corridor for reconstruction of I-74
and its bridges over the Mississippi River, VOCs were detected at 13 of 37
sites tested for VOCs. These sites included an elevator manufacturer, former
foundries, former and active machine shops, former and active gasoline

stations, and a former automobile dealer. PAHs above TACO Tier 1 residential standards were detected on an island in the Mississippi River. PCBs were detected at a former foundry and a control box for a railroad. Magnetic anomalies that might indicate the presence of USTs were detected in a park that formerly had been the site of a city garage, adjacent to a parking lot that formerly contained an automobile dealer, and at the sites of three former gasoline stations.

Conducting assessments of corridors for transportation facilities requires use of all available historical documents and government databases. In metropolitan areas, fire insurance maps have been found to be invaluable to determine past land use. Combined with aerial photographs and city directories, information contained on fire insurance maps, which have been revised infrequently, can be extended over periods between revisions and following the last fire insurance map available, which is often dated from the mid-1960s. Government databases can be related to the project area through information gathered from city directories and on-site visits. This complex project illustrates the importance of utilizing these multiple resources to screen large areas in a time- and cost-efficient manner. Completion of such corridor assessments has assisted the Illinois Department of Transportation to plan routes to avoid numerous heavily contaminated sites and to arrange excavations to reduce the amount of contaminated material that must be properly disposed of. These studies have helped IDOT to save millions of dollars in highway construction projects.

REFERENCES

ASTM (American Society for Testing and Materials). 2003. Standard Guide for Environmental Site Assessments: Phase II Environmental Site Assessment Process. ASTM Standard E1903-97(2002).

Anderson, R.C. 1980. Geology for pla nning in Rock Island County, Illinois. Illinois State Geological Survey Circular 510.

Chadwyck-Healey, Inc. 1886, 1892, 1898, 1906, 1912, 1950, and 1957. Sanborn fire insurance maps, Moline, Illinois.

Erdmann, A.L., Bauer, R.A., Bannon, P.L., and Schneider, N.P. 1996. A manual for conducting preliminary environmental site assessments for Illinois Department of Transportation highway projects. Illinois State Geological Survey Open File Series 1996 - 5.

Federal Emergency Management Agen cy (FEMA), National Flood Insurance Program (1980). Flood insurance rate map (FIRM). Moline panel 5, Rock Island County, Illinois.

Hansel, A.K., and Johnson, W.H. 1996. Wedron and Mason Groups: lithostratigraphic reclassification of de posits of the Wisconsin Episode, Lake Michigan Lobe area. Illinois State Geological Survey Bulletin 104.

Hickerson, W., and Anderson, R.C. 1994. Paleozoic stratigraphy of the Quad -Cities region, east-central Iowa, northwestern Illinois. Geological Society of Iowa Guidebook 59.

Illinois Pollution Control Board. 1997. Tiered Approach to Corrective Action Objectives. 35
    Illinois Administrative Code, Part 742.
University Publications of America. 1967. Sanborn fire insurance maps, Moline , Illinois.
Willman, H.B., Atherton, E., Buschbach, T.C., Collinson, C., Frye, J.C., Hopkins, M.E.,
    Lineback, J.A., and Simon, J.A. 1975. Handbook of Illinois stratigraphy. Illinois State
    Geological Survey Bulletin 95.

CHAPTER 32

TARGETED BROWNFIELDS ASSESSMENT OF A FORMER POWER PLANT USING THE TRIAD APPROACH

Barbara A. Weir[1], James P. Byrne[2], Robert Howe[3], Denise M. Savageau[4], and Kathy Yager[5]

[1]Metcalf & Eddy, Inc., 701 Edgewater Drive, Wakefiel d, MA 01880; [2]USEPA Region 1 Brownfields Team, Work Assignment Manager, 1 Congress Street, Suite 1100 (HIO), Boston, MA 02114-2023; [3]Tetra Tech, Inc., 4940 Pearl East Circle, Suite 100, Boulder, CO 80301; [4]Town of Greenwich, Conservatio n Director, 101 Field Point Road, Greenwich, CT 06830; [5]USEPA Office of Superfund Remediation and Technology Innovation, 11 Technology Drive, Chelmsford, MA 01863

Abstract: A Targeted Brownfields Assessment of a former power plant was conducted using the Triad approach, through the efforts of EPA Region I and Metcalf & Eddy, the Brownfields Technology Support Center, and the town of Greenwich, Connecticut. The Triad approach is an integrated strategy for managing decision uncertainty at hazardous waste sites that is being promoted by EPA, along with other Federal agencies. The Triad approach consists of three elements: systematic project planning, real-time measurement technologies, and dynamic work strategies (http://www.triadcentral.org). The Triad approach recognizes that overall decision uncertainty is g enerally governed more by sampling uncertainty (that is, the uncertainty that the samples collected adequately represent site contamination), than by analytical uncertainty (that is, the accuracy and precision of the analytical method) (Crumbling et al., 2001). The town plans to redevelop the nine -acre site as a waterfront park, and needed to decide whether this plan was feasible, and if so, what remedial measures might be necessary before the park could be constructed. Because coal ash was historically disposed at the site, it was considered probable that site surface soil throughout the entire nine acres would show concentrations of contaminants that exceed Connecticut residential direct exposure criteria . Historical information also suggested the possibility of localized petroleum and polychlorinated biphenyl (PCB) releases. The goal of the investigation was to obtain sufficient data, in one mobilization, to determine the nature and extent of surface soil contamination. Attaining this goal was judged to be infeasible using a traditional approach of soil sampling with analyses in off -site laboratories, and evaluation of results several weeks later. The Triad approach, with its focus on development of a conceptual site model, evaluation of decision uncertainty, use of field/real - time analytical methods, and field decision -making, was selected over a traditional approach. Field analytical te chniques included siteLAB [®] test kits for total petroleum hydrocarbons (TPH) and total polyaromatic hydrocarbons (PAHs), EPA's X-ray fluorescence instrument for metals , and the EPA mobile laboratory for PCBs. Soil samples were also analyzed by an off -site laboratory for arsenic with 48-hour turnaround. Sampling locations were added based on feedback from the field analyses. The use of field methods allowed for sampling using a random -grid approach, with many more locations sampled than would have been possible by relying solely on the more costly off-site laboratory analyses. The team located an area of PCB contamination that might have gone undiscovered if a random grid sampling approach had not been used. A correlation was developed between the TPH/PAH test-kit results and off-site laboratory results, which enabled use of the test-kit results to define the extent of soil contamination with TPH and PAHs. The results showed that site surface soil contains concentrations of arsenic, TPH, and PAHs at levels exceeding Connecticut residential direct exposure criteria. The observed concentrations are partia lly from the coal ash that is co-mingled with site soil. PCBs were also identified near an area where transformers had been located. The Triad approach yielded an estimated cost savings of approximately 35 percent, when compared with a traditional approach involving two mobilizations of 20 borings each, with locations selected judgmentally, and off -site laboratory analytical methods alone.

Key words: arsenic; Brownfields; coal ash; metals; PAHs; PCBs; petroleum; power plant; residential direct exposure criteria ; siteLAB; surface soil; Targeted Brownfields Assessment; test kit; Triad; XRF.

1.   INTRODUCTION

In 2002, EPA Region I awarded a Targeted Brownfields Assessment (TBA) grant to the town of Greenwich, Connecticut to help the town with its plans to remediate a former power plant and redevelop it as a waterfront park. The town was concerned that soil contamination might be present on the property, and needed to gain a greater understanding of site conditions to determine whether reuse as a park was feasible and if so, what cleanup might be required. The TBA program is designed to help municipalities with sites where perceived or actual contamination is an impediment to redevelopment. The program offers a grant of services that includes research into the site history and previous investigations, planning and execution of a site investigation, recommendations for additional investigation (if needed), and planning-level estimates for site cleanup (if needed). The typical value of the grant of services ranges from $50,000 to $100,000 depending on the complexity of the site.

EPA Region I also requested the assistance of the EPA's Brownfields Technology Support Center (BTSC) in applying the Triad approach to the site. The BTSC had been looking for a site in Region I to which the Triad approach might be applied, and the EPA Region I TBA project manager identified the Cos Cob site as a good candidate because of its size and potential for multiple release areas and contaminants of concern. It was felt that a collaborative approach to the investigation, combining the resources of the M&E TBA team, BTSC's experts in the Triad approach, and EPA Region I's mobile laboratory services, would be the most effective way to investigate the site.

The Triad approach is an integrated strategy for managing decision uncertainty at hazardous-waste sites that is being promoted by EPA, along with other Federal agencies. The Triad approach includes three primary elements: "(1) detailed and specific systematic project planning that begins by clearly defining desired project outcomes (*e.g.*, potential goals for site reuse), and exploring the uncertainties that stand in the way of achieving those outcomes; (2) dynamic work-planning strategies that can drastically save time and money over the project lifetime; and (3) real-time data generation and interpretation to support the real-time decision making of the dynamic work plan, while cost-effectively managing sampling uncertainty and data representativeness" (Crumbling *et al.*, 2003). The Triad makes use of new technologies when possible to cost-effectively increase sampling density so that the extent of contamination can be better characterized. The Triad also allows for adjustment of the sampling design in the field to reflect results from real-time analyses. Hence, data gaps can be addressed in the field, reducing the number of mobilizations to a site (Robbat, 1997 as cited in Crumbling *et al.*, 2003).

The BTSC is producing a case study, entitled *Case Study of the Triad Approach: Expedited Characterization of Petroleum Constituents and PCBs Using Test Kits and Mobile Chromatography Laboratory at the Former Cos Cob Power Plant Site, Greenwich, Connecticut* (USEPA, 2004). This paper is drawn from the case study, which is scheduled for printing in the Fall of 2004 and will be available to the public via the website http://www.triadcentral.org.

The Cos Cob Power Plant site is located on the west bank of the Mianus River in Greenwich, Connecticut. The portion of the property owned by the town is approximately 9 acres in area, and is bordered by transformer yards, an electrical substation, a condominium complex, and the Cos Cob Harbor. The property was deeded to the town by the state in 1987, with the understanding that the property would ultimately be open to all residents of the state. Possible uses of the land that the town is considering include walking trails, playing fields, picnic areas, and a boating facility.

The town and the state demolished the power plant during 1999 and 2000. There are currently no structures on the property owned by the town, but the town public works department makes use of the southern portion of the property for storage of construction materials. Access to the property is limited to town employees and is restricted by a fence along the northern and western boundaries.

1.1   Site History

The Cos Cob Power Plant was built in 1907 by the New York, New Haven, and Hartford Railroad, to provide power for electrification of the railroad. The coal-fired plant operated in this capacity until the 1960s. The plant was a multi-level concrete and metal building and housed boilers, PCB transformers, and other electrical generation and distribution equipment. Several additions were made to the building over the years, before its use was discontinued in the 1960s. In 1986, the idled plant was decommissioned by the Connecticut Department of Transportation, and the property was transferred to the town of Greenwich the following year. Plant equipment, such as transformers and boilers, was left intact but the plant itself, which had been essentially vacant for twenty years, continued to deteriorate and was vandalized. One instance of vandalism resulted in the release of transformer oils to the ground. The town reported the spill to the Connecticut Department of Environmental Protection (CTDEP) and contracted with Marin Environmental to remediate the release area and to conduct a limited investigation of site soil (Marin, 1998a).

In 1999, the town and the CTDEP became increasingly concerned about the potential for releases of asbestos-containing materials (ACM) from the deteriorating main building and smaller metal buildings. An inspection of the main building in 1999 by Osprey Environmental found ACM both inside the building and on its exterior. Osprey Environmental evaluated the potential for a release of ACM to the environment, and concluded that releases had likely occurred from several areas. Osprey recommended the removal of ACM, with the concurrence of the Connecticut Department of Public Health. The ACM was removed and the main building and smaller metal buildings were demolished during 1999 and 2000. The CTDEP assisted the town in accomplishing the demolition.

From examination of historic aerial photographs and conversations with town representatives, it was learned that the southernmost portion of the site, south of the former power house, is almost entirely composed of ash from the former power plant. A previous site investigation (TRC, 1988) documented the presence of ash to a significant depth in this area.

1.2   Areas and Contaminants of Potential Concern

Based on the review of background and historical information and previous site assessments, the contaminants of potential concern identified at the site were: metals, polyaromatic hydrocarbons (PAHs), asbestos-containing materials (ACM), total petroleum hydrocarbons (TPH), and polychlorinated biphenyls (PCBs).

Metals such as arsenic are primary contaminants of concern because of their presence in coal ash and the documented presence of ash used as fill on the site.   Arsenic has the lowest Connecticut residential direct exposure criterion (10 mg/kg), and is commonly present at levels above 10 mg/kg in soil impacted by coal ash.   PAHs are also of potential concern because of their presence in coal and coal ash.   Previous assessments (TRC, 1988) revealed low levels of PAHs (above Connecticut residential direct exposure criteria) in some surface soil samples.

The likelihood that surface soil is impacted by ACM was considered to be moderate, based on the evaluation conducted by Osprey Environmental in 1998.   Releases of ACM to the environment could have occurred before the ACM was abated and the buildings demolished in 2000.

Concentrations of TPH in excess of Connecticut residential direct exposure criteria were detected in surface soil by Marin (1998a). These samples were collected in areas suspected of being impacted by a release. Another Marin report (1998b) also documents the remediation of a PCB spill in the basement of the former power house.   Although soil samples collected for PCB analysis by TRC (1988) did not show detectable levels of PCBs, that investigation was limited in extent.

2.   SITE INVESTIGATION APPROACH

The primary objective of the TBA was to conduct a site investigation to assess the nature and extent of surface-soil contamination, sufficient to allow development of possible remedial strategies consistent with the town's plans for site reuse as a park.   It was considered probable that much of the site's surface soil (0 to 4 feet, consistent with the CTDEP definition) would show concentrations of coal ash-related contaminants that exceed the residential direct-exposure criteria established in the Connecticut Remediation Standard Regulations. The primary goal of the site investigation was to help delineate areas that exceed the criteria.

2.1   Triad and Traditional Approaches

During the early stages of planning, the Region I project team decided that a traditional approach to site investigation was unlikely to meet project goals within the time and funding available. The site was very large compared to the sites typically investigated under the TBA program, and it was considered likely that low levels of coal-ash-related contamination might be present in surface soil throughout the entire 9-acre site. Possible "hot spots" of PCB or TPH contamination were also considered likely because of past spills and incidents of vandalism at the site. A preliminary work plan was developed using a traditional approach, consisting of twenty judgmentally located soil borings with off-site laboratory analyses for the Connecticut-regulated metals, PCBs, TPH, and PAHs. The preliminary work plan included no provision for random grid sampling, field-based analyses, field-decision making, or collection of additional samples based on field results. It was realized that 20 borings would not provide sufficient coverage of the site, but under a traditional approach with reliance on more costly off-site laboratory analyses for all contaminants of concern, funding limitations would not allow for the addition of more "traditional" borings. The possibility of missing a release of PCBs or TPH was considered to be quite high with only 20 judgmentally-located borings distributed throughout the site. Also, it was considered very likely that evaluation of results from the 20 soil borings, after several weeks of waiting for off-site laboratory results, would almost certainly reveal data gaps that would not be addressable without a second full-blown "traditional" investigation.

A meeting was convened to decide how to re-structure the planned investigation using the Triad approach. Participants at the meeting included the EPA Region I TBA project manager, the M&E project manager, EPA's Brownfields Technology Support Center (BTSC) contractor Tetra Tech EM, Inc. (Boulder, Colorado), and the Region I Office of Environmental Measurement and Evaluation (OEME). The work plan was redesigned to focus on surface soil (defined by CTDEP as 0 to 4 feet below ground surface) and incorporate a random grid sampling strategy with a much greater density of samples than the traditional approach (approximately 100 surface soil sampling locations, as opposed to 20 "traditional" (that is, judgmentally-located) borings to the water table. The approach also made use of field-based analytical methods, and off-site laboratory analyses with 48-hour turnaround time, to allow on-site decision making and adjustment/addition of sampling locations based on the field and quick turnaround results. The analytical strategy was re-designed to focus on arsenic and carcinogenic PAHs, contaminants commonly associated with coal ash that also have low Connecticut residential direct exposure criteria.

Because the residential direct-exposure criterion for arsenic in soil, 10 mg/kg, is lower than can be detected by current field-based methods, it was decided that an off-site laboratory analysis with 48-hour turnaround would be used for arsenic. A test kit that measures TPH and total PAHs (siteLAB®) was selected to allow rapid field analyses for these contaminants. Finally, the OEME mobile laboratory was used to perform field-based analyses for PCBs using a modification of EPA Method 8082.

The Triad approach was selected for this site to save time, with the goal of fully delineating the extent of surface soil contamination in one mobilization. Cost savings was also a consideration, since it was not possible to increase the sampling density beyond 20 locations using only off-site laboratory analyses for all the contaminants of concern. The BTSC case study developed a cost comparison of Triad vs. traditional approaches for the Cos Cob site, by assuming that two "traditional" approaches would have been needed to obtain the level of site characterization that was obtained by using the Triad approach. A cost savings of 35% was estimated that takes into account the efforts of the BTSC to support the investigation, as well as the Demonstration of Methods Applicability described in Section 2.2.

Initial costs for using the Triad approach were higher than for a more traditional approach, because of the additional effort needed to plan the project (including the efforts of the BTSC and the Region I laboratory), efforts to identify and evaluate possible field analytical methods, and costs for performing the Demonstration of Methods Applicability (Section 2.2). Cost savings are projected over the length of the project, however, because a more complete site characterization was obtained during the TBA investigation than would have been obtained using the traditional approach originally envisioned.

2.2 Demonstration of Methods Applicability

Before the full site investigation was performed in February 2003, a limited (one day) sampling event was performed in December 2002 to collect samples to evaluate how the siteLAB test-kit results would correlate with off-site laboratory results for PAHs and TPH. Surface soil samples were collected from 17 locations in two areas of the site (Figure 1) and brought to the EPA Region I laboratory in North Chelmsford, MA. Selected samples were analyzed there using the siteLAB kits. Selected samples were also analyzed by off-site laboratories using EPA Method 8270 with Selected Ion Monitoring and the Connecticut method for Extractable Total Petroleum Hydrocarbons. The results indicated that multiple sources of PAHs and TPH might be present in site soil (coal ash and possible petroleum releases), that would complicate the interpretation of test-kit results. Broad decision criteria for when samples would be sent to off-site laboratories were established to account for the complexity of the PAH and TPH data observed during the demonstration of methods applicability. Within the limits of the analytical budget, it was decided that samples with total PAH concentrations (as reported by the test kit) of between 50 ppm and 1000 ppm would be sent to the off-site laboratory. For TPH test-kit results, a range of 100 ppm to 500 ppm was established. It was considered likely that samples with test-kit results below 50 ppm total PAHs and 100 ppm TPH would not exceed residential direct exposure criteria, while those with test-kit results above 1000 ppm total PAH and 500 ppm TPH would most likely exceed criteria for one or more analytes. It was decided that samples within the window of uncertainty would preferentially be selected for off-site laboratory analysis, with a limited number of samples outside the window also selected to allow development of more robust correlations once the main field event was completed.

3. SITE INVESTIGATION

The site investigation and data interpretation represented a collaborative effort among EPA Region I, M&E, the BTSC, and the town of Greenwich. Sampling locations were based on a 70-foot by 70-foot grid prepared by the BTSC and were surveyed by the town of Greenwich using their Global Positioning System (Figure 1). Locations within each grid were selected randomly. The former power-plant building footprint was not included in the grid, because according to town representatives, this location was filled in with clean fill when the power plant was demolished in 2000. Also, it was not completely accessible at the time because of its use for storage of soil by the town of Greenwich Department of Public Works. Other areas were also found to be inaccessible, because of soil piles, dense vegetation, or steep slopes. These inaccessible areas will most likely need to be investigated before a remedial action plan is prepared for the site, but it is anticipated that the level of soil contamination in these areas will be consistent with the areas that have been investigated.

The BTSC provided extensive support regarding use of field analytical methods and the Triad approach. The BTSC prepared decision trees for use in the field; provided several days of field support that included assistance with selection of samples for off-site laboratory analysis; and reviewed the siteLAB test-kit results and developed correlations with off-site laboratory results. The BTSC also performed statistical evaluation of the data and prepared figures that mapped the site contamination that was encountered. The BTSC case study (USEPA, 2004), due for release in the fall of 2004, should be consulted for further details regarding the work described in this paper, particularly with respect to the application of the Triad approach.

3.1   Arsenic in Surface Soil

Samples were collected at the locations identified on Figure 1 using a direct-push device with a four-foot acetate liner. The four-foot sample was segregated into separate one-foot intervals. The 0- to 1-foot interval was divided into several aliquots after mixing; one for arsenic analysis, one for TPH/PAHs analysis, one for PCBs analysis, and one for ACM analysis.

Samples from the 2- to 3- and 3- to 4- foot intervals were stored on site for possible later arsenic analysis. After review of the 0- to 1- and 1- to 2- foot results, however, it was decided that analysis of the deeper samples for arsenic would not be useful, because it was considered likely based on visual observations that the arsenic concentrations would not get lower with depth. The 0-to-1 and 1-to-2 foot samples were also analyzed in the mobile laboratory using the XRF instrument. Results for arsenic and other metals were reported for samples that were analyzed using the XRF instrument. The XRF instrument is not able to detect arsenic down to the CT residential direct exposure criterion of 10 mg/kg. For this reason, it was planned that any samples for which the XRF analysis indicated non-detected levels of arsenic (essentially all of the samples, in practice) would be sent to an off-site laboratory for arsenic analysis with 48-hour turnaround time (Severn Trent Laboratory-Connecticut, Shelton, CT). The results generated by Severn Trent Laboratory were faxed by the laboratory to M&E's Wakefield office, and used to decide whether additional sampling locations were warranted.

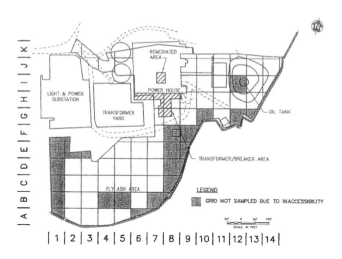

Figure 1. Sampling Locations

3.2   PCBs, TPH, and PAHs in Surface Soil

The 0- to 1-foot sample from every location was analyzed by the field team for TPH and PAHs using the siteLAB test kit.   Certain samples were selected for PCB analysis in the mobile laboratory, TPH analysis in an off-site laboratory, or PAHs in an off-site laboratory based on field-team judgment and a pre-established decision logic developed by the BTSC.   The decision logic for off-site laboratory sample selection was refined in the field, based on field observations, available analytical budget, and advice from the BTSC chemist who assisted in the field effort for two of the five field days.

PCBs. Selected soil samples from the 0- to 1-foot interval were analyzed for PCBs by the OEME mobile laboratory chemist.   Locations were pre-selected for PCB analysis based on past activities in the area (e.g., storage of PCB-containing electrical equipment) and proximity to areas where releases were considered likely based on past investigations and site history. Generally, if PCBs were detected by the mobile laboratory in the 0- to 1-foot interval, the 1- to 2-foot interval from the same location was also analyzed for PCBs, and so on until PCBs were no longer detected (up to a maximum of 4 feet below ground surface).   Also, if PCBs were detected at any location, the field team advanced additional borings around that location, as time allowed, to attempt to delineate the area of PCB contamination. Additional effort was focused towards delineating PCB releases because of the additional costs and regulatory issues that are generally involved when remediating PCB-contaminated soil.

The results of the mobile laboratory PCB analyses were used to select samples for off-site laboratory analysis. The OEME mobile laboratory chemist determined which samples to return to the OEME fixed laboratory for confirmation analysis. Any sample with possible PCB detections based on the field analyses was selected to be sent for confirmation analysis.

TPH and PAHs. The test-kit results were used to help determine whether a particular sample would be sent to an off-site laboratory for PAH analysis and/or TPH analysis. The off-site laboratory used specific extraction and extract cleanup procedures where necessary, to allow for quantitation of the PAHs at levels low enough for comparison to residential direct exposure criteria. The analytical method was GC/MS (SW-846 method 8270, with Selected Ion Monitoring). For TPH, the analytical method was the Connecticut method for Extractable Total Petroleum Hydrocarbons.

Samples for which the test-kit results indicated a total PAH concentration within the 50 to 1000 ppm range, and/or a TPH concentration within the 100 to 500 ppm range, were targeted for off-site laboratory analysis. It was also a goal to select some samples outside these ranges, to help develop more robust correlations between the test-kit results and off-site laboratory results. Because of analytical budget limitations, it was not possible to select as many samples for off-site laboratory analysis as would have been optimal. The project team relied on the decision logic, coupled with visual observations and the advice of the BTSC field chemist, to select an appropriate subset of samples for off-site laboratory analysis.

3.3 ACM in Surface Soil

No field analysis was conducted for ACM in soil. Selected sampling locations were submitted for ACM analysis by the OEME laboratory in North Chelmsford, MA. These locations were selected because of proximity to former site structures. Only the 0- to 1-foot interval was sent for ACM analysis, because any ACM was considered likely to have resulted from releases from the old power plant prior to completion of the ACM abatement and demolition project in 2000. Such releases would be expected to be confined to the surface. No burial of asbestos-containing material on site is suspected. No suspect bulk ACM was observed by the field team.

4.   RESULTS

It was considered probable that arsenic would be present in site soil at levels above its Connecticut residential direct exposure criterion (10 mg/kg), because of the documented presence of coal ash in site soil (TRC, 1988). Hence, the investigation focused on arsenic in terms of the number of off-site laboratory analyses performed. A large number of samples were also analyzed in the mobile laboratory for arsenic and other metals using the XRF instrument.   For arsenic, the reporting limit obtainable with the XRF instrument used in the mobile laboratory was generally greater than the arsenic concentration, and hence arsenic was not often detected by the mobile laboratory. Selected samples were submitted to the OEME off-site laboratory for analysis for a larger suite of metals, to check the assumption that arsenic would be the metal most likely found to exceed its residential direct exposure criterion (DEC), and therefore drive risk evaluations.

Results for the various metals analyses were consistent with the conceptual site model, which suggested that arsenic would be the only metal that would routinely exceed its residential DEC. Beryllium was found to exceed the residential DEC in one of the 15 soil samples submitted to the OEME off-site laboratory.   In this particular sample, the arsenic concentration was also elevated considerably above the residential DEC (120 mg/kg, vs. the residential DEC of 10 mg/kg for arsenic). No other detected metals concentrations were in excess of the residential DEC.

4.1   Arsenic in Surface Soil

The BTSC performed an analysis of the arsenic data and plotted the results on a site plan (Figure 2). Individual sampling grids are coded to display areas where samples were not collected, areas where the maximum arsenic result in the top 2 feet is below the residential DEC, areas where the maximum arsenic result in the top 2 feet exceeds the residential DEC, and areas where the maximum arsenic result in the top 2 feet exceeds two times the residential DEC. Regardless of what the calculated 95 percent upper-confidence limit (95UCL) is for a particular analyte, the CTDEP generally requires that no sample greater than two times the DEC be allowed to be left inplace without remedial action to limit exposure.

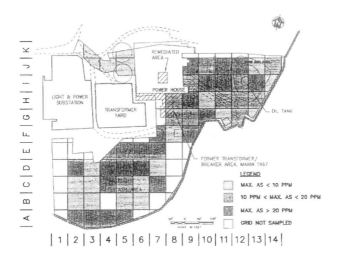

Figure 2. Arsenic Results Summary

Elevated levels of arsenic were found commonly at two times the residential DEC in the top 2 feet of soil across the site with higher concentrations generally being reported from the southern "fly ash" area. Further review of the data also reveals that 55 of the 112 total results (49 percent) for arsenic exceeded the residential DEC of 10 mg/kg. In addition, 32 of the 112 results (29 percent) exceeded two times the residential DEC, a value of 20 mg/kg.

The 95UCL for arsenic is approximately 30 mg/kg across the site. This type of contamination is likely present to a depth of almost 30 feet at the site based on the conceptual site model for disposal of coal ash at the site.

4.2   PAHs in Surface Soil

Correlations between the siteLAB results and the off-site laboratory results were developed by the BTSC using statistical methods, review of chromatograms, and professional judgment. Field results did not appear to have a simple relationship to the residential DECs for carcinogenic PAHs (cPAHs). This is not unusual considering the complexity of hydrocarbon chemistry and the distinctly differing TPH patterns observed across the site. To better understand the relationship between test-kit results and off-site lab results, cPAHs from the off-site laboratory were plotted in 2-dimensional scatterplots for comparison with field-based results. Total cPAH values were calculated by adding the sum of the five carcinogenic PAH compounds (benz[a]anthracene, benzo[b]fluroanthene, benzo[a]pyrene, indeno[1,2,3-

cd]pyrene, and dibenz[a,h]anthracene) for results reported from the fixed laboratory.

Results for the 19 comparison samples collected in February 2003 and the four sample results from the methods applicability study conducted in December 2002 were combined to make a 23 data point comparison (Figure 3). Total PAH results from the siteLAB field test kits were plotted on the y-axis, while cPAHs were plotted on the x-axis.

After reviewing the analytical results for cPAHs it was determined that the lowest total cPAH result where an individual cPAH compound exceeded the residential DEC was found in sample F-8 3-4 feet. The total cPAH result for this sample was 4,030 µg/kg, while individual cPAH results for benzo(a)anthracene (1,300 µg/kg) and benzo(a)pyrene (1,100 µg/kg) exceeded the residential DEC of 1,000 µg/kg. A conservative value of 4,250 µg/kg was then used in a multiple regression curve where the total PAH result from the siteLAB test kits was the dependent variable and the total cPAH result from the off-site laboratory was the independent variable. Using the best-fit line (Figure 3) and the multiple regression analysis, a value of 4,250 µg/kg for the total cPAH resulted in a corresponding siteLAB field test-kit value of 510,000 µg/kg. An additional 20 percent safety factor was built into the kit value and a value of 400,000 µg/kg was estimated as the kit value where corresponding cPAH compounds may begin to exceed the residential DEC of 1,000 µg/kg.

Field-based decision criteria for data from the siteLAB field test kits were then applied to the results of the February 2003 sampling event and used to prepare a map displaying soil samples whose concentration is expected to exceed the residential DEC. Figure 4 provides an overview of the site sampling grids and indicates locations where total PAH values from the siteLAB field test kits exceed 400,000 µg/kg. Elevated levels of total PAHs are indicated as being present in the top several feet of site soil across the site with higher concentrations found in the southern "fly ash" area. Further review of the data also reveals that of the 93 total PAH results from the siteLAB field test kits, 14 (15 percent) exceeded the decision criteria of 400,000 µg/kg and two of the 93 results (2 percent) exceeded two times the decision criteria, a value of 800,000 µg/kg. In addition, of the 23 comparison samples, seven (30 percent) had results from the siteLAB test kit that exceeded 400,000 µg/kg, while nine samples (39 percent) had individual cPAH values that exceeded the residential DEC of 1,000 µg/kg.

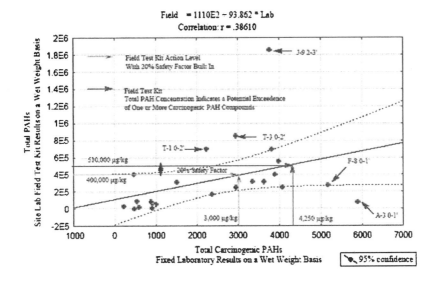

Figure 3. Laboratory Total cPAHs vs. Test-kit Total PAHs

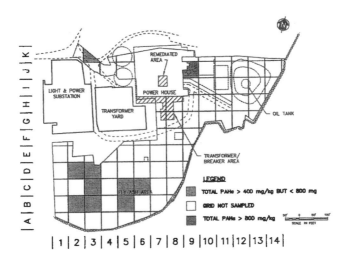

Figure 4. Total PAH Results Summary

4.3   TPH in Surface Soil

TPH results from the siteLAB field test kit and from the off-site laboratory were plotted in 2-dimensional scatterplots for comparison. Results for 8 of the comparison sample pairs collected in February 2003 and the 8 sample pairs from the methods applicability study conducted in December 2002 were combined to make a 16 data point comparison (Figure 5). TPH results from the siteLAB field test kits were plotted on the y-axis, while TPH from the off-site laboratory analyses were plotted on the x-axis.

After reviewing the analytical results for TPH it was determined that the lowest total TPH field result where the off-site laboratory result exceeded the residential DEC of 500 mg/kg was found in sample T-5 0-2 feet. A review of sample chromatograms from the off-site laboratory showed that there were problems with the quantitation of the TPH results for this sample, and also sample G-9 0-1 feet. Therefore, instead of using a value from the comparison results to estimate test-kit values where the corresponding off-site laboratory results would exceed the residential DEC of 500 mg/kg, a conservative value was chosen from the best fit line (Figure 5). The fixed laboratory value of 500 mg/kg was chosen as the residential DEC and then used in a multiple regression curve where the TPH from the field test kits was determined empirically. Using the best-fit line (Figure 5) and the multiple regression analysis, a value of 500 mg/kg for the off-site laboratory results indicated a corresponding siteLAB field test-kit value for TPH of 668 mg/kg. Given the good correlation (R= 0.78) between the off-site and field results, an additional 10 percent safety factor was built into the kit value and a value of 600 mg/kg was estimated as the field test-kit value where corresponding off-site laboratory TPH result may begin to exceed the residential DEC of 500 mg/kg.

Based on the field data collected using the siteLAB field test kits a map was prepared showing the estimated distribution of TPH above the residential DEC for TPH (Figure 6). Figure 6 provides an overview of the site sampling grids and indicates locations where TPH values from the siteLAB field test kits exceed 600 mg/kg. Elevated levels of TPH are distributed across the Site with higher concentrations found in the southern "fly ash" area. Further review of the data also reveals that of the 93 total TPH results from the siteLAB field test kits, 32 (34 percent) exceeded the field-based decision criteria of 600 mg/kg and 21 of the 93 results (23 percent) exceeded two times the decision criteria, a value of 1,200 mg/kg. Also, out of the 16 comparison sample pairs, four (25 percent) had results from the siteLAB test kit that exceeded 600 mg/kg, while six of the samples (38 percent) had off-site laboratory values that exceeded the residential DEC of 500 mg/kg.

4.4   PCBs in Surface Soil

PCBs were measured in site soil samples using the OEME mobile laboratory. The OEME mobile laboratory chemist selected a subset of the samples for confirmatory analysis at the OEME off-site laboratory. The mobile laboratory results were obtained on the same day or the day following the day of collection, and were used to identify additional sampling locations and sample depths for analysis to help define the extent of PCB contamination. Without the mobile laboratory, and the ability to add additional sample locations based on mobile laboratory results, it is considered likely that the PCB contamination in site soil might have been missed.

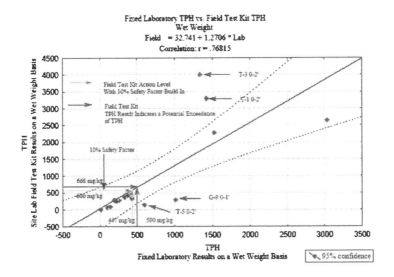

Figure 5. Laboratory TPH vs. Test-kit TPH

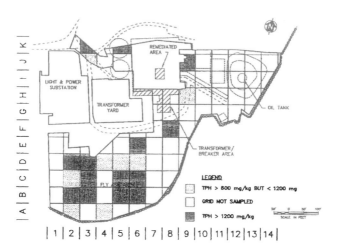

Figure 6. TPH Results Summary

At sampling grids F8, F9, and G11 where PCBs were detected at greater than 5 ppm in either the 0 to 1-foot or 1 to 2-foot intervals, the hotspots were further delineated vertically by sampling subsequent deeper 1-foot intervals until results were non-detected or the 3 to 4-foot interval was reached. The contamination was partially delineated horizontally by surrounding the location with an additional four boreholes approximately 10 feet from the original borehole. If the 10-foot step-out (horizontal) samples were analyzed and PCBs were detected then the distance from the original borehole was doubled and sampled again. This logic continued at the discretion of the M&E field team leader as time and budget permitted to try to delineate the approximate size of the identified PCB hotspots. The maximum total PCB concentrations (regardless of depth interval) are presented in Figure 7 for the F and G grids, which are the only grids where PCBs were detected in the mobile laboratory and confirmed by off-site laboratory analyses.

Fourteen samples were sent for off-site comparison analysis for PCBs. At locations D2, 0-1 foot, D5, 0-1 foot, and J12, 0-1 foot, samples analyzed in the field had elevated detection limits. Subsequent comparison analysis of samples from these locations in the off-site laboratory did not detect PCBs. At the other locations, off-site laboratory analyses confirmed the field laboratory results for both non-detected values and reported positive results.

4.5   ACM in Surface Soil

Selected soil samples from the 0- to 1-foot interval were submitted to the OEME off-site laboratory for analysis for asbestos-containing material (ACM), because the former power plant contained ACM. The intent of the sampling was to assess whether ACM may have been released to site soil, prior to the ACM abatement and demolition work conducted by the town in 2000. Grid locations near the former power plant structures were selected for ACM sampling and analysis.

For most samples, no ACM was observed. In certain samples, ACM was reported as being present at trace levels, but the concentration of fibers observed was less than the reporting limit of the testing method, which is one percent. Based on these results, surface soil contamination with ACM is not considered to be a concern at the site.

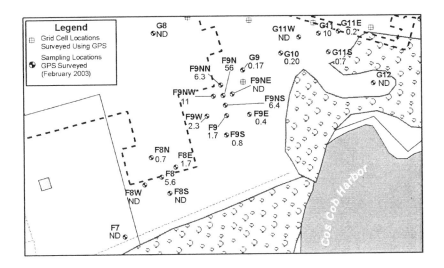

Figure 7. Total PCB Results Summary (Concentrations in mg/kg, wet weight)

5.   SUMMARY OF TBA FINDINGS

Detected concentrations of arsenic, and their locations throughout the site, are consistent with the conceptual site model of soil contamination with coal and coal ash. It appears that arsenic concentrations in the deeper interval (1 to 2 feet below ground surface) are somewhat greater than concentrations found in the 0- to 1-foot interval. This is also consistent with the conceptual site model, which indicates ash disposal throughout most of the site, and subsequent covering of the ash with soil. Previous investigations by TRC and Marin indicate that ash is present to a depth of up to 30 feet bgs in the "fly ash" area of the site. Remediation by excavation of the ash is not a practical measure, because of the great depth and lateral extent of the ash.

PAHs are present throughout site soil, with concentrations generally higher in the "fly ash" area of the site (Figure 4). The PAHs are in part from coal and coal ash, but petroleum releases are also possible sources. SiteLAB results for TPH (Figure 6) suggest exceedances of the residential DEC in multiple grid sectors, also mostly in the "fly ash" area.

An area of PCB contamination was detected within grid rows F and G (Figure 7). The area of PCB contamination may be limited to the F8, F9, and G11 grid sectors, but more sampling is needed to determine the full extent. The highest levels were noted at sampling location F9N. The F8 and F9 areas are located near the former transformer/circuit breaker area of the former power plant. It is not known when or how the PCBs came to be located in the soil, but the site history suggests that releases from PCB-containing electrical equipment may have taken place, either while the plant was still operating (prior to the 1960's), or possibly afterwards. Vandals are suspected to have caused the PCB release within the basement of the power plant, after the plant was abandoned but before it was demolished. That release was subsequently remediated.

Results from surface soil sampling (0 to 1 foot) for ACM suggest that ACM contamination of site soil is not a concern. Trace levels (less than one percent, the reporting limit of the test method) or non-detect levels were noted.

6.   OUTCOME AND NEXT STEPS

The Triad approach provided a characterization of site surface soil that has allowed the town of Greenwich to continue its plans for redevelopment of the site as a waterfront park.   Some inaccessible areas remain to be investigated, and further delineation of PCB contamination is needed, but the remaining investigation efforts are being combined with efforts needed to develop a remedial action plan. It is considered likely that at least two week-long site investigations using a traditional approach would have been necessary to obtain the same level of information as was obtained during the single week-long investigation using the Triad approach. Additional upfront planning and a Demonstration of Methods Applicability were needed to execute the Triad approach. However, the BTSC and M&E estimate that the savings from using the Triad approach for this site was on the order of 35%, including the additional planning and Demonstration of Methods Applicability, compared to two "traditional" investigations.

The town of Greenwich is pursuing further investigation and remediation of the site under the state of Connecticut's voluntary remediation program. Under this program, the property owner retains a Licensed Environmental Professional (LEP) to conduct additional investigation as needed, prepare a remedial action plan, supervise remediation, and submit a final remedial action report. Prior to commencement of remedial action, the town will submit the remedial action plan prepared by the LEP to the CTDEP.

Additional investigation to fill data gaps and define the extent of PCB contamination is scheduled to commence in the fall of 2004. A remedial action plan will then be developed, in consultation with a park designer, so that the remedial action can be coordinated with the park design. It is considered likely that the PCB-contaminated areas will be remediated by removal of PCB-contaminated soil. Once the PCB areas are remediated, contaminants such as arsenic and PAHs will remain in site surface soil at concentrations exceeding residential DEC. Additional soil and groundwater investigation will determine whether these remaining contaminants also present concerns with respect to pollutant mobility (i.e., migration to groundwater and surface water). It is considered probable that the contaminants will not be found to be mobile, and that a remedial action designed solely to limit direct exposure will be acceptable. In this case, a soil cap would be a protective remedy, when coupled with an Environmental Land Use Restriction to maintain cap integrity and prevent excavation without appropriate controls. Additional investigation and coordination is needed before an appropriate cap design can be developed. It may be possible to selectively cap certain areas and/or integrate the cap remedy with the town's site redevelopment plans, to reduce the amount of clean fill that would need to be purchased and brought to the site. Discussions have also been started with CTDEP regarding possible use of dredge spoils from Long Island Sound as capping material, which could substantially reduce the cost of capping.

ACKNOWLEDGEMENTS

This work was funded by the U.S. EPA under contract numbers 68-W6-0042 and 68-W-02-034. The authors wish to thank the field-team members from the EPA Region I Office of Environmental Measurement and Evaluation (Scott Clifford, Dick Siscanaw); Tetra Tech EM, Inc. (Stephen Dyment); Metcalf & Eddy (Cindy Castleberry, Robert Shoemaker, Stan

Hatfield, and Adam Burstein); and the town of Greenwich. We also wish to acknowledge Mr. Steve Greason from siteLAB for providing personal training on the use of the siteLAB test kits.

DISCLAIMER

Reference herein to any specific commercial product, process, or service by trade name, trademark, manufacturer, or otherwise, does not constitute or imply endorsement, recommendation, or favoring by the USEPA. The views and opinions of the authors expressed herein do not necessarily state or reflect those of the USEPA.

REFERENCES

Crumbling, D.M., Griffith, J., and Powell, D.M. 2003. Improving Decision Quality: Making the Case for Adopting Next -Generation Site Characterization Practices. Remediation, Spring 2003, 91-111. Available: http://www.triadcentral.org/tech/documents/spring2003 v13n2p91.pdf

Crumbling, D.M., et al. 2001. Managing Uncertainty in Environmental Decisions. Environ. Sci. Technol. October, 405A -409A. Available: http://www.triadcentral.org/tech/ documents /oct01est.pdf

Marin (Marin Environmental, Inc.). 1998a. Limited Soils Investigation : Cos Cob Power Plant. Report prepared for Town of Greenwich. March 1998.

Marin (Marin Environmental, Inc.). 1998b. Contaminated Soil Removal and Disposal: Cos Cob Power Plant. Report prepared for Town of Greenwich. July 6, 1998.

Robbat, A. 1997. A guideline for dynamic workplans and field analytics: The keys to cost - effective site characterization and cleanup , sponsored by the President's Environmental Technology Initiative, through the U.S. Environmental Protecti on Agency, Washington, DC. Available: http://clu -in.org/download/char/dynwkpln.pdf

TRC (TRC Environmental Consultants). 1988. Environmental Site Assessment Report: Cos Cob Power Plant. Report prepared for Town of Greenwich. April 1988.

USEPA (U.S. Environmental Protection Agency). 2004. Case Study of the Triad Approach: Expedited Characterization of Petroleum Constituents and PCBs Using Test Kits and a Mobile Chromatography Laboratory at the Former Cos Cob Power Plant Site, Greenwich, Connecticut. Office of Superfund Remediation and Technology Innovation, Brownfields Technology Support Center, Washington, DC 20460. EPA 542-R-04-008. To be published in the fall of 2004.

CHAPTER 33

CASE STUDY OF TCE ATTENUATION FROM GROUNDWATER TO INDOOR AIR AND THE EFFECTS OF VENTILATION ON ENTRY ROUTES

Alborz Wozniak and Christopher Lawless
Johnson Wright, Inc., 3687 Mt. Diablo Blvd, Suite 330, Lafayette, CA 94549

Abstract:　　An investigation of groundwater-to-indoor air vapor intrusion of trichloroethylene (TCE) was conducted at a commercial building to 1) identify vapor pathways; 2) measure TCE attenuation ; 3) evaluate whether building pressurization can reduce vapor intrusion; and, 4) evaluate seasonal effects on vapor intrusion. The investigation included a site assessment, both indoor and outdoor air sampling, and the modification of the existing heating, ventilation and air-conditioning system. The investigation found that preferential pathways, such as a floor crack, acted as conduits for vapor intrusion when the building was not properly ventilated. The investigation also found that air TCE concentrations were significantly atte nuated between ground surface and the breathing zone and proper building pressurization can effectively minimize vapor intrusion. Weather conditions at the site did not significantly affect the indoor air TCE concentrations.

Key words:　　vapor intrusion, ventilation, TCE, attenuation, preferential pathway, indoor air, building pressurization

1. INTRODUCTION

An investigation of groundwater-to-indoor air vapor intrusion was conducted at a Superfund site (the site) during 2003. Groundwater at the Site has elevated levels of trichloroethylene (TCE) and has been under cleanup by an groundwater extraction and treatment system since the 1980s. In 2002, the U. S. Environmental Protection Agency (USEPA) requested that the responsible parties at the site evaluate the health risks associated with vapor intrusion into their former facility. As a result, an indoor air sampling program was conducted between May and December 2003 to 1) identify vapor pathways; 2) measure TCE attenuation; 3) evaluate whether building pressurization can reduce vapor intrusion; and, 4) evaluate seasonal effects on vapor intrusion. The investigation involved gathering site-specific information, baseline air sampling, modifications of the facility's Heating, Ventilation, and Air Conditioning (HVAC) system, and two rounds of air sampling after the completion of HVAC repairs.

1.1   Site Conditions

The investigation was conducted at a single-storey commercial building constructed in 1965. The site building has a slab-on-grade concrete floor with a perimeter foundation. No information was available on the composition of the sub-slab material. The building has two independent office sections, each of which has its own separate HVAC system. Section 1 of the building was vacant during all of the sampling events. Section 2 of the building was occupied and had maximum daily work hours between 7:00 AM to 7:00 PM, Monday through Friday.

Groundwater depth was measured at four monitoring wells located within 30 feet of the site building. Groundwater-level measurements taken on May 22, August 28, and November 20, 2003, indicated that the depth of groundwater in the vicinity of the building ranged from 13.19 to 15.82 feet. Historical groundwater levels measured over the past 10 years are consistent with this range of depths. Groundwater samples collected from the two monitoring wells closest to the building had TCE concentrations of 77 and 270 parts per billion (ppb).

2.   MATERIALS AND METHODS

2.1   Site Screening

To evaluate the site conditions and aid in selecting appropriate sample locations, the following tasks were conducted prior to sampling:

- HVAC and other building construction as-built drawings were reviewed to obtain information about the HVAC system design and operation, building foundation construction details, and potential pathways for vapor intrusion

- An indoor air-quality building survey was completed by a facility employee to obtain information regarding site chemical usage and other facility-specific information that could affect indoor air quality
- A walk-through of the building with a photo ionization detector (PID) was conducted to monitor for the presence of volatile organic compounds
- A thorough search for outdoor TCE emission sources was conducted within a one-mile radius of the site

2.2   Sampling and Analysis Methods

Both representative and pathway air samples were collected within the building. Representative samples were collected in the breathing zone (i.e., 3.5 feet above ground surface). Representative samples were taken in offices, near cubicles and in a shop area (See Figure 1). Pathway samples were collected at ground surface at two potential preferential pathways above a bathroom floor drain and above a visible crack in the concrete floor. Outdoor samples were collected on the roof of the building at the HVAC unit intake for each section of the building to assess the potential contribution of TCE in the outdoor air to the indoor air TCE concentrations.

The sampling duration was based on maximum daily work hours was and lasted for twelve hours from 7:00 AM to 7:00 PM. The air samples were collected in 6-liter Summa canisters equipped with a flow controller set to allow for a steady flow of air over the 12-hour period.

All air samples collected during the investigation were analyzed for TCE, cis-1,2-dichloroethene, Freon-113, vinyl chloride, 1,1-dichloroethene and 1,1-dichlorethane using EPA Method TO-15, Selective Ion Monitoring (SIM) mode. All collection canisters were pre-cleaned and laboratory certified to meet the project-reporting limits. A state-certified laboratory analyzed all of the air samples.

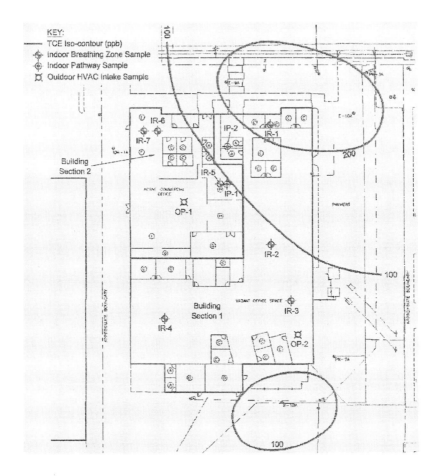

Figure 1. Air Sample Locations

2.3   Baseline Sampling

Two rounds of sampling were performed on May 6 and 13, 2003. During each of these sampling events, seven indoor representative samples, two indoor pathway samples, two outdoor samples, and one field duplicate were collected.

2.4   HVAC Modifications

The differential pressure between the inside and outside of the building was measured every two hours (seven times) during each of the sampling events using a Magnehelic Differential Pressure Gauge Model 2000-0. Differential pressure measurements taken during the initial May 2003 sampling events indicated that Section 2 of the building had negligble building pressurization. During the May sampling events, the HVAC units for Section 2 of the building appeared to be either inoperable or operating improperly. In addition, the tenants verbally stated that the HVAC system was not providing adequate air circulation and temperature control among various offices and throughout their portion of the building.

In July of 2003, an HVAC repair specialist examined and repaired the two HVAC units serving Section 2 of the building. The repairs included replacing all filters, changing fluids, replacing fan belts, and adjusting the dampers for outdoor air intake. Once the HVAC units were repaired, an HVAC testing and air balancing specialist measured the supply, return, exhaust, and outdoor air flow rates; examined the indoor air control settings; and recommended additional improvements for the ventilation systems. In particular, it was noted that one of the two HVAC units that service Section 2 had no means of introducing outdoor air into the space. Therefore, an outdoor air damper was installed in the HVAC unit to allow for outdoor air to be blended with the supply air.

Following HVAC repairs and improvements, the total air supply and outdoor air flow into the Section 2 of the building, along with the net positive pressure inside versus outside the building were measured. Those measurements indicated that the repaired HVAC system provided a positive pressure of approximately 0.02 inches of water inside versus outside the building. Such a building pressurization should significantly reduce the potential for subsurface vapor intrusion. In addition, the air exchange rate, defined as the number of times that outdoor air replaces the volume of air within the building per hour, was estimated at 1.2 air changes per hour (ACH) following the aforementioned HVAC repairs.

2.5   Post HVAC Repair Sampling

Follow completion of the HVAC improvements/modifications, sampling was conducted twice in September and once in December. Seven samples were collected during the September 4, 2003 sampling event: three representative samples, two outdoor samples, one pathway sample, and one field duplicate. Eight samples were collected during September 11, 2003 sampling event: four representative samples, two outdoor samples, one pathway sample, and one field duplicate. Only four samples from Section 2 of the building were collected during the December 23, 2003 sampling event: two representative samples, one pathway sample and one outdoor sample (sample locations IR-5, IR-6, IP-1 and OP-2; See Figure 1).

3.   RESULTS

3.1   Site Screening

The indoor air quality survey completed by the building Section 2 tenant indicated that there were no potential indoor sources of TCE either used or stored at the site. Section 1 of the building was vacant during all of the sampling events and no volatile organic compounds were detected with the PID during the site walk through.

3.2   Pre-HVAC Repair Sampling

The TCE concentrations measured during the air sampling investigation are shown in Figures 2 through 4. During the May 2003 sampling events, TCE concentrations in building Section 1 representative samples ranged from 0.14J to 0.35 $\mu g/m^3$. The maximum TCE concentration was detected in a sample collected from location IR-4 in an open area in the southwestern portion of the building (Figure 1). However, the representative air samples collected from building Section 2 during May 2003 had TCE concentrations ranging from 0.84 to 1.4 $\mu g/m^3$. The maximum detected TCE concentration was detected in a sample collected at location IR-6 in an open office area.

The May 2003 pathway samples collected in building Section 1 had concentrations of 0.37 and 0.5 $\mu g/m^3$. The pathway sample was collected at ground surface directly above a floor drain. Since the TCE concentrations in both of the pathway samples were not significantly above the representative and outdoor sample results, no additional pathway samples were collected during subsequent sampling events in building Section 1. In building Section 2, the pathway samples were collected at ground surface from location IP-2, directly above a crack in the concrete floor slab (Figure 1). The TCE concentrations detected in the samples collected from this location during the two May 2003 sampling events were 17 and 49 $\mu g/m^3$. By comparison, outdoor air TCE concentrations during the spring, summer and winter sampling events ranged from non-detect (i.e., less than 0.18 $\mu g/m^{3)}$ to 0.66 $\mu g/m^3$.

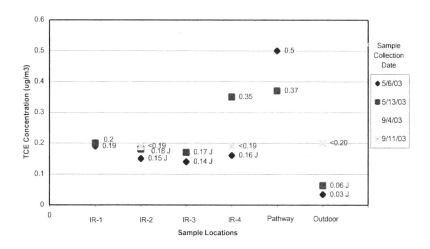

Figure 2. Building Section 1 - Trichloroethylene Air Sample Concentrations

Notes: IR = Indoor representative sample collected at 3.5 feet above ground surf ace; J = concentration is below the laboratory reporting limit

3.3   Post-HVAC Repair Sampling

During the September 2003 sampling events, all representative air samples collected in both building Sections 1 and 2 had TCE concentrations below laboratory-reporting limit. In the December 2003 sampling, the TCE concentrations detected in representative samples measured 0.27 and 0.3 $\mu g/m^3$. During the September 2003 sampling events, the pathway sample IP-1 had TCE concentrations of 0.23 $\mu g/m^3$ and non-detect, respectively. The TCE concentration in the December 2003 pathway sample was 0.62 $\mu g/m^3$.

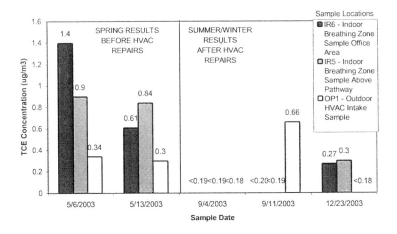

Figure 3. Building Section 2- Trichloroethylene Indoor Representative and Outdoor Air
Sample Concentrations

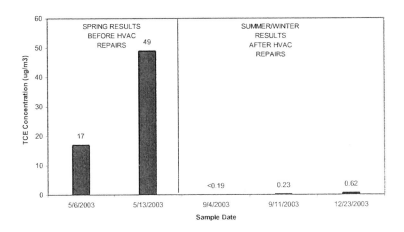

Figure 4. Building Section 2 - Trichloroethylene Concentrations in Pathway Air Sample IP -1

3.4 Quality Assurance

Air-sampling data collected in May were evaluated using the quality-control criteria for Precision, Accuracy, Representativeness, Completeness and Comparability. USEPA Region IX "Tier 3" Validation was conducted on three samples (10% of the total number) collected during the May 2003 sampling events. All of the May 2003 data met the quality assurance criteria for Precision, Accuracy, Representativeness, Completeness and Comparability with the exception of the field duplicate, which exceeded the relative percent difference criteria. Data collected during the September and December events were evaluated using the laboratory quality-assurance program. A total of five field blanks were collected during all of the air-sampling events and all samples were analyzed within 14 days of sampling.

4. DISCUSSION

4.1 Ventilation Events

During the May 2003 sampling events, building Section 2 had representative breathing zone air TCE concentrations that averaged 2 to 10 times higher than TCE concentrations in building Section 1. A review of HVAC operating conditions suggested that lack of adequate building pressurization and air circulation was the cause of this difference in indoor-air TCE concentrations, although other site conditions, such as occupancy and floor conditions, could have contributed.

During the May 2003 samplings, building Section 1 pathway sample TCE air concentrations were less than two times the maximum representative sample (See Figure 2). However, in building Section 2, the TCE pathway concentrations, 17 and 49 $\mu g/m^3$, were approximately 10 to 50 times greater than the value of the maximum representative sample, 1.4 $\mu g/m^3$. These data indicated that improperly operating HVAC systems had increased the potential for vapor intrusion. These data also indicated that the concentrations of TCE vapors that did enter the building through preferential pathways was significantly attenuated between ground surface and the breathing zone.

The building Section 2 TCE concentrations observed in the breathing zone before and after the HVAC system repairs are presented in Figure 3. As shown in Figure 3, after the HVAC system was repaired, all indoor air representative breathing zone samples were either below the laboratory-reporting limit (September) or very close to that limit (December). The building Section 2 TCE concentrations observed in pathway samples before and after the HVAC system repairs are presented in Figure 4. As shown in Figure 4, all indoor air-pathway sample TCE concentrations were significantly reduced after the HVAC system repair to less than 1 $\mu g/m^3$.

Differential pressure measurements taken prior to HVAC modifications in building Section 2 indicated that this section of the building was not under positive pressure. Differential pressure measurements taken after the HVAC repair indicated that the building was adequately pressurized.

During the September sampling events, the investigators turned on the HVAC fans the night before the sampling events, and these fans were left on for the duration of the sampling events. During the December sampling event, the investigators turned on the HVAC fans approximately 30 minutes prior to the sampling event and again the fans were left on for the duration of the events. This difference in HVAC operating conditions during sampling may account for the slight increase in the building Section 2 TCE air concentrations observed in the December sampling event (see Figures 3 and 4).

4.2   Seasonal Effects

Meteorological and building conditions during air sampling events are presented in Table 1. The weather conditions during the May, September, and December sampling events represent the expected range of annual weather at the site. During the September sampling event, outdoor temperatures reached 90°F, a value that is representative of a hot summer day. During the December sampling event, the highest temperature difference between indoors and outdoors was observed. The elevated temperatures observed indoors relative to outdoors can cause what is known as the "stack effect" in which higher indoor air temperatures tend to pull in air from outside the building, increasing the likelihood of vapor intrusion. Cloudy and rainy conditions, such as were present during the December sampling event, also represents the "worst-case" meteorological condition for vapor intrusion. The indoor and outdoor air samples collected in Sections 1 and 2 of the building were reviewed to evaluate the potential seasonal effects on vapor intrusion. All of the TCE concentrations from spring and summer samples collected from Section 1 of the building were at or below the laboratory reporting (Figure 2). Therefore, conclusions regarding the seasonal effects of vapor intrusion into Section 1 of the building could not be drawn.

Table 1. Meteorological and Building Condition Summary

Sample Date	Meteorological Conditions	Building Conditions
May 6, 2003	Cloudy, occasional rain Westerly Wind @ 5 mph T_{out} ~ 54-62°F	Poor ventilation system, stagnant air Doors periodically open T_{in} ~ 69-72 °F / ΔP <0.001" H_2O
May 13, 2003	Partly cloudy North-Northwest Wind @ 6.5 mph T_{out} ~ 59-67°F	Poor ventilation system, stagnant air Doors periodically open T_{in} ~ 69-72 °F / ΔP <0.001" H_2O
September 4, 2003	Sunny North-Northwest Wind @ 8.2 mph T_{out} ~ 65-74°F	Ventilation on full-time (24-hours/day) ΔP ~ 0.01" H_2O T_{in} ~ 73-75 °F
September 11, 2003	Sunny Wind from North-Northwest @ 3.9 mph T_{out} ~ 65-90°F	Ventilation on full-time (24-hours/day) ΔP ~ 0.02" H_2O T_{in} ~ 71-73 °F
December 23, 2003	Rain, heavy at times Wind from Southwest @ 8.2 mph T_{out} ~ 55-58°F	Ventilation on ½ hour before sampling ΔP ~0.02" H_2O T_{in} ~ 64-70 °F

Abbreviations:

T_{out} = Outside Temperature T_{in} = Inside Temperature   mph = miles per hour
°F = degrees Fahrenheit H_2O = water
ΔP = Pressure difference between inside and outside the building measured in inches of water
(" H_2O), where positive pressure means greater indoor air pressure.

The December and September sampling results from Section 2 of the building would be considered representative of indoor air conditions during winter and summer months. However, the relative difference in pathway indoor air TCE concentrations between September and December events was very small. Thus, seasonal weather conditions do not appear to noticeably affect the indoor pathway or breathing zone TCE concentrations at the site.

5.   CONCLUSIONS

The investigation found that TCE concentrations in groundwater of less than 300 ppb did result in indoor vapor intrusion when building ventilation/pressurization was not adequate. Preferential pathways, such as a floor cracks, acted as conduits for vapor intrusion when the building was not properly ventilated and pressurized. Sampling directly above a floor slab was identified as an advantageous way of identifying preferential pathways and assessing the TCE concentrations entering the building. The data collected above the floor crack in Section 2 of the building also showed that air TCE concentrations were significantly attenuated between ground surface and the breathing zone. Sampling conducted after HVAC modifications and repair indicated that proper building pressurization could effectively minimize vapor intrusion. Weather conditions at the site, however, did not significantly effect the indoor-air TCE concentrations.

Index